Interdisciplinary Biotechnological Advances

This series is an authoritative resource that explains recent advances and emerging directions in biotechnology, reflecting the forefront of research clearly and reliably, without excessive hype. Each volume is written by authors with excellent reputations and acknowledged expertise in the topic under discussion. The volumes span the entire field from an interdisciplinary perspective, covering everything from biotechnology principles and methods to applications in areas including genetic engineering, transgenic plants and animals, environmental problems, genomics, proteomics, diagnosis of disease, gene therapy, and biomedicine. The significance of these applications for the achievement of UN Sustainable Development Goals is highlighted. The series will be highly relevant for Master's and PhD students in Biotechnology, Nanochemistry, Biochemical Engineering, and Microbiology, medical students, academic and industrial researchers, agricultural scientists, farmers, clinicians, industry personnel, and entrepreneurs.

Jen-Tsung Chen • Sunny Ahmar
Editors

Plant Genome Editing Technologies

Speed Breeding, Crop Improvement and Sustainable Agriculture

Springer

Editors
Jen-Tsung Chen
Department of Life Sciences
National University of Kaohsiung
Kaohsiung, Taiwan

Sunny Ahmar
Institute of Biology, Biotechnology and Environmental Protection, Faculty of Natural Sciences
University of Silesia
Katowice, Poland

ISSN 2730-7069 ISSN 2730-7077 (electronic)
Interdisciplinary Biotechnological Advances
ISBN 978-981-99-9337-6 ISBN 978-981-99-9338-3 (eBook)
https://doi.org/10.1007/978-981-99-9338-3

This Springer imprint is published by the registered company Springer Nature Singapore Pte Ltd.
The registered company address is: 152 Beach Road, #21-01/04 Gateway East, Singapore 189721, Singapore

Paper in this product is recyclable

Preface

In the modern world, plant molecular biotechnology associated with crop breeding is expected to considerable enhance agricultural production, sustaining global food security in the face of changing climate and increasing human population. With the progress of biotechnology, the program of plant breeding can be advanced and accelerated to acquire future crops with high quality and yield as well as being stress-tolerant to achieve climate resilience. To support these goals, over the past years, plant biologists have explored advanced and promising molecular tools for uncovering the world of functional genomics, aiming to release the unlimited potential of plant genetic resources. Undoubtedly, genome editing is one of the cutting-edge tools that can be mediated by a number of systems via zinc finger nucleases (ZFNs), transcription activator-like effector nucleases (TALENs), or clustered regularly interspaced short palindromic repeats (CRISPR). Theoretically, these tools can modify crops to gain desired traits in an efficient and precise way and, therefore, lead to speed and precision breeding. Among them, currently, it is recognized that CRISPR/Cas-based genome editing has become the most popular tool and has been applied to produce a range of "CRISPR crops."

This book collects summaries organized by diverse and experienced experts in plant molecular biology and biotechnology, working on the exploration of emerging technologies for plant genome editing. It provides an overview of fundamental systems of genome editing by summarizing their molecular machinery and refining the general applications for crop improvement. In the way to obtain transgene-free or so-called non-genetically modified (non-GM) crops through genome editing, a part of the book content was designed to provide a detailed presentation of a series of techniques; additionally, the current protocols and future directions were discussed. The method based on CRISPR/Cas has been upgraded timely, and therefore, the following chapters introduce how to advance the genome editing system. It provides the current findings on the associations of Cas9, Cas12, and Cas13 proteins, with an emphasis on improving the resistance of crops for combating biotic stress/plant diseases. Also, the recent literature investigating plant abiotic/environmental stress tolerance achieved by genome editing technologies was summarized systematically, and interestingly, the potential involvement of CRISPR/Cas9-based base editor, prime editor, and the dual-sgRNA/Cas9 system to advance breeding of crops such as soybean was discussed. In horticultural crops including ornamental plants, vegetables, and fruits, genome editing technologies started to be applied to modify their quality such as flower characteristics, flavor, nutrition content, and yield. Recent achievements using CRISPR/Cas9 manipulated by various transgenic protocols involving *Agrobacterium*-mediated methods, particle bombardment, protoplast transformation, viral vector- and nanocarrier-mediated transformation, etc., were presented. Additionally, certain investigations showed successful genome editing events in woody plants, which have a related huge genome size, complex ploidy level, and a long generation time. A potent single-base editing protocol was described in this book, and researchers used this system to gain A to G base editing mediated by adenine deaminase/Cas9 fusion protein, which was proposed to be useful for directed genome changes that occur at low frequency. In vitro culture for the introduction of desired modifications is one of the major factors in conducting plant genome editing. This book provides a novel tissue culture protocol, namely fast-treated *Agrobacterium* co-culture (Fast-TrACC) which allows high-throughput delivery to reduce cost and labor. Nevertheless, a future direction has been proposed to gain a tissue culture-less regeneration system, eventually to achieve the highest efficiency.

Plant genome editing technologies are evolving rapidly, and novel technologies continue to be created such as "base editing" and "prime editing." Also, in the past years, non-GM genome-edited plants have been reported increasingly, and therefore, there is a high demand to summarize the current knowledge of their methods and applications and, additionally, to discuss the regulations, safety, and ethical implications in food and agricultural production. The knowledge of this book is essential for modern precision crop breeding and supports sustainable development goals (SDGs) by the United Nations, particularly for the goal of global zero hunger through achieving food security and sustainable agriculture. This book is, therefore, an ideal reference for students, teachers, and researchers in the research fields of plant science, plant physiology, plant molecular biology, and agricultural biotechnology, particularly for subtopics involving plant stress physiology, plant functional genomics, and crop breeding. As editors, we sincerely hope this book opens a window for students and young scientists to explore this critical and game-changing biotechnology in an "efficient and precise" manner, and after reading it, hopefully, new ideas can be inspired for their future research. In the end, we would like to thank all the authors for their invaluable contributions! The instruction and assistance from the editorial at Springer Nature are very grateful.

Kaohsiung, Taiwan — Jen-Tsung Chen
Katowice, Poland — Sunny Ahmar

Contents

Plant Genome Editing Technologies: An Updated Overview

Shreni Agrawal, Pradeep Kumar, Richa Das, Kajal Singh, Nancy Singh, Sakshi Singh, Amit Kumar Singh, Praveen Kumar Shukla, Vishnu D. Rajput, Tatiana Minkina, Indrani Bhattacharya, Sunil Kumar Mishra, and Kavindra Nath Tiwari

Abstract A genome is a representation of all the genetic material that forms an organism and is stored in its deoxyribonucleic acid (DNA). Genetic engineering, genome editing, and gene editing are all terms that relate to genetic operations in which genetic material is altered, deleted, or inserted in a living organism. Genome editing is now often employed across plant species to develop and investigate functional mutations and their effects on crop development. The development of novel plants with improved or desired traits has historically relied on challenging and time-consuming breeding techniques. With the advent of genome-editing (GE) technology, a new era of genetic engineering has emerged that makes it possible to quickly, precisely, and effectively engineer plant genomes. Several innovative gene editing methods have been discovered in the past few years, including zinc-finger nucleases (ZFNs), transcription activator-like effector nucleases (TALENs), and clustered regularly interspaced short palindromic repeats/Cas9 (CRISPR/Cas9). Targeted gene modification in living organisms enables study not just into the basic concepts underlying ecosystems but also to address a wide variety of aims in order to increase agricultural output and quality. This review gives detailed information

S. Agrawal (✉) · R. Das · I. Bhattacharya
Department of Biotechnology, Parul Institute of Applied Science, Parul University, Vadodara, Gujarat, India

P. Kumar · P. K. Shukla · K. N. Tiwari
Department of Botany, MMV, Banaras Hindu University, Varanasi, UP, India

K. Singh · N. Singh
Department of Biosciences, Galgotias University, Gautam Buddh Nagar, UP, India

S. Singh
Department of Bioscience and Biotechnology, Banasthali Vidhyapith, Tonk, Rajasthan, India

A. K. Singh · S. K. Mishra
Department of Pharmaceutical Engineering & Technology, Indian Institute of Technology, Banaras Hindu University, Varanasi, UP, India

V. D. Rajput (✉) · T. Minkina
Academy of Biology and Biotechnology, Southern Federal University, Rostov on Don, Russia

J.-T. Chen, S. Ahmar (eds.), *Plant Genome Editing Technologies*, Interdisciplinary Biotechnological Advances,
https://doi.org/10.1007/978-981-99-9338-3_1

about GE techniques and their applications to improve agriculture quality, yields, and resistance to biotic and abiotic conditions. In addition, this chapter has highlighted a number of potential applications of GE techniques for the prospects.

Keywords Genome-editing techniques · CRISPR/Cas9 · ZFN · TALENs

1 Introduction

In recent years, the natural environment has continuously been affected in a direction that is unfavorable for the growth of crops due to the rise of industry and the frequent occurrence of adverse weather conditions. Subsequently, abiotic conditions, including drought, salt, and severe temperatures, are only a few of the major constraints that seriously hinder the growth and productivity of crops. They cause crop plants to lose nearly 50% of their yield, which results in a $14 to $19 million economic loss per year globally (Oshunsanya et al. 2019). Moreover, biotic stressors, including diseases, insect pests, and weeds, may hinder crop yields ranging from a 17.2% loss in the production of potatoes to a 30% loss in the production of rice (Savary et al. 2019). The problems associated with agricultural outcomes are not only due to climate change or environmental stressors, but they are also due to the ever-increasing population worldwide. One study suggested that the worldwide population will rise from 7.8 billion to 8.3 billion by the year 2030. It is estimated that almost 800 million individuals globally are nutritionally deficient, with 98% of them residing in developing nations (Sinha et al. 2019). Additionally, according to the United Nations Children's Fund (UNICEF) (2021), about 340 million individuals have micronutrient deficiencies, including those in zinc, iodine, iron, and vitamin A (UNICEF 2021). Thus, maintaining sustainable agriculture is one of the main objectives of researchers for the control of climatic changes, environmental pressures in various crops, and the continuously growing world population.

Therefore, scientists and breeders concentrate on improving crop quality and productivity. Several techniques, including classical crossover breeding, radiation or chemical-assisted mutational cultivation, molecular-marker-mediated cultivation, and genetically modified cultivation, have been effectively used to enhance a variety of agricultural qualities (Wenefrida et al. 2013; Lusser et al. 2012; Ramesh et al. 2020; Chaudhary et al. 2019).

However, there are many drawbacks associated with conventional breeding techniques, such as the long period it takes to generate new cultivars with ideal agricultural qualities in any crop. This is influenced by multiple phases of crossing, selection, and testing during the breeding process, as well as the duration of the developing season and the stage of plant growth (particularly the long-period growers, like trees) (Ni et al. 2018). Furthermore, the results of chemical and physical mutation breeding are unreliable, considering the fact that essential genes have lower mutation rates than nonessential genes (Zhang 2022). In addition, extensive populations of mutagenized plants must undergo difficult and complicated screening and selection processes in order to find the desired characteristics (Ma et al. 2021). There is little question that transgenic technologies, which entail introducing gene coding for desirable features into top varieties, offer a substitute for combatting crop production damage (Sedeek et al. 2019). However, creating a genetically modified (GM) crop with desired features takes a lot of effort and money. The biggest disadvantage of this strategy is that GM products are not widely accepted, which leads to complex and strict health regulatory processes (Herman et al. 2019). Moreover, molecular deoxyribonucleic acid (DNA) markers have also made it possible to considerably shorten the time needed to develop new agricultural crop lines and types. But these methods also make it impossible to target the exact crop genome (Tuberosa 2012).

This growing evidence indicates a move toward genome-editing (GE) techniques for crop breeding, which efficiently and predictably modifies plant genomes. With the least likelihood of straying incorrect and no inclusion of extraneous genetic sequences, GE techniques can make estimated and genetically transmissible alterations in specific parts of the genome (Gaj et al. 2013). In plant research, genome-editing approaches involving sequence-specific nucleases (SSNs) have gained popularity. There are currently three SSNs: clustered regularly interspaced palindromic repeat (CRISPR)/CRISPR-associated (Cas) protein systems, transcription activator-like effector nucleases (TALENs), and zinc-finger nucleases (ZFNs). These methods enable an accurate selection and modification of specific genomic sequences in the following three stages: (1) an externally designed nuclease comprising a detection component and a nuclease region recognizes the desired DNA sequence and (2) adheres to the sequence to generate double-strand breaks (DSBs) at or close to the target location. (3) Nonhomologous end-joining (NHEJ) or homologous recombination (HR) is subsequently used to repair the DSBs. DSBs are accurately repaired by HR rather than NHEJ, which is more susceptible to errors and has a proper repair mechanism that commonly results in deletions and insertions (indel) alterations (Wada et al. 2020).

The FokI endonuclease motif and the zinc-finger DNA-binding motif are incorporated to produce ZFNs, which are the first kind of genome-editing nucleases (Kim et al. 1996), while TALENs are made up of a DNA-binding region and a FokI cleavage region from TALE proteins. When compared to ZFNs, the TALENs technique has greater target binding accuracy and a lower inaccurate possibility (Joung and Sander 2013). In 1987, The first incidence of CRISPR was found in *Escherichia coli*, which was described as an immune system defense against plasmid and viral DNA penetration (Ishino et al. 1987). CRISPR/Cas techniques have progressed to emerge as the most prevalent GE techniques in the past few years.

When employed in the field of agricultural development, GE can dramatically speed up the insertion of desirable characteristics and significantly save labor and additional expenditures. Genome editing holds out the possibility of producing novel crops more quickly and with a very low risk of adverse impacts. The fundamental advantage of these methods is that they can be used for any plant material in any type of laboratory setting, including those with complicated genomes that are challenging to breed using conventional methods (Abdallah et al. 2015). By utilizing these techniques, remarkable genetic alterations have been made to plants that enhance the physical features of plants and seeds, tolerance to biotic (pathogenic fungi, bacteria, or viruses) and abiotic conditions (salt, cold, drought), and nutrient quality; increase production, and crop quality; produce monoploid seeds; resist herbicides; and many other traits (El-Mounadi et al. 2020).

The industrialization of gene-edited crops, which is still a long way off, is one of the biggest issues related to these approaches. Additionally, not all of the conditions for modifying the plant genome have been satisfied by gene-editing techniques. Since multiple quantitative trait loci (QTLs) control a number of quality-related characteristics in plants, additional research will be needed to apply the CRISPR/Cas 9 technique, and changing a single gene might not significantly impact physiological characteristics. Therefore, creating new carrier materials would be a good idea. Other than them, another barrier to advances in plant breeding is the public's concern and the government's strict regulation of gene-editing technologies. Despite the obstacles that still require to be overcome, it is anticipated that GE techniques will become more prevalent and unavoidably play a significant role in agricultural production (Liu et al. 2021).

2 Genome-Editing Tools

Researchers now have the ability to quickly and economically insert changes within the genomes of a variety of cells and species in a specific sequence manner due to the recent development of highly adaptive genome-editing tools. The three main methods currently used to facilitate genome editing are zinc-finger nucleases (ZFNs), transcription activator-like effector nucleases (TALENs), and clustered regularly interspaced short palindromic repeat (CRISPR)-associated protein 9 (Cas9) (Gaj et al. 2016).

3 Zinc-Finger Nucleases (ZFNs)

Zinc-finger proteins with specifically designed DNA-binding domains are combined with FokI restriction endonucleases to construct ZFNs (Kim et al. 1996). These were the first programmable nucleases that were widely used (Porteus and Carroll 2005; Urnov et al. 2010). These nucleases work as dimers, where each monomer identifies a "half-site" of a sequence specifically. The monomers recognize nine to 18 base pairs in the target DNA. It is the DNA-binding region of the ZFN that identifies the target DNA sequence. ZFN protein dimerization is facilitated by FokI endonuclease. This cleavage domain splits DNA between the two zinc-finger domains by an intermediate sequence of five to seven base pairs (Smith et al. 2000). A single ZFN consists of three to four zinc fingers. Each zinc finger interacts with three nucleotides of the desired DNA by the use of about 30 residues

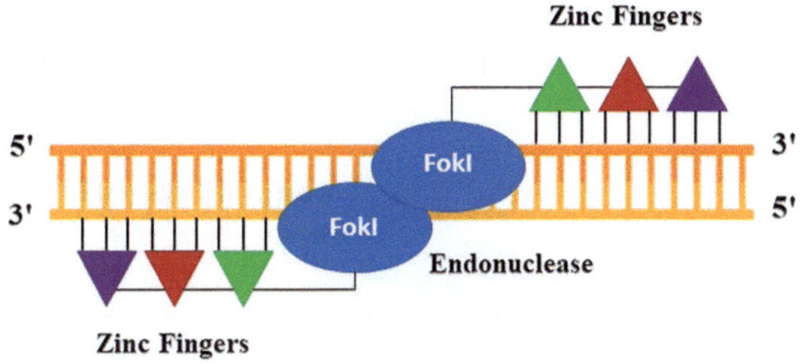

Fig. 1 Zinc-finger nuclease (ZFN) technique

of amino acids arranged in a ββα motif (Pavletich and Pabo 1991). The amino acid motifs in the α-helix region enhance this interaction (Wolfe et al. 2000). The FokI endonuclease identifies the nonpalindromic pentadeoxy-ribonucleotide 5′-GGATG-3′:5′-CATCC-3′. It cuts nine to 13 nucleotides away from the recognition site (Kim et al. 1994). During the dimerization of the cleavage domain, each ZFN binds with the forward and reverse strands of the DNA. A spacer sequence, which is five to seven base pairs long, separates both ZFNs (Zhang et al. 2017). The ZFNs create DSBs, generating cohesive overhangs at 5′-end. The DSBs generated can be corrected by homology-directed repair (HDR) or nonhomologous end joining (NHEJ). As a result, indels (insertions or deletions of nucleotides) are produced in the host DNA (Kamburova et al. 2021). The zinc-finger amino acid sequences, the number of zinc fingers, and how the FokI endonuclease interacts with the desired DNA all affect the specificity and recognition capabilities of ZFNs (Kamburova et al. 2021). Both domains of ZFNs are customizable. This gives researchers the freedom to construct new assemblies having better affinity and specificity for the target sequence (Li et al. 2020b). The indels caused by ZFNs are permanent. They integrate themselves rapidly into the locus (Ghosh et al. 2021). The phage display method can be employed (Choo and Klug 1994; Jamieson et al. 1994; Wu et al. 1995) to identify zinc-finger domains recognizing different DNA triplets (Segal et al. 1999; Dreier et al. 2001; Bae et al. 2003). A canonical linker peptide can be utilized to fuse these domains in tandem (Liu et al. 1997). To generate polydactyl zinc-finger proteins focusing a broad variety of genomic sequences (Beerli et al. 1998; Kim et al. 2009) (Fig.1).

4 Transcription Activator-like Effector Nucleases (TALENs)

TALEs (transcription activator-like effectors) are bacterial proteins. The code that is utilized by the TALE protein for the recognition of DNA was discovered in 2009 (Boch et al. 2009; Moscou and Bogdanove 2009). This discovery was a pioneer for the creation of custom TALENs that are capable of modifying a gene. TALE

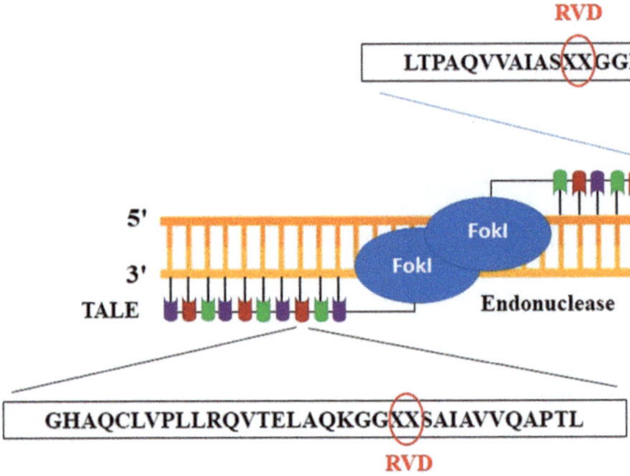

Fig. 2 Transcription activator-like effector nuclease (TALEN) technique

proteins, which bind to DNA, and FokI endonuclease, which cleaves DNA, are combined to construct TALENs. There are two units of TALE proteins, each bound to a cleavage domain. FokI dimerizes to generate DSBs on the target DNA, therefore activating the repair system to fix the indels (Bhardwaj and Nain 2021a). A spacer sequence that is 12–19 base pairs long separates the two units of TALE proteins (Miller et al. 2011). The TALE protein is constructed by repeat domains, where each repeat contains approximately 34 amino acids. The amino acids are substantially conserved, except in the 12th and 13th positions, which are known as repeat variable diresidues (RVDs) (Bhardwaj and Nain 2021a). In a 5′ to 3′ manner, each repeated sequence binds specifically to an individual nucleotide of the target DNA (Boch et al. 2009). According to biochemical research, the amino acid residue at the 13th position is in charge of uniquely recognizing a nucleotide in the desired DNA, which results in the development of a DNA-protein complex. This complex is stabilized by the amino acid residue at the 12th position (Stella et al. 2013; Deng et al. 2012; Mak et al. 2012). Different experiments have identified four frequent RVDs, NI, NG, HD, and NN, having an affinity for, A, T, C, and G accordingly, conferring target specificity (Mussolino and Cathomen 2012). When compared with ZFNs, TALENs have shown less toxicity and better specificity (Mussolino et al. 2014) (Fig. 2).

5 Clustered Regulatory Interspaced Short Palindromic Repeats (CRISPR/Cas9)

The CRISPR-Cas9 system is crucial for microbial adaptive immune responses (Horvath and Barrangou 2010; Marraffini and Sontheimer 2010). This system provides immunity to the bacteria by cleaving the genomes of invading plasmids and viruses with the help of proteins called Cas. The cleavage made here is a ribonucleic acid (RNA)-guided DNA cleavage (Wiedenheft et al. 2012; Sorek et al. 2013). The foreign DNA is incorporated into the locus of CRISPR and transcribed to crRNA (CRISPR RNA), which subsequently binds to tracrRNA (trans-activating crRNA), which triggers the Cas9 (CRISPR-associated protein 9) protein to degrade the foreign DNA when it binds to the specific sequence (Jinek et al. 2012). According to Doudna, Charpentier, and their colleagues, Cas9 needs a preserved PAM (protospacer-adjacent motif) sequence upward of the crRNA binding region in order to recognize the target sequence (Jinek et al. 2012). This structure is now composed of the Cas9 endonuclease and gRNA (single-guide RNA), which contains the two elements tracrRNA and crRNA (Gaj et al. 2016). The Cas9 endonuclease, tracrRNA, crRNA, and RNase III (ribonuclease III) make up the whole CRISPR/Cas9 complex. (El-Mounadi et al. 2020; Mohanta et al. 2017; Jansing et al. 2019). The assembly of tracrRNA and crRNA forms the sgRNA (Kumar et al. 2019). dsDNA is fragmented by the Cas9 protein (Li et al. 2020b). It consists of two domains, namely, His-Asn-His (HNH) and RuvC-like domains, taking part in cleaving the dsDNA at 3 bp upstream of the PAM sequence (5′ NGG or 5′-NAG) (Jiang and Doudna 2017; Hille and Charpentier 2016; Manghwar et al. 2019). The HNH motif splits the complementary strand of crRNA, whereas the RuvC-like motif breaks the strand opposite to dsDNA, resulting in DSBs that will be fixed later by NHEJ or HDR (Kumar et al. 2019; Jiang and Doudna 2017; Hille and Charpentier 2016; Manghwar et al. 2019; Hille et al. 2018; Liu et al. 2017). The sgRNA, which consists of crRNA and tracrRNA, is a 100-mer RNA sequence. The target DNA identification is mediated by a 20-base-pair-long guide sequence located at its 5′-end, which is followed by the PAM (Liu et al. 2017). The proper binding of the guide and target sequence is mediated by the loop-like structure present at its 3′-end. The generation of cleavage is achieved by RNP (ribonucleoprotein), which is formed by the assembly of sgRNA and Cas9 (Manghwar et al. 2019; Liu et al. 2017). An efficient DNA cleavage is achieved by RNP. The crRNA has a prominent role in identifying the target DNA and also helps locate RNP to the specific target site, therefore permitting it to bind with the target by forming an R-loop-like structure (Manghwar et al. 2019). The two Cas9 motifs, HNH and RuvC-like motifs, are triggered immediately as the R-loop has been generated, therefore causing cleavage and leading to the generation of blunt ends (Hille and Charpentier 2016). As the recognition of the target is completely mediated by gRNA, CRISPR-Cas9 is considered the most significant GE tool as it eliminates the requirement of designing novel proteins for every new target site. One of the constraints of this technique is the requirement that the PAM sequence for the recognition of Cas9 protein, which is necessary for splitting, must be situated just downward of the gRNA target region. The PAM sequence for *Streptococcus pyogenes* Cas9 is 5′-NGG-3′ (5′-NAG-3′ is occasionally permitted) (Hsu et al. 2013; Jiang et al. 2013a; Mali et al. 2013). Investigations have subsequently shed light on the molecular structure of DNA identified by Cas9, revealing that the formation of the heteroduplex between the gRNA and its opposite strand of DNA is contained within the protein among the

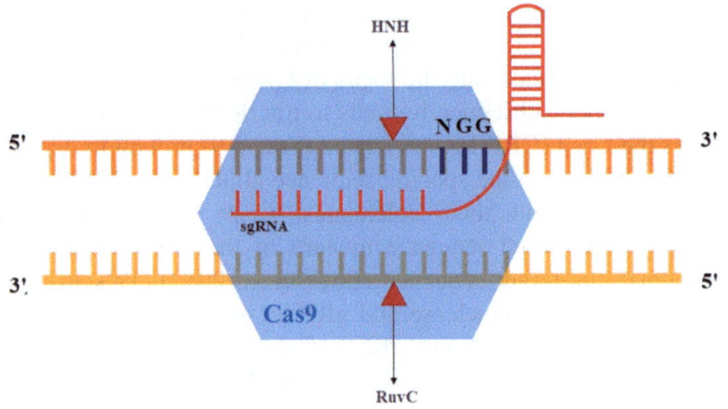

Fig. 3 Clustered regulatory interspaced short palindromic repeat (CRISPR/Cas9) technique

positively charged groove of the HNH and RuvC-like motifs (Nishimasu et al. 2014) and that the arginine-rich motif present in Cas9 facilitates the recognition of PAM (Anders et al. 2014). Since then, Doudna and associates have demonstrated that the displacement of the DNA strand leads to a conformational change in Cas9 that moves the nontarget DNA strand within the RuvC motif and positions the HNH domain close to the target DNA, allowing Cas9 to mediate the splitting of both DNA strands (Jiang et al. 2016) (Fig. 3).

6 Applications of Genome-Editing Techniques in Agriculture

Since GE techniques possess many different applications, it is currently being utilized more often to generate desirable and heritable features. These methods have been used by research teams to develop a number of agricultural systems for resistance to diseases, endurance to drought, salt tolerance, and thermotolerance. These methods support the development of crops that can withstand climatic change and enhance crop quality indicators, including appearance, palatability, nutritional value, and other desired features (Fig. 4 and Table 1).

7 Improving the Abiotic Stress Tolerance of Plants

Abiotic stressors are significant challenges to agricultural output, and it is predicted that climate change will make them much more dangerous in agricultural systems. The enhanced ability of genetically modified plants to function against abiotic conditions could potentially boost the sensitivity of resilient transgenes in altered varieties. Thus, scientists focused on genome-editing approaches to increase crop

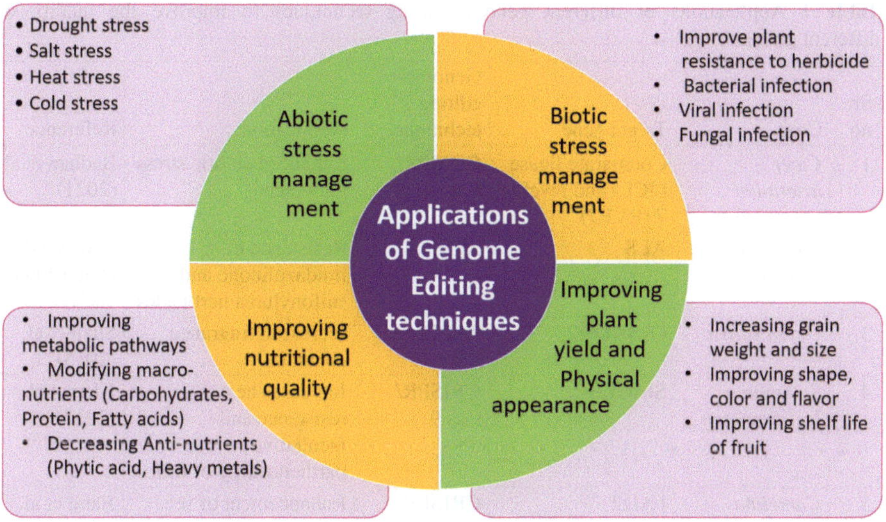

Fig. 4 Applications of genome-editing techniques for sustainable agriculture

resistance to a wide range of climatic conditions, such as drought, salt, cold, extreme temperatures, and other effects of climate change (Abdallah et al. 2021).

Drought is the most prevalent abiotic stress brought on by global warming that seriously harms the plant ecosystem. It is known that crops having a higher degree of ARGOS8 expression are more resilient to drought stress as it is a drought-resistant-reactive gene that has been altered by genome editing. Using CRISPR/Cas9, the GOS2 regulator was employed in maize to substitute the ARGOS8 regulator sequence, enhancing the gene's broad expression and the plant's resistance to drought stress (Shi et al. 2017). Additionally, the ability of *Arabidopsis thaliana* to withstand drought was improved by the CRISPR/Cas modification of the trehalase gene, a crucial component of trehalose metabolism (Nuñez-Muñoz et al. 2021). Moreover, it was reported that by employing the CRISPR/Cas9 technique, the effects of drought were greatly decreased by modifying the SlNPR1 gene in tomatoes (Li et al. 2019).

Another prominent abiotic stress that has significantly decreased the worldwide yield of several crops is salt stress (Sandhu et al. 2017). The plant's capacity to absorb nutrients via its roots and inhibit its development is compromised by the buildup of salts in the soil brought on by seawater drift, transpiration, or evaporation. Therefore, one of the greatest methods for helping plants tolerate salt stress is genome editing. Rice plants that had the OsRAV2 gene altered using CRISPR-Cas9 survived in environments with high salt concentrations (Liu et al. 2020). Under salt stress, the rice OsRR22 knockouts created with CRISPR/Cas9 performed better (Zhang et al. 2019). To increase the ability of rice to withstand salt and drought stress, the salt- and drought-resilient genes have been altered by utilizing CRISPR/Cas technology (Santosh Kumar et al. 2020).

Table 1 Applications of different genome-editing techniques to improve the quality of different plants

Sr. no.	Crop	Target gene	Genome-editing technique	Applications	Reference
1	*Cicer arietinum*	Coumarate ligase (4CL) and reveille 7 (RVE7)	CRISPR/Cas 9	Enhanced abiotic stress tolerance	Badhan et al. (2021)
2	*Nicotiana tabacum*	ALS	ZFN	Resistance to imidazolinone and sulfonylurea herbicides	Townsend et al. (2009)
3	*Oryza sativa*	OsBADH2	TALENs	Increased fragrance content	Shan et al. (2015)
4	*Solanum lycopersicum*	SlAGL6	CRISPR/Cas 9	Increases heat resistance and facultative parthenocarpy of fruit	Klap et al. (2017)
5	*Camelina sativa*	FAD2	CRISPR/Cas 9	Enhancement of seed oil	Jiang et al. (2017))
6	*Arabidopsis thaliana*	BSCTV genome elF4E	CRISPR/Cas 9	Beet severe curly top virus resistance	Ji et al. (2015)
7	*Triticum aestivum*	TaMLO-A1	CRISPR/Cas 9/TALENs	Powdery mildew resistance	Wang et al. (2016)
8	*Zea mays*	IPK1	ZFN	Herbicide tolerance	Shukla et al. (2009)
9	*Solanum tuberosum*	GBSS	CRISPR/Cas 9	High amylopectin starch	Andersson et al. (2017)
10	*Linum usitatissimum*	EPSPS	CRISPR/Cas9 TALEN	Glyphosate tolerance	Sauer et al. (2016)

Moreover, a temperature rise or drop can also significantly impair a plant's ability to grow since plants have an ideal temperature for their development (Fahad et al. 2017). By overexpressing heat shock transcription factors, such as the hsps gene, it is possible to grow plants that can withstand high temperatures while also minimizing cell protein degradation (Debbarma et al. 2019). The capacity of rice to tolerate heat stress is also increased by the protein kinase SAPK6 and the transcriptional factor OsbZIP46CA1 (Chang et al. 2017). Moreover, two rice genes, TIFY1a and TIFY1b, were suppressed using the CRISPR/Cas9 technique to determine their particular function in resistance to cold stress (Huang et al. 2017). Additionally, in the latest research, three rice genes—OsPIN5b, GS3, and OsMYB30—have been modified by using the CRISPR/Cas technique to boost their resistance to cold stress and increase crop yield (Zeng et al. 2020). Additionally, the molecular control of multiple genes implicated in the response to abiotic conditions has also been examined via genetic modification.

8 Improving the Biotic Stress Tolerance of Plants

Plants are vulnerable to a wide range of infections that lead to different ailments. As an alternative, various herbicides, insecticides, and fungicides are utilized, harming the environment both directly and indirectly. Thus, GE methods were employed to generate crops that are resilient to diseases brought on by bacteria, viruses, fungi, nematodes, insects, and weeds (Isman and Grieneisen 2014).

In agriculture, herbicides are frequently applied to prevent the development of weeds surrounding crops. However, there is also a significant risk that the growth of nontarget crops may be adversely affected by the application of pesticides during this time (Varshney et al. 2015). In order to improve plant resistance to herbicides, several genes have been modified by using various genome-editing methods. Among them, ALS and EPSPS are the prominent genes that have been reported to be inhibited by a variety of herbicides (Mazur et al. 1987). Thus, utilizing TALENs and CRISPR/Cas9 tools, the ALS gene was modified in a variety of crops, including maize, potatoes, and watermelon (Svitashev et al. 2015; Sun et al. 2016; Tian et al. 2018). Aside from ALS and EPSPS, there have been limited studies on using GE methods to change the acetyl-CoA carboxylase (ACCase) gene to produce pesticide tolerance in the desired plant (Dong et al. 2021). Moreover, utilizing CRISPR/Cas9 in rice plants has produced impressive outcomes in fighting against disease. The SWEET gene family in various plants encodes sucrose transporters, which are used by the majority of pathogens (Jiang et al. 2013b). To build resistance to bacterial leaf blight, CRISPR/Cas9 was used in two trials to target the promoter region of a few OsSWEET genes (Oliva et al. 2019). A tomato susceptibility gene called SlDMR6–1 was also modified using CRISPR/Cas9 to provide immunity against a variety of ailments caused by bacteria, oomycetes, and fungus (Thomazella et al. 2021).

Furthermore, CRISPR/Cas9-edited eIF4G, a eukaryotic elongation factor, in rice plants showed greater immunity against the rice tungro virus (Macovei et al. 2018). The begomoviruses, which cause cotton crops to develop the disease known as leaf curl, were also suppressed using CRISPR/Cas9 technology (Kis et al. 2019). Using a variant of the CRISPR/Cas9 system (Fncas9), immunity against the tobacco mosaic virus (TMV) and cucumber mosaic virus (CMV) was produced in the crops *Arabidopsis* and *Nicotiana benthamiana*. The virus-infected CRISPR-edited plants produced more yields compared to normal plants while leaving no traces of viral proteins (Zhang et al. 2018a). Targeting the mildew-resistance locus (MLO), along with additional homologous loci, also enhanced the susceptibility of other species to different fungal diseases. By concurrently targeting three MLO homologous, TaMLO-D, TaMLO-B, and TaMLO-A, using CRISPR/Cas9, wheat can become more immune against powdery mildew (Wang et al. 2014). Another example is the powdery mildew disease-resilient Tomelo tomatoes, which were produced using CRISPR/Cas9 by focusing on the SlMlo1 gene (Nekrasov et al. 2017).

9 Enhancing the Nutritional Quality of Plants

GE techniques play an important role in combating nutritional deficiencies by enhancing desired nutritional metabolites, decreasing antinutrients, and modifying macronutrient composition (Abdallah et al. 2021). Zhu et al. reported that food having an enhanced level of amylose and resistant starch (RS) lowers the incidence of significant noninfectious diseases and improves human health; however, RS-rich cereal crops are not commonly available (Zhu et al. 2012). The starch-branching enzyme (SBE) gene regulates the production of amylopectin, and studies have shown that crops with SBE mutations have higher concentrations of resistant starch (RS) and amylose, which can improve the nutritional quality of plants (Shimada et al. 2006). Li et al. further developed genetically modified wheat with enhanced starch formulation, structure, and characteristics using CRISPR/Cas9 and controlled the modification of the TaSBEIIa gene of both winter and spring wheat types (Li et al. 2021). Numerous studies have also demonstrated that crop quality is significantly impacted by the amount and proportion of protein content. CRISPR/Cas9 can be utilized to change the storage protein makeup of seeds to regulate their nutritional value. CRISPR/Cas9 technology was used by Li et al. to modify the D hordein gene in a spring wheat variety and produced two mutant lines with lower D hordein levels than the wild type, enhancing crop quality (Li and Xia 2020).

Currently, researchers have done significant work using CRISPR/Cas9 to improve the quality of oil, mostly focusing on a small number of oil crops, including rapeseed, soybean, camelina, etc. By concentrating on the BnaA.FAD2.a (FAD2_Aa) in *Brassica napus*, Okuzaki et al. (2018) were able to increase the concentration of oleic acid, which is ten times more auto-oxidizing stable and preferred for industrial application. Additionally, these genetically engineered oil crops can aid in lowering blood pressure and cholesterol levels in people (Okuzaki et al. 2018). Several studies showed that white rice is a common food in several countries, although it is deficient in provitamin A (mostly β-Carotene). By incorporating a 5.2 Kb carotenoid biosynthetic sequence made up of the coding regions of SSU-crtI and ZmPsy at two genomic-safe harbors in rice, Dong et al. (2020) successfully produced marker-free rice cultivars with an elevated amount of carotenoid in seeds (Dong et al. 2020). In addition, Zhang et al. (2018) focused on the LsGGP2 uORF start codon region. LsGGP2 is a crucial molecule in the production of vitamin C in lettuce, which improves the crop's ability to adapt to oxidative stress (Zhang et al. 2018b). A number of antinutritional factors (ANFs), including phytic acids (PAs), toxic elements, quinones, glycoalkaloids, and free asparagines, have adverse effects on a plant's ability to digest, absorb, and utilize nutrients. Using the CRISPR-Cas9 method, knocking out three of BnITPK's functional paralogs led to high levels of free phosphorus and low levels of PA (Sashidhar et al. 2020).

10 Improving Yield and the Physical Appearance of Plants

This can be determined by the perception that yield typically reflects a quantitative multigenic characteristic whose development is controlled by a number of quantitative trait loci (QTL) (Abdurakhmonov 2016). Additionally, QTL introduction across different varieties challenges standard yield-based choice, which is especially evident when there are closely spaced loci (Sedeek et al. 2019; Abdurakhmonov 2016).

GE is one method of increasing agricultural productivity by removing genes that have a negative impact on yield, such as those that limit grain weight and size (Song et al. 2016; Ma et al. 2016). Using CRISPR/Cas9, the genes responsible for four detrimental yield controllers (Gn1a, DEP1, GS3, and IPA1) were recently removed from the rice cultivar Zhonghua 11. Individual knockout mutants of DEP1, Gn1a, and GS3 all showed enhanced yield characteristics in the T2 stage (Li et al. 2016). Additionally, it has been demonstrated that the combinatorial CRISPR/Cas9 knockdown of the major inhibitory controllers of rice grain mass (TGW6, GW2, and GW5) allowed for a significant increase in grain mass (Xu et al. 2016). According to market preferences, the crops' size and shape can also be optimized using CRISPR/Cas9 technology. The rice and tomato crops supplied the greatest information on the control of fruit size and shape. Five species of japonica rice were essentially eliminated from the first QTL discovered for regulating grain length, GS3 (GRAIN SIZE 3). Across all genetic origins, the T1 progenitors' grain size has grown relative to the normal type (Shen et al. 2018; Yuyu et al. 2020).

Carotenoids, anthocyanins, and polyphenols are the typical components of plant pigments, which primarily influence a plant's color. Therefore, CRISPR/Cas9 may be used to disrupt genes active in the pigment production pathway, changing the color of fruits. For instance, by successfully knocking down SlMYB12, pink-colored tomatoes have been successfully cultivated (Yang et al. 2019). In addition, scientists generated purple and yellow tomatoes, respectively, by selectively focusing on PSY1 and ANT1 (Filler Hayut et al. 2017). The CRISPR/Cas9 method also holds great promise for improving the shelf life of fruits and vegetables, like tomatoes and bananas. Numerous organically generated altered genes, including Nr, alc, rin, nor, and Cnr, may also lengthen shelf life (Wang et al. 2020). One investigation revealed that alc mutation preserved fruit color and smell while significantly extending shelf life (Casals et al. 2011).

11 Challenges and Strategies Associated with Genome-Editing Techniques

For programmable nucleases to be used effectively, binding and cleavage at nonspecific loci must be prevented from causing off-site effects, a type of unintentional genome editing. Off-site cleavage may cause cell damage. When compared with TALENs, ZFNs exhibit higher off-site targeting because they can recognize shorter sequences. Several strategies have been employed to overcome this challenge. Different studies have reported that the activity and selectivity of a single ZFN half-enzyme are greatly improved by the presence of four to six zinc-finger domains. The length of the spacer sequence that maintains the distance between two target sites is another crucial factor. Spacer sequences longer than seven nucleotides reduce the specificity of ZFNs, and longer spacer sequences prevent ZFN dimerization, which causes off-site cleavage (Sood et al. 2013; Pattanayak et al. 2013). To improve ZFN selectivity, zinc-finger nickases (ZFNickases) were originally designed. These nickases cause a nick in an individual strand of the target DNA, resulting in homologous recombination rather than starting the allegedly more error-prone NHEJ healing procedure (Ramirez et al. 2012). Another approach to increasing the specificity of ZFNs is by creating obligate heterodimeric ZFNs. To prevent an unwanted homo-dimerization of FokI, it depends on charge-charge friction (Miller et al. 2007; Doyon et al. 2011), therefore minimizing off-target cleavage (Guo et al. 2010). Delivering ZFNs into cells as proteins is a promising strategy that shows significant promise for enhancing ZFN specificity. Due to the inbuilt cell-penetrating ability of zinc-finger domains, ZFN proteins may facilitate gene editing with less off-target consequences when delivered straight into cells as pure proteins as compared to being generated inside cells from genetic material (Gaj et al. 2012, 2014). Since then, altered ZFN proteins with improved cell penetration have been reported (Liu et al. 2015).

Compared to ZFNs, TALENs' sequence recognition code is more straightforward, which benefits targetability and development. However, TALENs have their own set of challenges (González Castro et al. 2021). Both specificity and binding affinity vary widely among individual RVDs; some exhibit a substantial binding affinity for a particular base but concurrent degradation, whereas others exhibit single-base selectivity but poor binding ability (Deng et al. 2019; Ousterout and Gersbach 2016). Customizable nucleases must exhibit strict specificity toward their DNA targets in order to be relevant for genetic analysis and application. However, complex genomes usually include several copies of the desired DNA target, which may result in inaccurate behavior and cell damage (Klap et al. 2017; Szczepek et al. 2007). In order to address this problem, new TALEN complexes with improved cleavage selectivity and reduced cytotoxicity have been created using structure- and selection-based methods (Jiang et al. 2017; Ji et al. 2015). Increased discrepancies between an on-target and off-target location reduce the potential of off-target events (Modrzejewski et al. 2020). A precise choice of the target sequence is said to be the primary factor in controlling off-target effects (Zhu et al. 2017). The proposed theory aims to estimate the probability of off-target effects based on the incidence of discrepancies between the on-target and off-target regions (Modrzejewski et al. 2020). Another limitation on the selection of TALEN nuclease locations is the necessity for T preceding the 5′-end of the desired sequence. However, the location is often selectable by varying the overall length of the spacer sequence (Nemudryi et al. 2014). Another significant drawback is the technically challenging assembling of several identical repeated sequences needed to create a TALEN array. This procedure is time-consuming and inefficient (Bhardwaj and Nain 2021b; Yamagata

2022). This issue can be solved by implementing tactics that facilitate quicker bespoke TALE assembly. Golden Gate molecular cloning (Cermak et al. 2011), connection-independent cloning approaches (Schmid-Burgk et al. 2013), and high-throughput solid-phase assembly are a few of the strategies (Reyon et al. 2012).

CRISPR/Cas techniques have progressed significantly over the past few years and have shown a lot of potential in a variety of life science research fields. This technique is now viewed as being more accurate, target-specific, and simple to utilize. Despite the great CRISPR developments, a number of problems remain that must be fixed in order to fully mature Cas technologies. The off-target cleavage of Cas9 is considered one of the primary issues associated with the CRISPR-Cas9 approach (Fu et al. 2013). There are several causes for the undesired cleavage of DNA, including an insufficient concentration of Cas9: sgRNA ratio, the existence of PAM sites, and insufficient Cas9 codon optimization (Song et al. 2016). Cleavage at off-target sites might not be a significant issue in plants because performing back-crossing repeatedly can eliminate undesirable mutations. However, backcrossing takes a lot of time and slows down crop improvement. Three distinct techniques, including double nick production, SpCas9, and SpCas9-FokI strategies, can be utilized to reduce off-target cleavage in plants having larger genome sizes but no genome sequence information (Ran et al. 2015; Tsai et al. 2014). In plants having their genomes fully sequenced, a truncated sgRNA approach can be employed to minimize off-target activity (Fu et al. 2014). Additionally, a number of computational tools, like, CRISPR Primer Designer, Cas-OFFinder, CRISPR-Plant, and CRISPR-P, have been applied to develop sgRNAs that are gene-specific with a low incidence of off-target activities (Bae et al. 2014). Finding off-target sites may be accomplished using a variety of next-generation sequencing methods, such as guide-seq, ChIP-seq, and digestome-seq. (Tsai et al. 2015). Targeting fidelity is improved by reducing the length of the gRNA spacer sequence to 17–18 base pairs. This also reduces off-target activity (Fu et al. 2014). Recently, three to four amino acid substitutions have been made to create high-fidelity variants of Cas9. These variants have proven to be quite useful for treating off-target challenges in plants (Kleinstiver et al. 2016). The lack of PAM sequence in the gene loci of interest is another significant problem. Cas12a and SpCas9 are two examples of Cas variants that are currently available and reduce PAM restrictions. These types of developments will give genome editing flexibility for the required precise targets (Kleinstiver et al. 2015). The size of the CRISPR-Cas9 system, which is roughly 4.3 kb, presents another difficulty. This significantly influences how it is used. Other Cas9 homologs from various species of bacteria were created to meet this challenge. One of these bacteria, *Staphylococcus aureus*, was utilized to create the unique CRISPR-Cas9 system known as SaCas9. This involves the binding of sgRNA at a location preceding a 5′-NNGRRT PAM sequence. The SaCas9 has been demonstrated to be efficient (Li et al. 2020a). SaCas9 has economically accessible reagents that enable direct site delivery as premade RNPs, much like other Cas9 homologs. The SaCas9 is only 3.3 kb in size. Fast substrate release prevents additional nuclease activity after the target DNA has been cleaved (Yourik et al. 2019). SaCas9, however, also comes with a drawback. The SaCas9 targeting range is constrained due to the requirement of a lengthy PAM sequence. Luna et al. effectively produced SaCas9-RL and SaCas9-NR, which are superior to the wild-type SaCas9 in terms of range and activity. These mutants have the potential to be useful as therapeutics (Luan et al. 2019).

12 Conclusion

The development of plant agriculture and food production using genome-editing technology may help feed the increasing global population. Plant genome engineering has been revolutionized, and its applications have become more widespread because of its effectiveness, engineering simplicity, and resilience. With the emergence of this genome-editing technology, scientists now can regulate crop-specific features more precisely and successfully. It should be highlighted that GE technology has enormous potential for creating novel crop varieties that are resilient to both biotic and abiotic stressors, along with enhancing food quality and productivity. In the field of plant engineering and molecular genomics, the CRISPR/Cas9 technique has become the most popular and adaptable technology. Furthermore, the results of this technique are readily accepted by humans since there is no insertion of foreign DNA and there are fewer regulatory constraints. In addition, these technologies have been explicitly applied to alter just a single gene for crop enhancement. However, the majority of gene-editing research aimed at crop enhancement is still at the stage of identifying regulatory mechanisms and genomic function. The commercialization of GE crops is also in its early stages. Additionally, not all of the conditions for modifying the plant genome have been satisfied by gene-editing techniques. Since multiple QTLs control a number of quality-related characteristics in crops, additional research will be needed to apply CRISPR/Cas, and changing just one gene might not significantly affect phenotypic characteristics. Therefore, introducing new carrier materials would be beneficial. GE technology is expected to gain popularity over the coming years and play an essential part in improving agricultural output despite the challenges that are still required to be resolved.

Acknowlegements The study was carried out with the financial support of the Ministry of Science and Higher Education of the Russian Federation (no. FENW-2023-0008) and the Strategic Academic Leadership Program of the Southern Federal University "Priority 2030".

References

Abdallah NA, Hamwieh A, Radwan K, Fouad N, Prakash C (2021) Genome editing techniques in plants: a comprehensive review and future prospects toward zero hunger. GM Crops Food 12(2):601–615

Abdallah NA, Prakash CS, McHughen AG (2015) Genome editing for crop improvement: challenges and opportunities. GM Crop Food 6(4):183–205

Abdurakhmonov IY (2016) Genomics era for plants and crop species—advances made and needed tasks ahead. In: Plant genomics. IntechOpen, London

Anders C, Niewoehner O, Duerst A, Jinek M (2014) Structural basis of PAM-dependent target DNA recognition by the Cas9 endonuclease. Nature 513(7519):569–573

Andersson M, Turesson H, Nicolia A, Fält AS, Samuelsson M, Hofvander P (2017) Efficient targeted multiallelic mutagenesis in tetraploid potato (Solanum tuberosum) by transient CRISPR-Cas9 expression in protoplasts. Plant Cell Rep 36:117–128

Badhan S, Ball AS, Mantri N (2021) First report of CRISPR/Cas9 mediated DNA-free editing of 4CL and RVE7 genes in chickpea protoplasts. Int J Mol Sci 22(1):396

Bae KH, Do Kwon Y, Shin HC, Hwang MS, Ryu EH, Park KS et al (2003) Human zinc fingers as building blocks in the construction of artificial transcription factors. Nat Biotechnol 21(3):275–280

Bae S, Park J, Kim JS (2014) Cas-OFFinder: a fast and versatile algorithm that searches for potential off-target sites of Cas9 RNA-guided endonucleases. Bioinformatics 30(10):1473–1475

Beerli RR, Segal DJ, Dreier B, Barbas CF III (1998) Toward controlling gene expression at will: specific regulation of the erbB-2/HER-2 promoter by using polydactyl zinc finger proteins constructed from modular building blocks. Proc Natl Acad Sci 95(25):14628–14633

Bhardwaj A, Nain V (2021a) TALENs—an indispensable tool in the era of CRISPR: a mini review. J Genetic Eng Biotechnol 19(1):1–10

Bhardwaj A, Nain V (2021b) TALENs—an indispensable tool in the era of CRISPR: a mini review. J Genet Eng Biotechnol 19(1):1–10

Boch J, Scholze H, Schornack S, Landgraf A, Hahn S, Kay S et al (2009) Breaking the code of DNA binding specificity of TAL-type III effectors. Science 326(5959):1509–1512

Casals J, Cebolla-Cornejo J, Roselló S, Beltrán J, Casañas F, Nuez F (2011) Long-term postharvest aroma evolution of tomatoes with the alcobaça (alc) mutation. Eur Food Res Technol 233:331–342

Cermak T, Doyle EL, Christian M, Wang L, Zhang Y, Schmidt C et al (2011) Efficient design and assembly of custom TALEN and other TAL effector-based constructs for DNA targeting. Nucleic Acids Res 39(12):e82–e82

Chang Y, Nguyen BH, Xie Y, Xiao B, Tang N, Zhu W et al (2017) Co-overexpression of the constitutively active form of OsbZIP46 and ABA-activated protein kinase SAPK6 improves drought and temperature stress resistance in rice. Front Plant Sci 8:1102

Chaudhary J, Alisha A, Bhatt V, Chandanshive S, Kumar N, Mir Z et al (2019) Mutation breeding in tomato: advances, applicability and challenges. Plan Theory 8(5):128

Choo Y, Klug A (1994) Toward a code for the interactions of zinc fingers with DNA: selection of randomized fingers displayed on phage. Proc Natl Acad Sci 91(23):11163–11167

Debbarma J, Sarki YN, Saikia B, Boruah HPD, Singha DL, Chikkaputtaiah C (2019) Ethylene response factor (ERF) family proteins in abiotic stresses and CRISPR–Cas9 genome editing of ERFs for multiple abiotic stress tolerance in crop plants: a review. Mol Biotechnol 61(2):153–172

Deng P, Carter S, Fink K (2019) Design, construction, and application of transcription activation-like effectors. Methods Mol Biol 1937:47–58

Deng D, Yan C, Pan X, Mahfouz M, Wang J, Zhu JK et al (2012) Structural basis for sequence-specific recognition of DNA by TAL effectors. Science 335(6069):720–723

Dong H, Huang Y, Wang K (2021) The development of herbicide resistance crop plants using CRISPR/Cas9-mediated gene editing. Genes 12(6):912

Dong OX, Yu S, Jain R, Zhang N, Duong PQ, Butler C et al (2020) Marker-free carotenoid-enriched rice generated through targeted gene insertion using CRISPR-Cas9. Nat Commun 11(1):1178

Doyon Y, Vo TD, Mendel MC, Greenberg SG, Wang J, Xia DF et al (2011) Enhancing zinc-finger-nuclease activity with improved obligate heterodimeric architectures. Nat Methods 8(1):74–79

Dreier B, Beerli RR, Segal DJ, Flippin JD, Barbas CF (2001) Development of zinc finger domains for recognition of the 5′-ANN-3′ family of DNA sequences and their use in the construction of artificial transcription factors. J Biol Chem 276(31):29466–29478

El-Mounadi K, Morales-Floriano ML, Garcia-Ruiz H (2020) Principles, applications, and biosafety of plant genome editing using CRISPR-Cas9. Front Plant Sci 11:56
Fahad S, Bajwa AA, Nazir U, Anjum SA, Farooq A, Zohaib A et al (2017) Crop production under drought and heat stress: plant responses and management options. Front Plant Sci 8:1147
Filler Hayut S, Melamed Bessudo C, Levy AA (2017) Targeted recombination between homologous chromosomes for precise breeding in tomato. Nat Commun 8(1):15605
Fu Y, Foden JA, Khayter C, Maeder ML, Reyon D, Joung JK, Sander JD (2013) High-frequency off-target mutagenesis induced by CRISPR-Cas nucleases in human cells. Nat Biotechnol 31(9):822–826
Fu Y, Sander JD, Reyon D, Cascio VM, Joung JK (2014) Improving CRISPR-Cas nuclease specificity using truncated guide RNAs. Nat Biotechnol 32(3):279–284
Gaj T, Gersbach CA, Barbas CF (2013) ZFN, TALEN, and CRISPR/Cas-based methods for genome engineering. Trends Biotechnol 31(7):397–405
Gaj T, Guo J, Kato Y, Sirk SJ, Barbas CF III (2012) Targeted gene knockout by direct delivery of zinc-finger nuclease proteins. Nat Methods 9(8):805–807
Gaj T, Liu J, Anderson KE, Sirk SJ, Barbas CF III (2014) Protein delivery using Cys2–His2 zinc-finger domains. ACS Chem Biol 9(8):1662–1667
Gaj T, Sirk SJ, Shui SL, Liu J (2016) Genome-editing technologies: principles and applications. Cold Spring Harb Perspect Biol 8(12):a023754
Ghosh D, Kumar A, Sinha N (2021) Targeted genome editing: a new era in molecular biology. In: Advances in animal genomics. Academic, Cambridge, MA, pp 75–89
González Castro N, Bjelic J, Malhotra G, Huang C, Alsaffar SH (2021) Comparison of the feasibility, efficiency, and safety of genome editing technologies. Int J Mol Sci 22(19):10355
Guo J, Gaj T, Barbas CF III (2010) Directed evolution of an enhanced and highly efficient FokI cleavage domain for zinc finger nucleases. J Mol Biol 400(1):96–107
Herman RA, Fedorova M, Storer NP (2019) Will following the regulatory script for GMOs promote public acceptance of gene-edited crops? Trends Biotechnol 37(12):1272–1273
Hille F, Charpentier E (2016) CRISPR-Cas: biology, mechanisms and relevance. Philos Trans R Soc Lond B Biol Sci 371(1707):20150496
Hille F, Richter H, Wong SP, Bratovič M, Ressel S, Charpentier E (2018) The biology of CRISPR-Cas: backward and forward. Cell 172(6):1239–1259
Horvath P, Barrangou R (2010) CRISPR/Cas, the immune system of bacteria and archaea. Science 327(5962):167–170
Hsu PD, Scott DA, Weinstein JA, Ran FA, Konermann S, Agarwala V et al (2013) DNA targeting specificity of RNA-guided Cas9 nucleases. Nat Biotechnol 31(9):827–832
Huang X, Zeng X, Li J, Zhao D (2017) Construction and analysis of tify1a and tify1b mutants in rice (Oryza sativa) based on CRISPR/Cas9 technology. J Agric Biotechnol 25(6):1003–1012
Ishino Y, Shinagawa H, Makino K, Amemura M, Nakata A (1987) Nucleotide sequence of the iap gene, responsible for alkaline phosphatase isozyme conversion in Escherichia coli, and identification of the gene product. J Bacteriol 169(12):5429–5433
Isman MB, Grieneisen ML (2014) Botanical insecticide research: many publications, limited useful data. Trends Plant Sci 19(3):140–145
Jamieson AC, Kim SH, Wells JA (1994) In vitro selection of zinc fingers with altered DNA-binding specificity. Biochemistry 33:5689–5695
Jansing J, Schiermeyer A, Schillberg S, Fischer R, Bortesi L (2019) Genome editing in agriculture: technical and practical considerations. Int J Mol Sci 20(12):2888
Ji X, Zhang H, Zhang Y, Wang Y, Gao C (2015) Establishing a CRISPR–Cas-like immune system conferring DNA virus resistance in plants. Nat Plants 1(10):1–4
Jiang W, Bikard D, Cox D, Zhang F, Marraffini LA (2013a) RNA-guided editing of bacterial genomes using CRISPR-Cas systems. Nat Biotechnol 31(3):233–239
Jiang F, Doudna JA (2017) CRISPR–Cas9 structures and mechanisms. Annu Rev Biophys 46:505–529

Jiang WZ, Henry IM, Lynagh PG, Comai L, Cahoon EB, Weeks DP (2017) Significant enhancement of fatty acid composition in seeds of the allohexaploid, Camelina sativa, using CRISPR/Cas9 gene editing. Plant Biotechnol J 15(5):648–657

Jiang F, Taylor DW, Chen JS, Kornfeld JE, Zhou K, Thompson AJ et al (2016) Structures of a CRISPR-Cas9 R-loop complex primed for DNA cleavage. Science 351(6275):867–871

Jiang W, Zhou H, Bi H, Fromm M, Yang B, Weeks DP (2013b) Demonstration of CRISPR/Cas9/sgRNA-mediated targeted gene modification in Arabidopsis, tobacco, sorghum and rice. Nucleic Acids Res 41(20):e188

Jinek M, Chylinski K, Fonfara I, Hauer M, Doudna JA, Charpentier E (2012) A programmable dual-RNA–guided DNA endonuclease in adaptive bacterial immunity. Science 337(6096):816–821

Joung JK, Sander JD (2013) TALENs: a widely applicable technology for targeted genome editing. Nat Rev Mol Cell Biol 14(1):49–55

Kamburova VS, Salakhutdinov IB, Shermatov SE, Abdurakhmonov IY (2021) Using of genome editing methods in plant breeding. In: Plant breeding-current and future views. IntechOpen, London

Kim YG, Cha J, Chandrasegaran S (1996) Hybrid restriction enzymes: zinc finger fusions to Fok I cleavage domain. Proc Natl Acad Sci 93(3):1156–1160

Kim HJ, Lee HJ, Kim H, Cho SW, Kim JS (2009) Targeted genome editing in human cells with zinc finger nucleases constructed via modular assembly. Genome Res 19(7):1279–1288

Kim YG, Li L, Chandrasegaran S (1994) Insertion and deletion mutants of FokI restriction endonuclease. J Biol Chem 269(50):31978–31982

Kis A, Hamar É, Tholt G, Bán R, Havelda Z (2019) Creating highly efficient resistance against wheat dwarf virus in barley by employing CRISPR/Cas9 system. Plant Biotechnol J 17(6):1004

Klap C, Yeshayahou E, Bolger AM, Arazi T, Gupta SK, Shabtai S et al (2017) Tomato facultative parthenocarpy results from Sl AGAMOUS-LIKE 6 loss of function. Plant Biotechnol J 15(5):634–647

Kleinstiver BP, Prew MS, Tsai SQ, Topkar VV, Nguyen NT, Zheng Z et al (2015) Engineered CRISPR-Cas9 nucleases with altered PAM specificities. Nature 523(7561):481–485

Kleinstiver BP, Tsai SQ, Prew MS, Nguyen NT, Welch MM, Lopez JM et al (2016) Genome-wide specificities of CRISPR-Cas Cpf1 nucleases in human cells. Nat Biotechnol 34(8):869–874

Kumar R, Kaur A, Pandey A, Mamrutha HM, Singh GP (2019) CRISPR-based genome editing in wheat: a comprehensive review and future prospects. Mol Biol Rep 46(3):3557–3569

Li J, Jiao G, Sun Y, Chen J, Zhong Y, Yan L et al (2021) Modification of starch composition, structure and properties through editing of TaSBEIIa in both winter and spring wheat varieties by CRISPR/Cas9. Plant Biotechnol J 19(5):937–951

Li M, Li X, Zhou Z, Wu P, Fang M, Pan X et al (2016) Reassessment of the four yield-related genes Gn1a, DEP1, GS3, and IPA1 in rice using a CRISPR/Cas9 system. Front Plant Sci 7:377

Li R, Liu C, Zhao R, Wang L, Chen L, Yu W et al (2019) CRISPR/Cas9-mediated SlNPR1 mutagenesis reduces tomato plant drought tolerance. BMC Plant Biol 19(1):1–13

Li A, Tanner MR, Lee CM, Hurley AE, De Giorgi M, Jarrett KE et al (2020a) AAV-CRISPR gene editing is negated by pre-existing immunity to Cas9. Mol Ther 28(6):1432–1441

Li S, Xia L (2020) Precise gene replacement in plants through CRISPR/Cas genome editing technology: current status and future perspectives. aBIOTECH 1:58–73

Li H, Yang Y, Hong W, Huang M, Wu M, Zhao X (2020b) Applications of genome editing technology in the targeted therapy of human diseases: mechanisms, advances and prospects. Signal Transduct Target Ther 5(1):1

Liu J, Gaj T, Wallen MC, Barbas CF (2015) Improved cell-penetrating zinc-finger nuclease proteins for precision genome engineering. Mol Ther Nucleic Acids 4:e232

Liu Q, Segal DJ, Ghiara JB, Barbas CF III (1997) Design of polydactyl zinc-finger proteins for unique addressing within complex genomes. Proc Natl Acad Sci 94(11):5525–5530

Liu X, Wu D, Shan T, Xu S, Qin R, Li H et al (2020) The trihelix transcription factor OsGTγ-2 is involved adaption to salt stress in rice. Plant Mol Biol 103:545–560

Liu X, Wu S, Xu J, Sui C, Wei J (2017) Application of CRISPR/Cas9 in plant biology. Acta Pharm Sin B 7(3):292–302

Liu Q, Yang F, Zhang J, Liu H, Rahman S, Islam S et al (2021) Application of CRISPR/Cas9 in crop quality improvement. Int J Mol Sci 22(8):4206
Luan B, Xu G, Feng M, Cong L, Zhou R (2019) Combined computational–experimental approach to explore the molecular mechanism of SaCas9 with a broadened DNA targeting range. J Am Chem Soc 141(16):6545–6552
Lusser M, Parisi C, Plan D, Rodríguez-Cerezo E (2012) Deployment of new biotechnologies in plant breeding. Nat Biotechnol 30(3):231–239
Ma L, Kong F, Sun K, Wang T, Guo T (2021) From classical radiation to modern radiation: past, present, and future of radiation mutation breeding. Front Public Health 9:768071
Ma X, Zhu Q, Chen Y, Liu YG (2016) CRISPR/Cas9 platforms for genome editing in plants: developments and applications. Mol Plant 9(7):961–974
Macovei A, Sevilla NR, Cantos C, Jonson GB, Slamet-Loedin I, Čermák T et al (2018) Novel alleles of rice eIF4G generated by CRISPR/Cas9-targeted mutagenesis confer resistance to Rice tungro spherical virus. Plant Biotechnol J 16(11):1918–1927
Mak ANS, Bradley P, Cernadas RA, Bogdanove AJ, Stoddard BL (2012) The crystal structure of TAL effector PthXo1 bound to its DNA target. Science 335(6069):716–719
Mali P, Aach J, Stranges PB, Esvelt KM, Moosburner M, Kosuri S et al (2013) CAS9 transcriptional activators for target specificity screening and paired nickases for cooperative genome engineering. Nat Biotechnol 31(9):833–838
Manghwar H, Lindsey K, Zhang X, Jin S (2019) CRISPR/Cas system: recent advances and future prospects for genome editing. Trends Plant Sci 24(12):1102–1125
Marraffini LA, Sontheimer EJ (2010) CRISPR interference: RNA-directed adaptive immunity in bacteria and archaea. Nat Rev Genet 11(3):181–190
Mazur BJ, Chui CF, Smith JK (1987) Isolation and characterization of plant genes coding for acetolactate synthase, the target enzyme for two classes of herbicides. Plant Physiol 85(4):1110–1117
Miller JC, Holmes MC, Wang J, Guschin DY, Lee YL, Rupniewski I et al (2007) An improved zinc-finger nuclease architecture for highly specific genome editing. Nat Biotechnol 25(7):778–785
Miller JC, Tan S, Qiao G, Barlow KA, Wang J, Xia DF et al (2011) A TALE nuclease architecture for efficient genome editing. Nat Biotechnol 29(2):143–148
Modrzejewski D, Hartung F, Lehnert H, Sprink T, Kohl C, Keilwagen J, Wilhelm R (2020) Which factors affect the occurrence of off-target effects caused by the use of CRISPR/Cas: a systematic review in plants. Front Plant Sci 11:574959
Mohanta TK, Bashir T, Hashem A, Abd Allah EF, Bae H (2017) Genome editing tools in plants. Genes 8(12):399
Moscou MJ, Bogdanove AJ (2009) A simple cipher governs DNA recognition by TAL effectors. Science 326(5959):1501–1501
Mussolino C, Alzubi J, Fine EJ, Morbitzer R, Cradick TJ, Lahaye T et al (2014) TALENs facilitate targeted genome editing in human cells with high specificity and low cytotoxicity. Nucleic Acids Res 42(10):6762–6773
Mussolino C, Cathomen T (2012) TALE nucleases: tailored genome engineering made easy. Curr Opin Biotechnol 23(5):644–650
Nekrasov V, Wang C, Win J, Lanz C, Weigel D, Kamoun S (2017) Rapid generation of a transgene-free powdery mildew resistant tomato by genome deletion. Sci Rep 7(1):1–6
Nemudryi AA, Valetdinova KR, Medvedev SP, Zakian SM (2014) TALEN and CRISPR/Cas genome editing systems: tools of discovery. Acta Naturae 6(3 (22)):19–40
Ni Z, Han Q, He YQ, Huang S (2018) Application of genome-editing technology in crop improvement. Cereal Chem 95(1):35–48
Nishimasu H, Ran FA, Hsu PD, Konermann S, Shehata SI, Dohmae N et al (2014) Crystal structure of Cas9 in complex with guide RNA and target DNA. Cell 156(5):935–949
Nuñez-Muñoz L, Vargas-Hernández B, Hinojosa-Moya J, Ruiz-Medrano R, Xoconostle-Cázares B (2021) Plant drought tolerance provided through genome editing of the trehalase gene. Plant Signal Behav 16(4):1877005

Okuzaki A, Ogawa T, Koizuka C, Kaneko K, Inaba M, Imamura J, Koizuka N (2018) CRISPR/Cas9-mediated genome editing of the fatty acid desaturase 2 gene in Brassica napus. Plant Physiol Biochem 131:63–69

Oliva R, Ji C, Atienza-Grande G, Huguet-Tapia JC, Perez-Quintero A, Li T et al (2019) Broad-spectrum resistance to bacterial blight in rice using genome editing. Nat Biotechnol 37(11):1344–1350

Oshunsanya SO, Nwosu NJ, Li Y (2019) Abiotic stress in agricultural crops under climatic conditions. In: Sustainable agriculture, forest and environmental management. Springer, Cham, pp 71–100

Ousterout DG, Gersbach CA (2016) The development of TALE nucleases for biotechnology. Methods Mol Biol 1338:27–42

Pattanayak V, Lin S, Guilinger JP, Ma E, Doudna JA, Liu DR (2013) High-throughput profiling of off-target DNA cleavage reveals RNA-programmed Cas9 nuclease specificity. Nat Biotechnol 31(9):839–843

Pavletich NP, Pabo CO (1991) Zinc finger-DNA recognition: crystal structure of a Zif268-DNA complex at 2.1 Å. Science 252(5007):809–817

Porteus MH, Carroll D (2005) Gene targeting using zinc finger nucleases. Nat Biotechnol 23(8):967–973

Ramesh P, Mallikarjuna G, Sameena S, Kumar A, Gurulakshmi K, Reddy BV et al (2020) Advancements in molecular marker technologies and their applications in diversity studies. J Biosci 45:1–15

Ramirez CL, Certo MT, Mussolino C, Goodwin MJ, Cradick TJ, McCaffrey AP et al (2012) Engineered zinc finger nickases induce homology-directed repair with reduced mutagenic effects. Nucleic Acids Res 40(12):5560–5568

Ran FA, Cong L, Yan WX, Scott DA, Gootenberg JS, Kriz AJ et al (2015) In vivo genome editing using Staphylococcus aureus Cas9. Nature 520(7546):186–191

Reyon D, Tsai SQ, Khayter C, Foden JA, Sander JD, Joung JK (2012) FLASH assembly of TALENs for high-throughput genome editing. Nat Biotechnol 30(5):460–465

Sandhu M, Sureshkumar V, Prakash C, Dixit R, Solanke AU, Sharma TR et al (2017) RiceMetaSys for salt and drought stress responsive genes in rice: a web interface for crop improvement. BMC Bioinformatics 18:1–11

Santosh Kumar VV, Verma RK, Yadav SK, Yadav P, Watts A, Rao MV, Chinnusamy V (2020) CRISPR-Cas9 mediated genome editing of drought and salt tolerance (OsDST) gene in indica mega rice cultivar MTU1010. Physiol Mol Biol Plants 26:1099–1110

Sashidhar N, Harloff HJ, Potgieter L, Jung C (2020) Gene editing of three BnITPK genes in tetraploid oilseed rape leads to significant reduction of phytic acid in seeds. Plant Biotechnol J 18(11):2241–2250

Sauer NJ, Mozoruk J, Miller RB, Warburg ZJ, Walker KA, Beetham PR et al (2016) Oligonucleotide-directed mutagenesis for precision gene editing. Plant Biotechnol J 14(2):496–502

Savary S, Willocquet L, Pethybridge SJ, Esker P, McRoberts N, Nelson A (2019) The global burden of pathogens and pests on major food crops. Nature Ecol Evol 3(3):430–439

Schmid-Burgk JL, Schmidt T, Kaiser V, Höning K, Hornung V (2013) A ligation-independent cloning technique for high-throughput assembly of transcription activator–like effector genes. Nat Biotechnol 31(1):76–81

Sedeek KE, Mahas A, Mahfouz M (2019) Plant genome engineering for targeted improvement of crop traits. Front Plant Sci 10:114

Segal DJ, Dreier B, Beerli RR, Barbas CF III (1999) Toward controlling gene expression at will: selection and design of zinc finger domains recognizing each of the 5′-GNN-3′ DNA target sequences. Proc Natl Acad Sci 96(6):2758–2763

Shan Q, Zhang Y, Chen K, Zhang K, Gao C (2015) Creation of fragrant rice by targeted knockout of the Os BADH 2 gene using TALEN technology. Plant Biotechnol J 13(6):791–800

Shen L, Wang C, Fu Y, Wang J, Liu Q, Zhang X et al (2018) QTL editing confers opposing yield performance in different rice varieties. J Integr Plant Biol 60(2):89–93

Shi J, Gao H, Wang H, Lafitte HR, Archibald RL, Yang M et al (2017) ARGOS 8 variants generated by CRISPR-Cas9 improve maize grain yield under field drought stress conditions. Plant Biotechnol J 15(2):207–216

Shimada T, Otani M, Hamada T, Kim SH (2006) Increase of amylose content of sweetpotato starch by RNA interference of the starch branching enzyme II gene (IbSBEII). Plant Biotechnol 23(1):85–90

Shukla VK, Doyon Y, Miller JC, DeKelver RC, Moehle EA, Worden SE et al (2009) Precise genome modification in the crop species Zea mays using zinc-finger nucleases. Nature 459(7245):437–441

Sinha P, Davis J, Saag L, Wanke C, Salgame P, Mesick J et al (2019) Undernutrition and tuberculosis: public health implications. J Infect Dis 219(9):1356–1363

Smith J, Bibikova M, Whitby FG, Reddy AR, Chandrasegaran S, Carroll D (2000) Requirements for double-strand cleavage by chimeric restriction enzymes with zinc finger DNA-recognition domains. Nucleic Acids Res 28(17):3361–3369

Song G, Jia M, Chen K, Kong X, Khattak B, Xie C et al (2016) CRISPR/Cas9: a powerful tool for crop genome editing. Crop J 4(2):75–82

Sood R, Carrington B, Bishop K, Jones M, Rissone A, Candotti F et al (2013) Efficient methods for targeted mutagenesis in zebrafish using zinc-finger nucleases: data from targeting of nine genes using CompoZr or CoDA ZFNs. PLoS One 8(2):e57239

Sorek R, Lawrence CM, Wiedenheft B (2013) CRISPR-mediated adaptive immune systems in bacteria and archaea. Annu Rev Biochem 82:237–266

Stella S, Molina R, Yefimenko I, Prieto J, Silva G, Bertonati C et al (2013) Structure of the AvrBs3–DNA complex provides new insights into the initial thymine-recognition mechanism. Acta Crystallogr D Biol Crystallogr 69(9):1707–1716

Sun Y, Zhang X, Wu C, He Y, Ma Y, Hou H et al (2016) Engineering herbicide-resistant rice plants through CRISPR/Cas9-mediated homologous recombination of acetolactate synthase. Mol Plant 9(4):628–631

Svitashev S, Young JK, Schwartz C, Gao H, Falco SC, Cigan AM (2015) Targeted mutagenesis, precise gene editing, and site-specific gene insertion in maize using Cas9 and guide RNA. Plant Physiol 169(2):931–945

Szczepek M, Brondani V, Büchel J, Serrano L, Segal DJ, Cathomen T (2007) Structure-based redesign of the dimerization interface reduces the toxicity of zinc-finger nucleases. Nat Biotechnol 25(7):786–793

Thomazella DPDT, Seong K, Mackelprang R, Dahlbeck D, Geng Y, Gill US et al (2021) Loss of function of a DMR6 ortholog in tomato confers broad-spectrum disease resistance. Proc Natl Acad Sci 118(27):e2026152118

Tian S, Jiang L, Cui X, Zhang J, Guo S, Li M et al (2018) Engineering herbicide-resistant watermelon variety through CRISPR/Cas9-mediated base-editing. Plant Cell Rep 37(9):1353–1356

Townsend JA, Wright DA, Winfrey RJ, Fu F, Maeder ML, Joung JK, Voytas DF (2009) High-frequency modification of plant genes using engineered zinc-finger nucleases. Nature 459(7245):442–445

Tsai SQ, Wyvekens N, Khayter C, Foden JA, Thapar V, Reyon D et al (2014) Dimeric CRISPR RNA-guided FokI nucleases for highly specific genome editing. Nat Biotechnol 32(6):569–576

Tsai SQ, Zheng Z, Nguyen NT, Liebers M, Topkar VV, Thapar V et al (2015) GUIDE-seq enables genome-wide profiling of off-target cleavage by CRISPR-Cas nucleases. Nat Biotechnol 33(2):187–197

Tuberosa R (2012) Marker-assisted breeding in crops. In: Encyclopedia of sustainability science and technology. Springer, New York, pp 6425–6448

UNICEF (2021) 2021: A year marked by conflict, COVID and climate change. UNICEF, New York

Urnov FD, Rebar EJ, Holmes MC, Zhang HS, Gregory PD (2010) Genome editing with engineered zinc finger nucleases. Nat Rev Genet 11(9):636–646

Varshney S, Khan MIR, Masood A, Per TS, Rasheed F, Khan NA (2015) Contribution of plant growth regulators in mitigation of herbicidal stress. J Plant Biochem Physiol 3(2):4

Wada N, Ueta R, Osakabe Y, Osakabe K (2020) Precision genome editing in plants: state-of-the-art in CRISPR/Cas9-based genome engineering. BMC Plant Biol 20:1–12
Wang Y, Cheng X, Shan Q, Zhang Y, Liu J, Gao C, Qiu JL (2014) Simultaneous editing of three homoeoalleles in hexaploid bread wheat confers heritable resistance to powdery mildew. Nat Biotechnol 32(9):947–951
Wang R, Lammers M, Tikunov Y, Bovy AG, Angenent GC, de Maagd RA (2020) The rin, nor and Cnr spontaneous mutations inhibit tomato fruit ripening in additive and epistatic manners. Plant Sci 294:110436
Wang F, Wang C, Liu P, Lei C, Hao W, Gao Y et al (2016) Enhanced rice blast resistance by CRISPR/Cas9-targeted mutagenesis of the ERF transcription factor gene OsERF922. PLoS One 11(4):e0154027
Wenefrida I, Utomo HS, Linscombe SD (2013) Mutational breeding and genetic engineering in the development of high grain protein content. J Agric Food Chem 61(48):11702–11710
Wiedenheft B, Sternberg SH, Doudna JA (2012) RNA-guided genetic silencing systems in bacteria and archaea. Nature 482(7385):331–338
Wolfe SA, Nekludova L, Pabo CO (2000) DNA recognition by Cys2His2 zinc finger proteins. Annu Rev Biophys Biomol Struct 29(1):183–212
Wu H, Yang WP, Barbas CF 3rd (1995) Building zinc fingers by selection: toward a therapeutic application. Proc Natl Acad Sci 92(2):344–348
Xu R, Yang Y, Qin R, Li H, Qiu C, Li L et al (2016) Rapid improvement of grain weight via highly efficient CRISPR/Cas9-mediated multiplex genome editing in rice. J Genet Genomics 43(8):529–532
Yamagata M (2022) Programmable proteins: target specificity, programmability and future directions. SynBio 1(1):65–76
Yang T, Deng L, Zhao W, Zhang R, Jiang H, Ye Z et al (2019) Rapid breeding of pink-fruited tomato hybrids using the CRISPR/Cas9 system. J Genet Genomics 46(10):505–508
Yourik P, Fuchs RT, Mabuchi M, Curcuru JL, Robb GB (2019) Staphylococcus aureus Cas9 is a multiple-turnover enzyme. RNA 25(1):35–44
Yuyu C, Aike Z, Pao X, Xiaoxia W, Yongrun C, Beifang W et al (2020) Effects of GS3 and GL3. 1 for grain size editing by CRISPR/Cas9 in rice. Rice Sci 27(5):405–413
Zeng Y, Wen J, Zhao W, Wang Q, Huang W (2020) Rational improvement of rice yield and cold tolerance by editing the three genes OsPIN5b, GS3, and OsMYB30 with the CRISPR–Cas9 system. Front Plant Sci 10:1663
Zhang J (2022) Important genomic regions mutate less often than do other regions. Nature 602:38–39
Zhang A, Liu Y, Wang F, Li T, Chen Z, Kong D et al (2019) Enhanced rice salinity tolerance via CRISPR/Cas9-targeted mutagenesis of the OsRR22 gene. Mol Breed 39:1–10
Zhang H, Si X, Ji X, Fan R, Liu J, Chen K et al (2018b) Genome editing of upstream open reading frames enables translational control in plants. Nat Biotechnol 36(9):894–898
Zhang H, Zhang J, Lang Z, Botella JR, Zhu JK (2017) Genome editing—principles and applications for functional genomics research and crop improvement. Crit Rev Plant Sci 36(4):291–309
Zhang T, Zheng Q, Yi X, An H, Zhao Y, Ma S, Zhou G (2018a) Establishing RNA virus resistance in plants by harnessing CRISPR immune system. Plant Biotechnol J 16(8):1415–1423
Zhu C, Bortesi L, Baysal C, Twyman RM, Fischer R, Capell T et al (2017) Characteristics of genome editing mutations in cereal crops. Trends Plant Sci 22(1):38–52
Zhu L, Gu M, Meng X, Cheung SC, Yu H, Huang J et al (2012) High-amylose rice improves indices of animal health in normal and diabetic rats. Plant Biotechnol J 10(3):353–362

Navigating the Path from Lab to Market: Regulatory Challenges and Opportunities for Genome Editing Technologies for Agriculture

Mayla Daiane Correa Molinari, Renata Fuganti Pagliarini, Lilian Hasegawa Florentino, Rayane Nunes Lima, Fabrício Barbosa Monteiro Arraes, Samantha Vieira Abbad, Marcelo Picanço de Farias, Liliane Marcia Mertz-Henning, Elibio Rech, Alexandre Lima Nepomuceno, and Hugo Bruno Correa Molinari

Abstract Over recent decades, an array of molecular tools has been applied in plant genome engineering, including TALENs (transcription activator-like effector nucleases), ZFNs (zinc-finger nucleases), and CRISPR/Cas systems (clustered regularly interspaced short palindromic repeats). At present, CRISPR/Cas systems have caught significant industry attention owing to their cost-effectiveness and precision in genomic modulation, thereby serving as a potent tool in plant science research. Importantly, plants subjected to genome editing via CRISPR/Cas systems might not be classified as genetically modified organisms (GMO), which could streamline their acceptance worldwide. Originally discovered as a defense mechanism against plasmids and invading viruses in bacteria and archaea, the CRISPR/Cas system includes two components: the CRISPR ribonucleic acid (crRNA) and the Cas protein. The crRNA guides the Cas protein to a specific (DNA) target sequence. Once there, the protein cleaves the sequence, thereby impeding replication. In relation to plant genome editing, researchers have modified the crRNA to

Mayla Daiane Correa Molinari and Renata Fuganti Pagliarini contributed equally to the chapter.

M. D. C. Molinari · F. B. M. Arraes · S. V. Abbad · M. P. de Farias · H. B. C. Molinari (✉)
Sempre AgTech/WIN, São Paulo, São Paulo (SP), Brazil
e-mail: hm@sempre.agr.br

R. F. Pagliarini
Picolla Scientific Consulting, Saskatoon, SK, Canada

L. H. Florentino · R. N. Lima · E. Rech
Embrapa Genetic Resources and Biotechnology/National Institute of Science and Technology in Synthetic Biology, Distrito Federal, Brasília, Brazil

L. M. Mertz-Henning · A. L. Nepomuceno
Embrapa Soybean, Londrina, PR, Brazil

J.-T. Chen, S. Ahmar (eds.), *Plant Genome Editing Technologies*, Interdisciplinary Biotechnological Advances,
https://doi.org/10.1007/978-981-99-9338-3_2

target distinct genome sequences, and the Cas protein has been manipulated to function as either an endonuclease or a base editor. The most frequently used enzymes from the Type II CRISPR/Cas system are CRISPR/Cas9 and CRISPR/Cas12a (Cpf1). CRISPR/Cas systems and other genome editing tools harbor immense potential to revolutionize plant breeding and biotechnology. Nevertheless, their use must undergo stringent regulation to ensure safe and responsible application. The future holds promise for plant genome editing, with safety being a paramount concern for crop gene editing. As such, it is vital to perpetuate research and development in this field to fully exploit its potential advantages for plant science and agriculture because, as this technology advances and new tools emerge, it becomes crucial for governments to keep abreast of cutting-edge scientific progress. This awareness allows for a balance between gene editing benefits and the associated safety and ethical considerations.

Keywords CRISPR/Cas, crop breeding, food safety, genetically modified organisms

1 Introduction

In recent years, genome editing technologies, also known as new breeding technologies (NBTs), have emerged as flexible tools that allow researchers to modify deoxyribonucleic acid (DNA) sequences in a rapid and cost-effective manner. The CRISPR/Cas (clustered regularly interspaced short palindromic repeats) methodology has revolutionized genome edition by enabling easy targeting of DNA sequences, promoting rapid advances in fields such as synthetic biology, gene therapy, and agricultural sciences (Gaj et al. 2016).

Concerning plant science, the use of NBTs offers new possibilities for developing cultivars that can overcome the limitations of classical breeding techniques. These technologies eliminate or reduce the risks associated with the first generation of tools used to create genetically modified organisms (GMOs) through transgenic techniques. Although many applications of NBTs are still in the research phase, some products are already commercially available. As new scientific information is disseminated, disinformation and ideological barriers are being overcome, allowing the benefits of products derived from NBTs to reach producers, supermarket shelves, and consumers.

A notable advantage of products resulting from NBTs is the potential to bypass the laborious and costly regulation process that current GMOs are subject to. Several countries, including Brazil, are discussing and implementing new legislation to regulate NBTs. Under these new terms, depending on the type of genetic edition performed, most of these products will be considered nongenetically modified (non-GM), which simplifies and reduces the cost of the deregulation process. This fosters competition in the scientific landscape, allowing new players beyond large multinational companies to emerge and accelerate the time it takes for NBT-derived products to reach producers and consumers.

NBTs are used in distinct sectors for genome modification. In the public sector, NBTs are applied to basic research in crop cultivation, aiming to expand knowledge about environment and crop interactions (Julkaisut n.d.). Public research institutes, such as Embrapa (Brazilian Agricultural Research Corporation), are also involved in introducing and evaluating market-oriented traits.

In the private sector, companies have intensified their studies by incorporating genome editing as an additional tool alongside classical breeding techniques (Jorasch 2020). Larger Research and Development (R&D) companies have diversified their approaches to different cultivars, considering regulations and consumer acceptance of genetically edited products to make them more suitable for specific markets. This has led to the emergence of several start-ups, such as Tropic Bioscience, Benson Hill, Pairwise, Plantedit, Calyxt, SolEdits, and Inari, which focus on applying genome editing technology to enhance cultivars of different species, thereby driving the economies of their home countries and the global market (Europa.eu n.d.). Developing countries like Brazil are rapidly adopting NBTs to enhance their agricultural sector, with domestic private companies like SEMPRE AgTech/WIN investing in and developing innovative technologies aimed at transforming agriculture. These technologies are designed to help farmers overcome unique challenges such as disease outbreaks, water scarcity, and climate change effects. Increased investment and implementation of NBTs are expected to enhance productivity and sustainability in Brazilian agriculture while generating significant economic benefits for society.

This chapter will discuss the main genome editing techniques and provide examples of their application in economically important crops in both research and commercialization phases. It will also address regulatory aspects related to NBT-derived products worldwide, public perception, and future expectations for this technology.

2 CRISPR/Cas Gene Editing

The CRISPR/Cas system is a feature of bacteria's adaptive immunity. Within these organisms, the Type II CRISPR system serves as a safeguard against the intrusion of viral DNA and plasmids by cleaving DNA guided by Cas proteins (Wiedenheft et al. 2012; Sorek et al. 2013). Foreign DNA segments are incorporated into the CRISPR site, which is transcribed into CRISPR ribonucleic acid (crRNA), which attaches to the transactivating crRNA (tracrRNA). This process results in the Cas protein inducing a guided breakdown of the harmful DNA sequence (Jinek et al. 2012). In 2012, a team including Emmanuelle Charpentier and Jennifer Doudna made a discovery that target identification by Cas9 demands a distinct sequence in crRNA, along with a neighboring protospacer-adjacent motif (PAM) preceding the crRNA binding site. From that point, this characteristic has been streamlined for genome alteration and currently only necessitates the Cas variant and a single guide

RNA (gRNA), which comprises elements from both crRNA and tracrRNA (Gaj et al. 2016).

In summary, the CRISPR/Cas9 system involves two key molecules that introduce DNA changes. Operating like "molecular scissors," the Cas9 enzyme snips double-stranded DNA at a particular location, facilitating the insertion or deletion of DNA segments. It operates in conjunction with a predesigned gRNA (guide RNA), approximately 20 bases in length, which contains complementary bases to the genomic target sequence. This gRNA is embedded within a longer RNA sequence that guides Cas9, ensuring precise cutting at the desired genomic site. Hence, the cell recognizes the damaged DNA and attempts to repair it. Therefore, researchers can harness the cell's DNA repair machinery to insert alterations into genes within the target cell's DNA, thus enabling precise editing.

The remarkable versatility of these technologies stems from their capacity to efficiently promote unidirectional double-stranded breaks (DSBs) in DNA. Such interruptions prompt the cellular mechanisms for DNA repair, enabling the implantation of particular mutations at the intended site. This method is employed to execute gene knockout by introducing insertions and/or deletions (INDELs) via nonhomologous end joining (NHEJ). Alternatively, given the availability of a donor template bearing resemblance to the target DNA site, the integration of genes or base correction can be achieved through the homology-directed repair (HDR) pathway (Gaj et al. 2016).

The straightforward nature of programming CRISPR/Cas, along with its precise DNA cleavage mechanism, target recognition capabilities, and the availability of numerous system variants, has led to remarkable developments. This cost-effective and user-friendly method enables precise targeting, editing, modification, regulation, and labeling of genomic loci in various cells and organisms (Doudna and Charpentier 2014).

3 Examples of Application of New Breeding Technologies (NBTs)

NBTs are gaining momentum, and the presence of products developed with these technologies in the agricultural market and on supermarket shelves is expected to increase. The use of NBTs is clearly irreversible, and to ensure the success of these products, reliable regulation and accurate labeling are essential to provide consumers with precise and effective information about edited crops and their related products. Below are some notable examples of successful applications of NBTs in agricultural product development in both research and commercial phases.

As recorded in April 2023, an exploration of the EU-SAGE database (https://www.eu-sage.eu/genome-search) revealed 614 plants that have been modified utilizing the CRISPR/Cas method, with most of these modifications achieved through site-directed nuclease 1 (SDN1) mutagenesis. NBTs have been applied to over 60 distinct plant species. Notably, the number of publications utilizing NBTs has significantly increased since 2015, with a substantial rise in subsequent years (Europa.eu n.d.; Dima et al. 2022; Eu-Sage.eu n.d.). EU-SAGE is a continuously updated database that includes new studies published in scientific journals.

Global data until 2023 indicate that market-oriented applications have encompassed cultivated and ornamental plants, fruits, and flowers (Eu-Sage.eu n.d.). The most extensively edited crops include rice, tomato, corn, wheat, and soybean. However, other crops with smaller cultivation areas, such as canola, potato, tobacco, and barley, as well as various fruit species (strawberry, kiwi, orange, grapefruit, cocoa), and flowers (dandelion, morning glory, orchid, torenia, petunia), have also been targeted for gene editing. Overall, most modified traits fall into agronomic categories, such as growth performance, increased yield, product quality, industrial applications, color and flavor enhancement, herbicide tolerance, extended shelf life, resistance to biotic stresses (fungi, bacteria, viruses), and abiotic stresses (heat, salinity), among others. These traits have been achieved through a combination of approaches (Eu-Sage.eu n.d.; Menz et al. 2020). The CRISPR/Cas9-based tool has been the predominant gene editing methodology used, while other enzymes, like Cas13 and Cas12a (Cpf1), have been employed to a lesser extent. In terms of the quantity of gene-edited items, China and the United States are at the forefront, with several European countries (France, Germany, the Netherlands, Sweden, the United Kingdom, Spain, Hungary, Italy, and Belgium) also actively involved. Brazil, Israel, Saudi Arabia, South Korea, New Zealand, Kenya, Russia, and Turkey are also conducting research on NBTs (Eu-Sage.eu n.d.; Menz et al. 2020; European C et al. 2021).

More specifically, some of the characteristics targeted in economically important crops include early flowering in rice with larger grain size and number, enriched with carotenoids, and disease resistance. Camelina is being modified to alter oil composition, while soybean is being engineered to have higher oil and protein content. Maize is being enhanced for increased grain productivity under both normal and water-deficit conditions. Sugarcane is being developed with higher sucrose content and improved biomass saccharification process. Tomato varieties are being created with early flowering, high lycopene content, and extended shelf life. Fiber-rich wheat is being engineered to have low gluten content to reduce allergenic factors and resistance to fungi. Mustard is being improved for better taste and peanuts for altered oil composition. Potatoes are being modified to have reduced glycoalkaloid content and altered amylopectin levels. Fruit species are also being targeted, with grapevines engineered for fungus resistance and the removal of banana streak virus and strawberries designed to flower multiple times (Eu-Sage.eu n.d.).

Despite research efforts, applications in the precommercial and market stages are still limited (European C et al. 2021). This may be due to the relative novelty of new genomic techniques, particularly those based on CRISPR, and regulatory uncertainties surrounding these techniques in various countries. However, it is expected that many more products will emerge in the future, benefiting end consumers (Europa.eu n.d.; Metje-Sprink et al. 2020).

In the United States, a commercially available soybean cultivar with high oleic acid content in its oil is already on the market. This variety has higher levels of oleic fatty acids and lower levels of linolenic acids, resulting in the increased oxidative stability of the oil when exposed to heat (Demorest et al. 2016). In Japan, a tomato variety with elevated levels of gamma-aminobutyric acid (GABA) was introduced to the market in September 2021. Developed by Sanatech Seed's Sicilian Rouge and approved for commercialization in December 2020, this Japanese tomato claims to offer health benefits as GABA intake is believed to help lower blood pressure, promote relaxation, and enhance concentration (Europa.eu n.d.; Nonaka et al. 2017; Waltz 2022). In Japan, approval was also granted in 2021 for the sale of two fish species edited using CRISPR: tiger pufferfish and goldfish. These edited species, developed by the start-up Regional Fish Institute in Kyoto in collaboration with Kyoto and Kindai Universities, have been enhanced to grow larger and faster than conventional fish. In tiger pufferfish, researchers successfully disrupted the leptin receptor gene, leading the fish to eat more and gain weight more quickly. The edited fish grow 1.9 times faster than their conventional counterparts, reaching market size earlier. In goldfish, researchers deactivated the myostatin protein, allowing the fish to grow approximately 1.2 times larger while maintaining the same amount of food. These developments aim to reduce production costs for fish cultivated in aquaculture systems (Anon 2022).

One interesting fact is that researchers have successfully edited up to 35 alleles simultaneously using CRISPR/Cas9 in wheat. In a study by Sánchez-León et al. (2018), they edited 35 out of 45 copies of the α-gliadin gene. Gliadin is a component of wheat gluten proteins that triggers celiac disease, an autoimmune disorder, in genetically predisposed individuals. By mutating these 35 copies, wheat immunoreactivity was reduced by 85% (Sánchez-León et al. 2018).

In Brazil, several genetically edited organisms have gained approval for commercial use, including economically important crops (Produtos Avaliados n.d.). This includes three soybean cultivars (two developed by GDM Genética do Brasil S.A. and one by Embrapa Soybean), two sugarcane varieties developed by Embrapa Agroenergy, and a maize line by Corteva. The GDM soybean cultivars have lower levels of raffinose and increased drought tolerance. By editing the raffinose gene, a 75% reduction in raffinose and a 50% reduction in stachyose were achieved in the seeds, adding value to the soybean production chain. The use of these edited cultivars in poultry and swine farming is expected to improve the nutritional quality of animal feed, reduce fattening costs, and promote faster weight gain in animals (Produtos Avaliados n.d.).

Embrapa Soybean has also focused on manipulating antinutritional factors in soybeans, given their importance in the feed industry. They have used CRISPR/Cas9 technology to deactivate lectin, one of the antinutritional factors, in an edited cultivar. These antinutritional factors hinder the digestion of soybeans by monogastric animals with reduced stomach capacity, causing gas, intestinal discomfort, and reduced weight gain. Conventional soybeans for animal feed require thermal processing to inactivate these factors, leading to increased production costs. The edited event has the potential to reduce these costs and improve the utilization of soybeans in animal feed. Additionally, it is worth noting that these edited cultivars are considered nontransgenic, allowing for faster commercial release, reduced costs, and easier market access with assured biosafety (Produtos Avaliados n.d.).

Embrapa Agroenergy has developed two varieties of sugarcane through the silencing of the genes *BAHD1* and *BAHD5* (acyl transferases) using the ribonucleoprotein (RNP) method via CRISPR/Cas9. These edited varieties exhibit enhanced saccharification of plant biomass and higher accumulation of sucrose, leading to increased yields in conventional second-generation (2G) ethanol production, animal feed, and the production of valuable chemical compounds (Produtos Avaliados n.d.).

Similarly, DuPont Pioneer (now Corteva) utilized CRISPR/Cas technology to delete a portion of the waxy gene (*Wx1*) in dent corn lines. This resulted in the waxy corn phenotype, which reduces the production of functional enzymes and disrupts the amylose synthesis pathway, ultimately leading to starch with nearly 100% amylopectin content. Waxy corn offers industrial advantages such as improved uniformity, stability, and texture in various food products, as well as enhanced binding properties in paper manufacturing, the textile industry, and adhesive production (GM Waxy Maize n.d.).

Looking ahead, it is expected that more genetically edited cultivars will become available to farmers, providing them with a broader range of options. Furthermore, new products with diverse traits will reach consumers. For instance, in 2023, a nutrient-enriched variety of mustard is anticipated to be introduced to the American market (Europa.eu n.d.).

CISRPR-Cas technology has led to the emergence of distinct start-ups in the agricultural industry. These companies are developing innovative solutions for improving crop yields, disease resistance, improved flavor, improved nutritional content, longer shelf life, and crops that are resistant to specific regional conditions, such as heat, drought, salinity, and other environmental stresses. Moreover, the development of plant genome editing also presents new business opportunities for start-ups in the area of intellectual property. As CRISPR systems become more widely used, there will be an increasing demand for patents and licenses for new genetic sequences and modifications. Start-ups that specialize in intellectual property management and licensing could play an important role in this emerging market. As CRISPR technology continues to develop, we can expect to see more start-ups entering the field and developing innovative products and services for the agricultural and biotech industries.

4 Ensuring the Biosafety of Products Generated by New Breeding Techniques (NBTs)

The regulation of products generated by new breeding techniques (NBTs) remains a highly debated topic globally. The central issue revolves around whether genetically edited organisms (GEOs) should be subjected to the same level of scrutiny as genetically modified organisms (GMOs). This section will delve into the fundamental principles of regulation for edited crops in various countries.

In a groundbreaking decision in 2023, the highest court in the European Union, the European Court of Justice (ECJ), ruled that genetically modified plants are not subject to the same regulatory framework as GMOs. This legally binding decision, based on scientific evidence, aligns with the existing legislation in leading countries like Brazil and the United States. This milestone denotes a significant advancement in European and global plant biotechnology, opening new market opportunities in the agricultural sector. It represents a major stride for the scientific community, revolutionizing the way we cultivate our food and contributing to global food security (Reuters.Com n.d.).

Ongoing efforts are being made to establish unified legislation for NBT products. Two general approaches are under consideration: the first involves creating a distinct regulatory framework for GMOs, defining a GMO based solely on the genetic description of the claimed new characteristic. The methodology employed would not be a determining factor; rather, other genetic and epigenetic alterations resulting from the process would be evaluated based on the characteristics of the product rather than its development history. The second approach aims to develop a specialized regulatory system for GMOs that contributes to sustainability goals. Under this approach, GMOs labeled as "sustainable" would undergo a streamlined evaluation process, while "nonsustainable" GMOs would not receive authorization (Objective 2030 n.d.).

In the United States, the regulation of new plants produced by NBTs falls under various agencies: the Environmental Protection Agency (EPA) oversees plant-incorporated protectants in accordance with the Federal Insecticide, Fungicide, and Rodenticide Act; the Animal and Plant Health Inspection Service (USDA APHIS), a part of the United States Department of Agriculture, administers the risk of plant pests under the Plant Protection Act; and the Food and Drug Administration (FDA) manages food safety as per the Federal Food, Drug, and Cosmetic Act.

These three agencies regulate the characteristics of the products themselves rather than the process used to develop them. The document titled "SECURE" (Sustainable, Ecological, Consistent, Uniform, Responsible, Efficient), compiled by APHIS in 2020, would exempt (not regulate) plants edited with genes that could have been developed through conventional methodologies, subjecting them to the same regulations as existing GMOs. The intent of these exemptions is to bring the oversight of edited plants more in line with the guidelines applicable to traditionally bred crops. Even though these are viewed as "low risk," they carry risks considered to be "manageable by accepted standards" (Crispr-Gene n.d.).

In Canada, the duty of regulating genetically modified plants falls under the jurisdiction of the Canadian Food Inspection Agency (CFIA), known as Plants with Novel Traits (PNTs), to ensure environmental safety. The country's health department, Health Canada, oversees food and feed safety. It evaluates and approves PNTs intended for human or animal consumption and notifies the public before they can be marketed. The CFIA conducts evaluations of each product independently to ascertain if it exhibits new traits, prioritizing the end product over the technique used in its development. Any plants, foods, or feeds that have new traits undergo environmental and safety assessments to obtain approval. Although Canada is moving toward a less restrictive regulation of genetically edited crops, there remains uncertainty regarding which types and how many genetically edited products will require supervision and the level of that supervision. In this context, both agencies are currently proposing new "regulatory guidance" documents focused on PNTs developed through gene-modifying technology (Smyth 2019). In May 2023, Marie-Claude Bibeau, the Minister of Agriculture and Agri-Food, announced that the seed guidelines of the Canadian Food Inspection Agency (CFIA) now permit certain modified plants. The agency has officially published updated guidance for Part V (5) of the Seed Regulations to make it clear which plants require assessment from the CFIA before being released into the environment. So long as the resulting plants do not contain transgenic DNA sequences, CFIA does not require additional assessment (Realagriculture n.d.).

In January 2022, the Chinese Ministry of Agriculture and Rural Affairs introduced new regulations regarding field trials of plants obtained through new breeding techniques (NBTs). These regulations aim to streamline crop improvement efforts for enhanced food safety. Notably, the initial gene editing guidelines specify that upon the successful completion of pilot tests involving edited plants, a production certificate can be obtained, eliminating the need for extensive field trials typically required for genetically modified (GM) plants (ISAAA 2019).

Moreover, it is worth highlighting that Shandong Shunfeng Biotechnology has recently obtained safety approval for the first gene-edited soybeans in China. This significant achievement underscores the increasing adoption of gene editing technology in agriculture to bolster food production. The gene-edited soybean developed by this company features two modified genes that significantly enhance the level of healthy oleic acid in the plant. The safety certificate for this gene-edited soybean has been granted for a period of 5 years, starting from April 21, 2023 [source: Reuters]. This approval represents a remarkable advancement in crop innovation and exemplifies the potential of gene editing in addressing food security challenges.

Japan has chosen a more adaptable approach to its regulations related to edited crops, which are mandated to be registered but do not need safety or environmental assessments unless the plant includes foreign DNA. Cultures and food products originating from NBTs undergo evaluation on an individual basis and necessitate government notification, which includes details about the technique utilized and the genes modified. Moreover, four separate ministries supervise the regulation of genetically modified/edited crops and food in Japan: the Ministry of Agriculture, Forestry and Fisheries (MAFF); the Ministry of Education; the Ministry of Environment (MOE), Culture, Sports, Science and Technology (MEXT); and the Ministry of Health, Labour and Welfare (MHLW). The Food Safety Commission (FSC), an independent entity, conducts safety risk assessments for food and feed on behalf of MHLW and MAFF. Moreover, local governments retain the authority to establish additional regulatory requirements for edited crops. Nevertheless, the recommendations do not encompass labeling requirements for food derived from NBTs (Crispr-Gene n.d.).

In Latin American countries, there is no consensus. Countries like Belize, Bolivia, Mexico, Peru, and Venezuela have strict restrictions on GEOs, while Brazil, Argentina, Chile, Paraguay, Colombia, Ecuador, Guatemala, and Honduras have lower levels of restriction. The establishment of regulatory rules for these products is currently underway in Panama and El Salvador. On the other hand, Costa Rica, the Dominican Republic, Nicaragua, Trinidad and Tobago, and Uruguay do not have defined regulations, while there is no available information on the status of regulation for products obtained through NBTs in French Guiana, Guyana, and Suriname.

In Brazil, all guidelines and technical prerequisites for presenting inquiries to CTNBio about products resulting from NBTs are detailed in Normative Resolution No. 16 (RN16), published on January 15, 2018.

In essence, through an individualized examination, this document establishes whether a product derived from NBTs should be categorized as a GMO or not. For this evaluation, the originating institution must supply information about the original organism and the product, which includes all steps of the methodology and its molecular analysis. In practical terms, products created through directed random mutation at the nonhomologous end-joining site (nonhomologous end joining (NHEJ)), regarded as SDN-1 (site direct nuclease 1 – direct nuclease mutation site), or directed homologous repair (homologous repair directed (HRD)) at the site involving a few nucleotides (SDN-2) that conform to the conditions stipulated in RN16 could be deemed non-GMO (Produtos Avaliados n.d.; Correa et al. n.d.).

Conversely, transgene insertions directed at the site (SDN-3) will most probably be classified as GMOs. If the product is marked as a GMO, the developer must adhere to all biosafety protocols and will only receive approval after a risk assessment by CTNBio. If the product is not categorized as a GMO, it can be registered following the existing procedures.

5 Minimizing the Impact of the Term GMO

According to Tagliabue (2018), the term "GMO" has a detrimental impact and misleads the public. Much of the public's rejection of GM products and by-products stems from the argument that GMOs are associated with a "Frankenfood" trend that can have unintended consequences. However, a closer examination of this disapproval reveals that misinformation is the driving force behind it. First question: Who holds these concerns? Answer: Some members of the public, but certainly not most scientists, are involved in the development of these products. Second question: Who amplifies these concerns? Answer: Antibiotechnology groups that profit from spreading alarmism, along with the media, tend to create a false understanding by equating consistent scientific research results with unfounded alarmism. Third question: What is the basis for these concerns? Answer: There is none as they arise from a major epistemological error.

Furthermore, as stated by the same author, the controversy surrounding GMOs arises from an excessive focus on the processes used to develop the new plant, microbial, and animal varieties, rather than considering environmental and safety impact assessments and the phenotypic characteristics of the resulting products (Tagliabue 2017). In fact, unsatisfactory results can also occur with traditional techniques as numerous non-GMO attempts have been discarded. In general, those who demand endless testing for GMOs and now apply the same pressure to products obtained through NBTs are unable to provide a robust scientific basis for their arguments (Anon 2018).

In the field of agrifood biotechnology, the most frequent objection is that the results of techniques used to obtain GM products and now those derived from NBTs are unpredictable. However, critics always fail to acknowledge what science already knows: there is no crystal ball to predict the outcomes of these technologies. Nonetheless, science relies on previous experiments, studies, analyses, and knowledge, as well as subsequent evaluations, to determine potential flaws and implications and to formulate solutions. In fact, unpredictability is inherent in any creative process (Anon 2018).

Despite the challenge of substantiating the "risks" associated with GMOs and products derived through NBTs, the hurdle of public opinion persists and must be addressed. Nevertheless, the future seems promising. A 2022 study that involved 2000 Americans indicated that an individual's propensity to consciously consume or evade gene-edited food (GEF) is primarily influenced by their views on food, science and technology, trust, and food awareness. The findings also revealed that those who are receptive to gene-edited foods and display a readiness to eat them are generally supportive of technology, possess knowledge about GEFs, and trust the government and industry to monitor the products as they hit the market. Conversely, those skeptical of GEFs, who refuse to consume them and aim to steer clear of GEFs, are more likely to be driven by food ethics and beliefs and have the propensity to hold strong religious beliefs and identify as more politically conservative (Cummings and Peters 2022).

This group, for the most part, distrusts the industry and the government to oversee GEFs and, instead, places more trust in environmental organizations that better represent their values regarding food. An important conclusion of this study was the need to improve consumer-facing communication. The authors believe that as people become more familiar with the development processes, more products enter the commercial market, better communication reaching broader audiences is implemented, and consumer acceptance of products obtained through NBTs will be ensured (Cummings and Peters 2022).

Another study examined 59 manuscripts related to consumer attitudes and willingness to pay for food generated from NBTs and found similar results. The data suggested that large segments of consumers, but not all, are willing to consume and pay for food derived from NBTs, especially if they possess characteristics that consumers perceive as beneficial. The authors believe that due to the still limited commercialization of food derived from NBTs, ongoing research should be conducted as new products become widely available at the retail (Beghin and Gustafson 2021).

In this context, it becomes clear that public acceptance of NBT-derived food is likely to become less contentious, possibly with the launch of a wide range of products derived from GMOs in the consumer market, whether due to increased access to information or the lack of media hype (Fig. 1). Either way, this scenario will make the results of these technologies more readily accessible to the public and producers, contributing to the production, food safety, and economies of producing countries.

6 Regulation of CRISPR Methods

6.1 Introduction

Nowadays, world agriculture has been facing deep challenges; among them, we can highlight the commonly called nine billion people problem (The Economist 2011), which predicts that in 2050, the growth of the global population will demand the

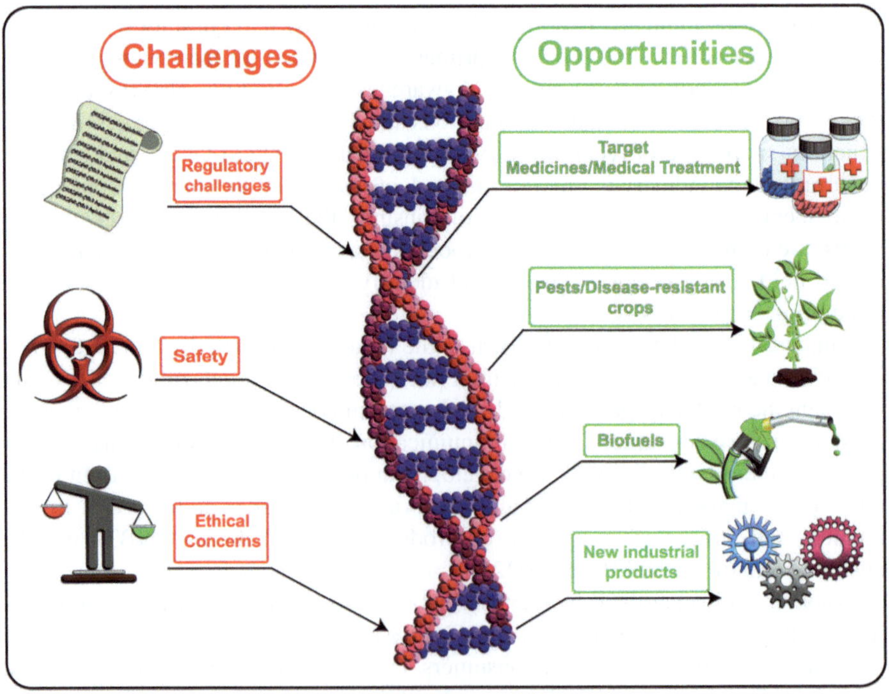

Fig. 1 Challenges and opportunities for both GMO and CRISPR technology public acceptance

duplication of the food production capability (Bullock et al. 2019; Kanchiswamy et al. 2015). The use of biotechnology is a great component of agricultural innovation that contributes to the great success of modern agriculture. Products derived from biotechnology are quickly becoming the biggest global agricultural commodities, which are used to produce human and livestock food, clothes, and fuel for eco-friendly cars.

At present, we are witnessing a significant surge in the cultivation of genetically modified organism (GMO) crops globally, and the confidence in these products has correspondingly increased, with them being deemed safe for both consumption and the environment. Over the past two decades, the number of GM crops has witnessed the most rapid expansion in human history, with an impressive 100-fold increase from 1.7 to 185.6 million hectares of cultivated land between 1996 and 2020 (Brookes 2022). Over the past 2 years, merely 26 countries have accounted for the cultivation of biotech crops. Industrialized countries, including the United States, Australia, Spain, Canada, and Portugal, grow 46% of biotech crops. Among the developing nations, which are responsible for 54% of the biotech crop cultivation area, Argentina, Brazil, and India are among the top five countries with the largest areas for GM crop cultivation. Consequently, only a small number of diverse GM crops have been developed and commercially released, which include insect-resistant cotton, maize, soybean and eggplant, herbicide-tolerant soybean, maize, canola, cotton, alfalfa and sugar beet, and virus-resistant papaya and squash (Brown et al. 2014).

This exponential growth has been taking place despite the troubled regulatory design around the world, from the prohibition of genetically modified products to the statute that includes conventional and biotechnological products under the same regulatory framework (Turnbull et al. 2021). The responsibility for the regulation of biotechnological crops/products has fallen on the legislators, in the last 40 years. So following the exponential growth of technology, there must be a development in the assessment of the risk associated with these innovations (Aven 2016). The risk evaluation standards applied to all agricultural products intended for consumption and usage, including those developed through traditional plant breeding techniques, relate to their safety for human consumption, animal feeding, and environmental protection (Turnbull et al. 2021).

The Cartagena Protocol on Biosafety, edited by Christoph Bail in 2002 under the United Nations (UN), served as an international guideline to direct states and their governments in determining their countries' respective biosafety legislations, establishing risk assessment and management strategies for the release and commercialization of GMO crops. As per the Cartagena Protocol (Article 3 (i)), a GMO or GM crop is defined as a "living modified organism" (LMO). To qualify as a GMO, a plant must adhere to the following criteria: (1) it must contain a novel combination of genomic material, and (2) this material must be inserted via modern biotechnology techniques. These techniques are characterized as the use of either in vitro nucleic acid methods (inclusive of recombinant DNA and the direct injection of nucleic acid into cells or organelles) or the fusion of cells beyond taxonomic families (as per the Cartagena Protocol on Biosafety) (Turnbull et al. 2021; Mackenzie et al. 2003).

In the early 2000s, when the Cartagena Protocols were written, the legal definition of "biotechnologies techniques" had a well-established definition that separates the "Modern" from the more traditional techniques, such as plant breeding, selection, and conventional mutagenesis techniques (Mackenzie et al. 2003). Despite fitting in the first rule of the GMO definition, plant development based on mutation breeding was excluded from the GMO legal charges because it began to be used before the establishment of recombinant DNA methods (Bado et al. 2015). The advent of what are referred to as new breeding techniques (NBTs), such as plant genome editing methods, and the subsequent blurring of lines between naturally occurring and artificially induced genetic modifications pose fresh challenges for regulatory authorities worldwide (Dederer and Hamburger 2019). A summarized timeline of technological and regulatory advances in CRISPR biology and genome editing is presented in Fig. 2.

6.2 Regulating Genetically Modified (GM) Crops Versus Genetically Edited (GE) Crops

The majority of nations adopt regulations based on the Cartagena Protocol's GMO definition or a similar variation within their legislation governing GM crops. Identifying a GMO or a new crop or product activates national regulations, which include relevant risk assessment and management strategies contingent on the product's intended use. Following this strategy, a crop or product intended for human consumption falls under the Regulatory Framework for Food, whereas products meant for animal feed abide by the Regulatory Framework for Animals. Furthermore, crops intended for cultivation comply with the Regulatory Framework for Agriculture and/or Environment (Huesing et al. 2016). Generally, when setting regulations for biotech crops, a distinction is made between approvals for the cultivation of GM plants, for import and export, and for the consumption of GM food and feed products. This differentiation is based on the varying risks associated with cultivation, trade, and consumption, which necessitate distinct regulatory approaches that involve multiple official sectors.

Normally, GM regulations are categorized as product oriented or process oriented (Eckerstorfer et al. 2019; Medvedieva and Blume 2018). In general, process-based regulations for GMOs are more stringent than product-based ones, necessitating additional time to secure regulatory approval (Davison 2010; Ramessar et al. 2008). For instance, US regulation defines GMOs as organisms and products that have been modified or created through genetic engineering. On the other hand, the Canadian Act determines a GMO as a food sourced from an animal, plant, or microorganism that has undergone genetic engineering to exhibit changed characteristics (Araki and Ishii 2015).

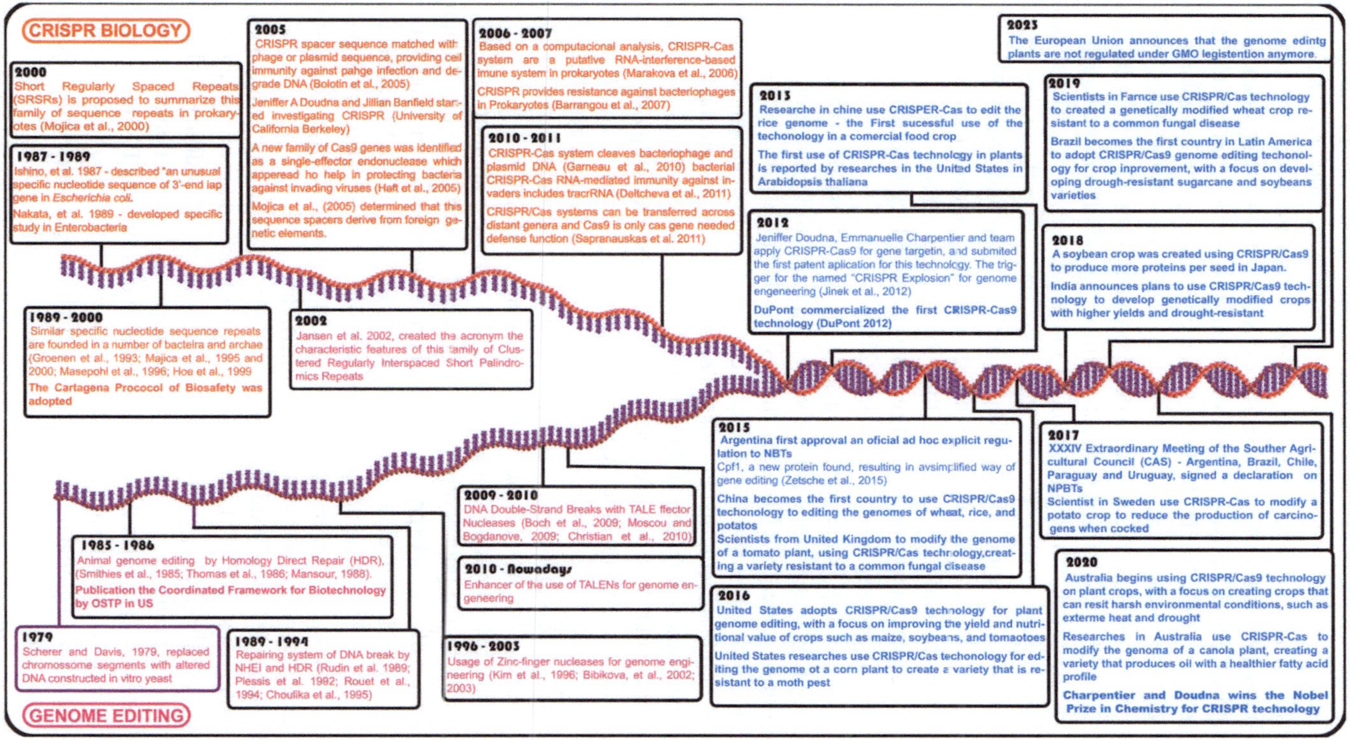

Fig. 2 Key milestones of development and regulation of genome editing techniques and CRISPR biology

Regulations that are process oriented view GM technologies as a novel technique relative to traditional methods, thereby prompting the application of specific legislation. The focus is on the process used to derive the new product, not on the product itself. In such scenarios, GM crops are subjected to regulatory scrutiny, including an extensive procedure based on the scientific evaluation of risks to human health and the environment. Specifically, the EU adopted this regulatory approach to GM crops under Regulation (EC) No 1829/2003. Other countries following this regulatory approach include New Zealand, under the Hazardous Substances and New Organisms Act 1996, and Australia, under the Gene Technology Act 2000 (Araki and Ishii 2015). On the contrary, product-oriented regulations emphasize the unique attributes of the product in comparison to those yielded by traditional breeding methods (McHughen 2016). Certain countries, including Argentina under the National Biosafety Framework, Canada under the Food and Drugs Act, and the USA under 7 CFR Part 340, evaluate health and environmental risks linked to a GMO based on the end product rather than the processes involved. Up until now, Canada is the sole country that has entirely based its GM legislation on the product as opposed to the process (Turnbull et al. 2021).

Globally, regulatory bodies often legislate the evaluation of GMOs, endorsing the scientific viewpoint that the plant phenotype produced from a biotechnological procedure should undergo safety verifications (Wolt et al. 2016).

Worldwide, numerous authoritative bodies partake in the evaluation process for the approval of release requests. For instance, in the United States, the final product type could be examined by the USDA (United States Department of Agriculture), EPA (Environmental Protection Agency), or FDA (Food and Drug Administration). Around the world, governmental regulatory agencies primarily aim to implement rules that safeguard their communities, citizens, and ecosystems. Similarly, legislation concerning plants and crops designated for human and animal consumption and industrial purposes aligns with these goals (Turnbull et al. 2021).

Among various countries, Canada stands out for its evaluation of GM crops based primarily on the phenotype (product-oriented regulation) when defining the regulatory status of a plant expressing a novel trait in terms of environmental release (Turnbull et al. 2021; Wolt et al. 2016; Smyth and Phillips 2014). When the focus is on product versus process approaches, Plant Novel Traits (PNTs) developed from traditional breeding, mutagenesis, transgenesis, or genome editing will undergo the same regulatory approval process (Wolt et al. 2016).

The landscape drastically shifted with the advent of what is known as new plant breeding techniques (NPBTs), causing a fusion of boundaries between naturally and artificially triggered genetic modifications. Genome editing technologies, which consist of a range of newly formulated methods for exact genome modifications in organisms, epitomize the most advanced and promising PNBTs today. Their by-products have rapidly transitioned from research labs to commercial markets. Owing to the surge in global research activities utilizing genome editing techniques, there is a tremendous proliferation in the creation of crops bearing traits coveted by the market. Consequently, the first wave of commercially available genome-edited crops has already been launched by businesses (Waltz 2022). These fast changes give regulators around the world the bigger challenge that they faced in recent years – redefine, or at least, elucidate the scope of regulatory framework application with respect to genome-edited organisms (GEOs) or GE crops and/or products. As per Wasmer's 2019 definition, the phrase "Genome-Edited Organism" (GEO) is used to describe an organism that has been modified through these techniques, resulting in a specimen that is indistinguishable from a conventionally bred variety or a variant that occurs naturally (Wasmer 2019).

The foundation of the genome editing system rests on the use of site-directed nucleases (SDNs) for precision breaks in the targeted DNA region. Currently, there are five genome editing tools in use: (1) oligonucleotide-directed mutagenesis (ODM), (2) zinc-finger nucleases (ZFNs), (3) meganucleases, (4) transcription activator-like effector nucleases (TALENs), and (5) clustered regularly interspaced short palindromic repeat (CRISPR) systems. Generally, regulators categorize these technologies broadly as SDN-1, SDN-2, and SDN-3, a classification first introduced by Lusser and his team (Turnbull et al. 2021; Lusser et al. 2012). To put it succinctly, these categories delineate the following modifications in the DNA: SDN-1 techniques steer the enzyme to a specific genome site to initiate a single double-strand break (DSB), inducing single-point mutations or InDels, or two DSBs to excise a portion of the local DNA. The SDN-2 techniques involve Indels using an external DNA-template sequence, with small donor DNA templates employed to guide the genomic DNA's repair, generating the desired modification. SDN-3 refers to the insertion of longer strands of foreign or native sequences, utilizing a considerably lengthier donor DNA template that is inserted at the target site, similar to traditional recombinant DNA technology (Turnbull et al. 2021; Podevin et al. 2013).

This categorization (SDN1/2/3), or similar, is widely legally used around the world, but just a few countries have adopted novel regulation frameworks specifically for GEO or GE (genome editing) crops (and related technologies). Some countries have already changed their actual regulations, although still most countries are stuck in debates about whether and how to regulate GE and other new techniques that can edit genomes (Menz et al. 2020). Despite these modifications in the legal issues, technological development is so fast that the recently released regulations are already out of date. So the great challenge is the creation of a regulatory regime flexible enough to follow the technology evolution and guarantee legal security for all developers, producers, traders, and consumers (Sprink et al. 2022).

There is a great deal of uncertainty regarding the regulatory frameworks for genome-edited organisms (GEOs), and lively discussions are taking place among politicians, farmers, corporations, seed developers, environmentalists, and other civil society actors about the feasibility of the appropriate regulations. The major question surrounding the regulation of new plant biotechnologies is how legislators will categorize new breeding techniques (NBTs), such as CRISPR/Cas methods, as "genetic modifications" capable of generating GMOs or not. A small number of countries, primarily in the Americas, have adopted approved legislation or issued guidelines that support the use of genome modification strategies. For other countries, the path forward is under discussion, either because the legal categorization is ambiguous or due to a consensus being challenged by the reinvigorated debate on the scope of GMOs.

Genome editing techniques using CRISPR/Cas9 have a great potential to transform agriculture, but the concretion of this potential depends in large part on the regulatory issues' evolution. According to the quick technological progress in molecular biology, and the widely contrasting interests of producers, global agriculture faces an urgent need for the government's new handling of genetically engineered crops and animals. Essentially, this new regulation should stimulate technological development and simultaneously keep governmental control and differentiate between GMOs and non-GMOs (Bratlie et al. 2019; Globus and Qimron 2018).

Over the last 10 years, there has been a significant technological advancement worldwide, which has led to extensive public discourse on whether nontransgenic plants should be viewed as conventional lines and, consequently, whether edited plants should be classified as GMOs (Buchholzer and Frommer 2023). There is a lack of uniformity among various countries regarding these three potential regulatory frameworks. The regulation of gene-edited crops has been the subject of extensive debate within the scientific community and among regulatory bodies (Metje-Sprink et al. 2020; Lusser et al. 2012; Podevin et al. 2013; Ahmad et al. 2021; Friedrichs et al. 2019; Lassoued et al. 2019; Lusser et al. 2011; Lusser and Davies 2013; Smyth 2022). This framework is crucial for fostering open and proactive discussions among the public, researchers, and regulatory agencies about the societal acceptance of GE crops. These crops are recognized for their potential to satisfy the requirements of breeders, stakeholders, and consumers; enhance worldwide food security; and ensure environmental safety. Thus, the global expansion of GE crops is hindered not by technological considerations but rather by societal acceptance (Araki and Ishii 2015).

6.3 Regulatory Approaches in Different Countries

The regulatory frameworks governing genetically modified or gene-edited plants/animals, authorized for commercial use in 34 nations, are largely shaped by two factors: (1) the adherence to the Cartagena Protocol on Biosafety and (2) a categorization model focused on either process or product. Despite the disparities, these two frameworks, namely process oriented and product oriented, often intersect and shape one another. The process-oriented approach tailors regulations based on the techniques employed in the creation of new plant species. Conversely, product-oriented regulation models base their criteria on the unique characteristics of biotech crops, comparing them to traditional varieties. Nevertheless, they both employ an individual assessment method, with overlapping elements, making the legal characterization rather intricate. In addition to these national "regulatory classes," an international law principle – the Cartagena Protocol on Biosafety – is recognized by the countries that have ratified it. This protocol, which has received approval from over 173 countries in the United Nations, was signed in 2000 and came into force on September 11, 2003 (Mackenzie et al. n.d.; Rozas et al. 2022).

The trailblazing countries that first issued regulations, perspectives, or guidelines on genome editing include Chile, Argentina, the United States, and Canada. Subsequently, nations such as Colombia, Brazil, and Paraguay have established regulatory resolutions concerning genomic alterations, taking inspiration from Argentina's and Chile's approaches (Menz et al. 2020). Many legal systems explicitly categorize genome editing techniques that can lead to a GMO separately, which makes it more convenient for plant breeders and producers. Nonetheless, numerous countries still harbor uncertainties regarding the legal position of GE crops or are engaged in ongoing deliberations. Nations like Norway, Switzerland, Russia, India, and several African countries are yet to finalize their regulations on GE crops. Up until 2020, both Sudan and South Africa were cultivating genetically modified plants, with South Africa starting to investigate the legal framework associated with GEOs. In recent times, Nigeria, Burkina Faso, and Ghana have initiated the cultivation of GM crops, whereas Uganda continues discussions on GMO legislation without well-established rules for genome editing. The Africa Biennial Bioscience Communication (ABBC) Symposium in 2019 issued a declaration emphasizing that GEO regulations should pave the way for easier access to genome editing tools (Anon 2019). Subsequently, India too published a document concerning GEOs, suggesting a stratified risk approach toward the regulation of products resulting from genome editing (Anon 2020). The current overview of the status of the global regulatory framework is shown in Fig. 3.

6.3.1 The United States

First, the United States stands out because even its approach toward GM crops differs from that of others – using a unique and selective trigger for GMO regulation. In principle, GMOs could potentially evade the entire regulatory system, provided they do not significantly align with the conditions that activate the GMO regulatory frameworks. The basement purely in a product-based trigger for food regulation, under an informal consultation procedure, is another distinguishing mark in the US regulatory regime. At least between the countries studied, this characteristic of the US regulatory framework is unique. Around the world, GM crops are regulated as either process or product oriented. Europe, for example, regulates GEOs as a process, which has resulted in strict crop regulations in the past. The United States, Canada, and numerous other countries have historically focused on the final product and whether novel characteristics are created (Buchholzer and Frommer 2023; Anon 2023). Within the existing US regulatory structure, the regulatory stance on GEOs remains somewhat ambiguous. This uncertainty stems from the "product-based" methodology, wherein the presence of "products" with specific traits can initiate the enforcement of existing regulations. Consequently, GEOs can bypass these regulations if they do not fit within any of the categories of regulated "products" (Buchholzer and Frommer 2023).

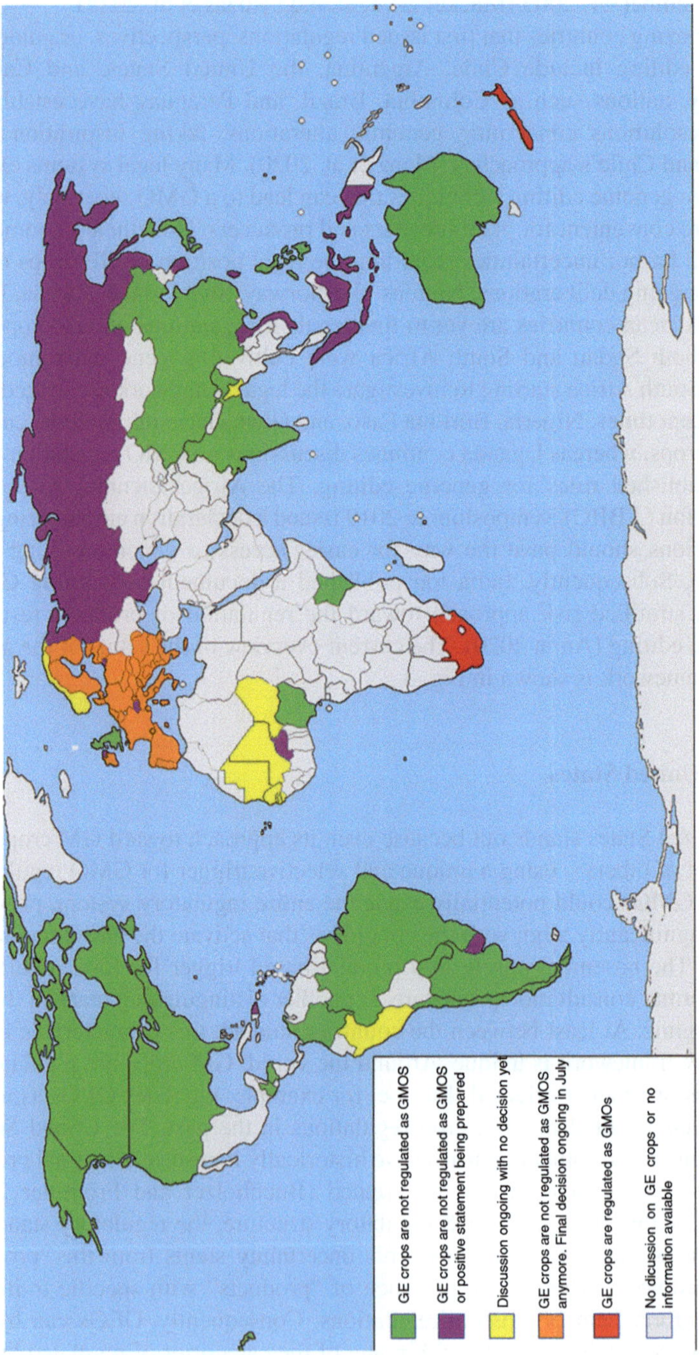

Fig. 3 Overview of the current genome-edited (GE) crops' regulatory landscape. Adapted from ref. (Sprink et al. 2022)

This way, the United States has no formal legislation for GE crops, products, and food; therefore, with no specific federal laws in place, the regulation of GMOs and GEOs falls under three governmental agencies (USDA, FDA, and EPA) (Anon 2023). In 1986, the US Office of Science and Technology Policy (OSTP) developed a piece of legislation intended for the research and products of biotechnology, known as the Coordinated Framework for Regulation of Biotechnology (Ostp 1986). According to this framework, federal regulatory oversight should be based on risk and concentrate on the attributes of the new product rather than the genetic modification technique used for its development. This OSTP framework underwent an update, resulting in the 2017 version of the Coordinated Framework (Us 2017).

Modifications to incorporate the innovative products derived from advanced biotechnologies, such as genome editing, into this Coordinated Framework and the US regulatory structure were achieved through two distinct government initiatives. The initial measure was the release of the "Memorandum on Modernizing the Regulatory System for Biotechnology Products" by the Executive Office of the US President in July 2015 (Holdren et al. 2015). This Memorandum stipulated that the key regulatory bodies for agricultural biotechnology products (USDA, EPA, and FDA) must revise regulatory norms and responsibilities under the Coordinated Framework for the Regulation of Biotechnology. Furthermore, they should devise an adaptable approach to ensure that biotechnology regulations encapsulate future biotechnological products. In line with this objective, the "National Strategy for Modernizing the Regulatory System for Biotechnology Products" was issued in 2016, and in 2017, the "Update to the Coordinated Framework for the Regulation of Biotechnology" was concluded (Us 1986, 2017).

In June 2019, the US President issued the "Executive Order (EO) on Modernizing the Regulatory Framework for Agricultural Biotechnology Products." This order called upon regulatory agencies to review their guidelines concerning the regulation of GEOs (Executive Office of the USPEO 2019). This directive specifically encourages agricultural innovation and regulatory streamlining. It instructs agencies to exercise their existing statutory authority, as deemed appropriate, to exclude low-risk agricultural biotechnology products from unnecessary and excessive regulatory systems (Entine et al. 2021). As a result, the USDA revised its procedures and released a Final Rule for its Biotechnology Regulations (known as the SECURE rule – Sustainable, Ecological, Consistent, Uniform, Responsible, and Efficient – on the Movement of Certain Genetically Engineered Organisms). Simultaneously, the EPA proposed a rule for "Exemptions of Certain Plant-Incorporated Protectants (PIPs) Derived from Newer Technologies" (Hoffman 2021; Anon n.d.-a).

The USDA's Animal and Plant Health Inspection Service (APHIS) is the agency entrusted with safeguarding US agriculture, the environment, and the economy from pests and diseases. It holds regulatory authority over certain biotechnological plants and plant products under the Plant Protection Act (PPA), a mandate requiring the USDA to ensure plant health (Entine et al. 2021). The SECURE excuses (does not regulate) GE plants that otherwise may be created through conventional breeding because the US government considers the changes obtained from gene editing – adding or deleting genes from plant genome – equivalent to conventional breeding.

Despite these regulatory frameworks still relying on the process of genetic engineering as a trigger for regulatory studies, it confirms a focus on the legislation of GE plants. These approaches were the target of many social critics; in response, APHIS says that these exceptions have as a goal to bring a more flexible regulation, which would be aligned with the guidelines for conventionally bred crops, where the concept is not "risk-free" but risks determined to be "manageable by accepted standards."

6.3.2 Canada

Since 1995, Canada has been at the forefront of adopting GM crops, with two cultivars of herbicide-tolerant canola being the first GM crops approved for commercial release. This approval was soon followed by the release of corn and soybean varieties in 1996 and 1997, respectively. As GM crop technologies swiftly moved from labs to open-field trials, regulatory frameworks were simultaneously evolving, reaching standardization when the first GM varieties received approval in the United States and Canada in 1994 and 1995, respectively. Throughout this period, Canada continued to adhere to scientific principles in regulating plants with novel characteristics (Dederer and Hamburger 2019).

Canada distinguishes itself globally with a unique regulatory framework that introduces a novel term to describe a new plant variety, "plant with novel traits" (PNTs). For a new variety to be recognized as a PNT, it must exhibit specific distinguishing features compared to traditional cultivars, irrespective of the technique used to create it, whether it is transgenesis, conventional breeding, or new breeding techniques (NBTs). Consequently, a new crop that is classified as a GMO globally can be termed as a PNT in Canada (Rozas et al. 2022). Beyond this definitional criterion, Canadian regulations incorporate a safety evaluation system. This system emphasizes allergenicity, toxicity, impact on field release, and even effects on organisms other than the PNT. These evaluations are conducted through the Canadian Food Inspection Agency (CFIA) (Smyth 2022; Rozas et al. 2022).

As of now, all commercialized GM crops have undergone safety evaluations and are deemed as PNTs. However, as per the CFIA, PNTs are not solely developed through genetic modification techniques but can arise from various other techniques, such as mutagenesis and somaclonal variation, among others, which some countries might classify as "traditional" breeding. Similar to the United States, the emergence of genome editing techniques has not led to changes in Canada's regulatory frameworks. However, thanks to their product-oriented policy, the system retains flexibility and can adapt to all plants, irrespective of their breeding methods (Menz et al. 2020).

In this regulatory framework, all PNTs are subjected to the same regulatory scrutiny, irrespective of whether they were produced through biotechnology or conventional breeding, including traditional mutagenesis. Every product is considered on a case-by-case basis, according to the novelty of product characteristics, which might vary from other regulated products. Even though there is a somewhat unclear definition of novelty in this product-based legislation, there is a general rule that designates a product as novel if it has a 20% difference in a specific trait compared to a previously established reference product (Menz et al. 2020).

For Canadian regulation, gene editing techniques are treated no differently than other "older" technologies. As such, whenever a new product based on these modern technologies is developed, the Canadian PNT regulations are invoked, which results in an additional regulatory burden. However, despite the situation where it is still premature for CRISPR-developed crops to complete a full field trial and the necessary tasks for compiling a regulatory submission package are not yet compiled, they are expected to reach this stage swiftly. Further information is limited because it is treated as confidential business information by technology development companies. Typically, it takes about 3 years of field studies, from evaluating a greenhouse variety line to collecting the agronomic data required for informing regulators.

The Canadian PNT system does not differentiate among any gene editing technologies. Consequently, submissions are assessed on a case-by-case basis. However, governmental authorities maintain that the PNT system is adequately flexible to handle the increasing number of genome editing crop requests (Dederer and Hamburger 2019). This regulatory approach is similar to the legislative review undertaken by South Africa's Academy of Science, aimed at aligning regulations with submissions for gene-edited varieties (Academy of Science of South Africa 2016).

A noteworthy point is Canada's pioneering attitude toward gene editing research as it is among the first countries to initiate field trials globally. Canada embarked on field trials with genetically engineered flax and canola as early as 1986, thus boasting 37 years of experience in regulating innovative plant breeding technologies (Smyth and McHughen 2008). The initial genetically engineered crops to secure regulatory approval were two herbicide-tolerant canola varieties, granted in March 1995 (Cibus n.d.). Over the following 28 years, Canada has evaluated the risks of genetically engineered crops and has commercially approved 123 crop varieties (Canadian Food Inspection Agency C n.d.). These approvals, which are based on scientific evidence, validate the flexibility and robustness of the Canadian regulatory system. They establish that there is no additional risk associated with the production of genetically engineered crops, which can be different from the risk related to producing other conventional and/or GMO crop varieties (Entine et al. 2021).

Currently, breeders from private companies are more inclined to utilize genome editing techniques, such as CRISPR, compared to public breeders, with rates of 74% and 60%, respectively. Canada stands as one of the countries that extensively employ genome editing technologies. In 2020, it was discovered that 66% of plant breeders were either using CRISPR technologies or planned to use them in the near future (Gleim et al. 2020). The primary reason for choosing CRISPR technology to develop new crop varieties, according to 90% of the surveyed breeders, was the simpler route to regulatory approval. This large percentage of Canadian plant breeders using or planning to use a genome editing technology like CRISPR believe that the PNT regulatory framework needs updating to better represent the status of the new plant breeding techniques (Smyth et al. 2020). The current Canadian PNT regulatory framework is perceived to negatively impact plant breeding as one-third of the researchers surveyed in Canada reported having terminated research projects before determining whether the target variety is a PNT or not. This discussion holds greater importance for the public sector due to the high cost of maintaining dual efforts in breeding programs to determine if a crop is classified as a PNT (Entine et al. 2021).

The international commodity community has expressed its unwillingness to accept the global commercialization of genetically modified crops, rejecting the import of GM products without stringent and rigorous testing studies (Gleim et al. 2020). Other significant commodity-producing nations, such as Brazil, Argentina, and the United States, have embraced the policy that if a crop variety does not contain foreign DNA, it will not be regulated as a GMO variety (Smyth et al. 2020). In this context, Canada's primary challenge is the demand from some commodity organizations for import approval for key commodity export markets for new varieties commercialized in Canada. This process increases the cost of commercialization, necessitating extensive outreach and communication with key markets by certain commodity groups. They aim to illustrate that the regulation of plants with novel traits (PNTs) cannot be the same as for GMOs due to the differing risks involved (Gleim et al. 2020; McDougall 2011).

Another challenge Canadian trade entities face is the fact that most private-sector organizations developing crops function in both Canada and the United States, necessitating equivalent regulatory frameworks. The regulatory disparity between Canada's plant with novel trait (PNT) approach and the US's gene editing (GE) regulation, which is based on the absence of foreign DNA in the final crop, could discourage companies. They may think that the time and cost associated with gaining approval in Canada do not warrant the investment, considering the relatively smaller size of the Canadian seed market (Entine et al. 2021).

According to research conducted by Lassoued et al. (2019) among plant breeders and regulatory experts, the cost and time to develop a gene-edited (GE) crop variety, from the laboratory to the market, could be as much as $24.5 million over a period of 14 years if it is regulated similarly to genetically modified organism (GMO) crop varieties. Conversely, if the GE crop is regulated like conventional varieties, it would cost approximately $10.5 million and take about 5 years to develop (Lassoued et al. 2019).

A delay of 9 years in the commercialization of gene-edited (GE) varieties represents a significant cost and could deter investments. As S. J. Smyth et al. noted in their 2014 study, regulatory delays of just 6 years are enough to significantly reduce the return on investment, to the point where the private sector might decide to stop investing in the development of new varieties (Entine et al. 2021). In response, the last 25–30 years have been marked by Canada and the United States working together to harmonize their regulations, now allowing the same data needed for risk assessment to be submitted to both Canadian and United States regulators, regardless of where genetically edited crops were grown (Entine et al. 2021).

6.3.3 Brazil

Following the United States, Brazil is the second largest country in terms of area cultivated with genetically modified (GM) crops, it being the major exporter of soybeans, maize, and cotton. In 1995, the Brazilian National Congress approved the first biosafety law addressing the controversy surrounding GMOs. In the same year, the Congress also established the "National Biosafety Technical Commission" (CTNBio) tasked with formulating policies for the production, handling, release, and commercialization of GMO crops and their derivative products, overseeing the development, experimental activities, and field trials of GMOs (Gupta et al. 2021). The National Biosafety Law N° 11.105, established on 24 March 2005 and initially crafted in 1995 before being revised in 2015, sets the rules for the management of genetically engineered crops. Its foundations are aimed at promoting progress in various sectors of biosafety and biotechnology, ensuring the protection of life, and human health, along with plant and animal welfare, and preserving the environment, all while aligning with the standards set by the Cartagena Protocol on Biosafety (Entine et al. 2021). In Brazil, CTNBio's management of genetically engineered organisms hinges on risk evaluation. However, the National Biosafety Council can amend these decisions due to socioeconomic factors (Mackelprang and Lemaux 2020).

Originally, new breeding techniques (NBTs) were not encompassed in the earlier legislation. However, a committee was assembled in 2015 to scrutinize the matter of genome modification technologies and bring the existing law up to date. As a result, Brazil's regulatory system became a complex of four institutions – International Biosafety Committees (CIBio), the National Biosafety Council (CNBS), the CTNBio, and appropriate supervisory and registration entities (Gupta et al. 2021). The primary body of this legislation uses the phrase "Precision Breeding Techniques" (PBTs), similar to the early European NBTs. Consequently, Brazil's regulatory structure lacks a clear list of sanctioned NBTs and does not specify the characteristics or methodologies that qualify for non-GMO crop designation. This strategy aligns with the vagueness and initial case-by-case scrutiny seen in Argentina. While a detailed enumeration may lend greater transparency and foresight for current technologies, it could also introduce hurdles for integrating novel technologies (Dederer and Hamburger 2019).

During the XXXIV Extraordinary Meeting of the Southern Agricultural Council (CAS) in 2017, aiming to adjust their regulatory infrastructure to the expansive emergence of new plant breeding technique (NPBT) crops, the administrations of Brazil, Argentina, Chile, Uruguay, and Paraguay all inked a declaration on NPBTs (Turnbull et al. 2021; Benítez Candia et al. 2020). The proclamation stipulated the need to foster regional collaboration and information sharing to prevent discrepancies in the authorization process throughout the region. It also aimed to stimulate the growth of technical expertise in NBT and establish a dynamic regulatory system to ease the development, experimentation, farming, and commerce of genetically engineered crops and products (Rozas et al. 2022). Eight of the 12 Latin American nations consented to a legislative approach involving case-by-case evaluation, which resulted in certain genetically engineered products being exempt from stringent GM regulations (Ahmad et al. 2021; Gatica-Arias 2020). A shared trait among these jurisdictions is the implementation of case-by-case and product-centric assessments for food biosafety and field release safety, in accordance with the authorities designated for these responsibilities in each individual country (Turnbull et al. 2021; Rozas et al. 2022).

Falling under the purview of the Brazilian Biosafety Law and utilizing a case-by-case analysis framework, the CNTBio ratified Normative Resolution N° 16 (RN16) unanimously on 15 January 2018. In addition, the formulation of this Normative drew upon the experiences of other nations in establishing a swifter and more efficient regulatory system that could encompass NBTs (Gupta et al. 2021). RN16 continues to employ a case-by-case analysis methodology to ascertain whether a crop produced by NBTs should be classified as a GMO. This document, founded on various criteria, concludes that products derived from NBTs may not be labeled or regulated as GMOs. However, this classification only applies in situations where there is no recombinant DNA/RNA in the offspring when radiation or chemical exposure induces naturally occurring Indels, and when there are genetic elements present that could have been acquired through conventional breeding (Gupta et al. 2021; Eriksson et al. 2019). RN16 is universally applicable to all categories of organisms, regardless of their developmental stage (Entine et al. 2021).

As per RN16, practically speaking, genetically engineered crops achieved through a site-directed random mutation that connects nonhomologous ends (SDN1 mutation) or based on homologous repair involving one or a few nucleotides (SDN2 mutation) are labeled as non-GMOs and can be documented following conventional product registration procedures (Tripathi et al. 2022). On the other hand, products related to site-directed transgene insertions (SDN3 mutation) are governed as GMOs and must adhere to biosafety prerequisites. They can only gain approval after a comprehensive risk evaluation (Gupta et al. 2021). The entity developing the genetically engineered crop is required to provide details on the method used to create the crop and the procedure implemented for modification to the relevant authorities for proper evaluation. The progressive policy encapsulated in CTNBio's RN16 facilitated the rise of new businesses (start-ups) and bolstered medium to large national corporations in the pursuit of biotechnological products and solutions for agribusiness, industrial sectors, and animal/human health (Hua et al. 2019; Zhao et al. 2019; Li et al. 2020).

In June 2018, the first genetically engineered organism, a *Saccharomyces cerevisiae* strain designed to enhance alcohol production from sugarcane, was assessed and approved under the RN16 framework. This yeast underwent mutation in four genes from another *S. cerevisiae* strain with the use of the CRISPR/Cas9 technique, leaving only mutation in the final product. This method was recognized as non-GMO, given that such outcomes can be achieved by traditional breeding or older mutagenesis techniques, albeit with significantly less precision. In December 2018, the waxy maize genotype was deemed non-GMO and received approval for commercial sale in Brazil. The development of this crop also involved a CRISPR/Cas9 technique to deactivate a specific metabolic pathway responsible for amylose production (Gupta et al. 2021). By 2020, a genome-edited hornless cow was also deemed for regulation as a conventional organism, alongside a second genetically engineered yeast meant for bioethanol production (Gatica-Arias 2020).

6.3.4 Argentina

Argentina is part of an ensemble of six nations that, in 1996, were among the first globally to simultaneously approve the commercialization of GM crops. Since then, Argentina has risen to become the third largest cultivator of GM crops, boasting 24.5 million hectares (Gupta et al. 2021). As a result, Argentina swiftly and proactively implemented a regulatory framework to cover genetically engineered crops and their derivatives, establishing a straightforward yet groundbreaking regulation. This led Argentina to become the third largest producer of GE crops, following the USA and Brazil. In 2015, Argentina was the inaugural nation to introduce an official, dedicated regulation addressing NBTs (inclusive of GE crops), formally stating that GE crops would not be governed under GMO biosafety laws, provided the crop product does not contain foreign DNA/RNA (Resolution N° 173/2015) (Dederer and Hamburger 2019; Schmidt et al. 2020). This regulation is procedural in nature rather than substantive and does not specify rules on risk assessment, authorization, or labeling. This approach, globally known as the "Argentina model," necessitates that a comprehensive report on gene-edited products is submitted on a case-by-case basis to the Argentine Biosafety Commission (CONABIA) to ascertain the exemption. CONABIA considers three factors: (1) the techniques used in the process, (2) genetic modification in the final product, and (3) the absence of a transgene in the final product.

In 2018, a consortium of 13 nations, comprising Argentina, Canada, Australia, Brazil, and the USA, declared their commitment to "support policies that enable agricultural innovation, including genome editing." In 2019, Argentina's Ministry of Production and Work – Directorate of Biotechnology instituted a "Form for the presentation of instances of prior consultation for plants, animals, and microorganisms obtained through NBTs," which accelerated CONABIA's process for determining whether new microorganisms, animals, and crops produced via gene editing techniques should be classified as GMOs (Apps.Fas.Usda.Gov n.d.).

On October 7, 2020, Argentina made history as the first country to approve a genetically engineered (GE) wheat crop when the government authorized HB4 Eco Wheat, developed by the Argentine company Bioceres Crop Solutions, known for its drought resistance. This comprehensive approval (encompassing cultivation, food safety, processing, commercial sale, etc.) is closely tied to its approval in Brazil (Argentina's primary wheat market). Brazil only granted partial approval (for flour derived from wheat) for this crop in November 2021, amid concerns from Brazilian producers. In Argentina, numerous farmers are apprehensive about how this approval could affect the international wheat market of Argentina. The status of HB4 wheat in Argentina remains in flux, owing to Brazil's nonapproval of its importation and grain sale. In response, the Government of Argentina has declared new rules to monitor the cultivation of HB4 wheat in the country (Apps.Fas.Usda. Gov n.d.).

Throughout 2020 and 2021, the Government of Argentina (GOA) persistently updated its biotechnology regulatory framework to enhance alignment with international protocols, strengthening its ties to the Cartagena Protocol on Biosafety. It specifically revamped the combined assessment of GM crops, which presently applies to these crops. In the past, every GM event in a given crop had to be individually analyzed. This remains true, but the process has been streamlined by eliminating irrelevant information from the final assessment. Consequently, combined traits are now evaluated on a case-by-case basis for each new event (Apps.Fas.Usda. Gov n.d.).

6.3.5 European Union

The European Union (EU) upholds the strictest legislation regarding the cultivation and consumption of GMOs within its borders, as evidenced by the fact that less than 1.5% of its agricultural land is used for GM crop cultivation (Davison 2010). Only one transgenic Bt maize event (MON810) has been authorized for commercial cultivation in the EU, and only in Spain and Portugal. Hence, the EU is the largest import market for transgenic soybeans and corn used for animal feed (Dederer and Hamburger 2019; Vieira et al. 2021). The European Union's regulatory framework rigorously adheres to the "Precautionary Principle." This principle dictates that if an action could potentially lead to irreversible environmental or public harm and there is a lack of absolute scientific consensus, the burden of proof falls on the party planning to carry out the potentially harmful act or action. The existing GMO legislation in the EU is primarily based on Directive 2001/18 and Regulations 1829/2003 and 1830/2003.

The EU embraces a process-oriented or precautionary approach toward worldwide regulation. This regulatory process begins with a notification to the appropriate authorities of the country, detailing the specific genetic modification and the method used to achieve it, its intended use (for food, feed, and/or cultivation), which outlines the specific regulatory pathway, as well as safety information (Mackelprang and Lemaux 2020). Next, the European Food Safety Authority (EFSA) provides an opinion based on a risk assessment, which is then passed to the European Commission (EC). The EC's decision is submitted to a committee of representatives from each EU Member State for voting. This is followed by a reconsideration by an Appeal Committee comprised of Member State representatives. At any of these stages, they cannot contest the EFSA's risk assessment without valid scientific evidence. If a deadlock arises, the decision falls back on the EC. Nevertheless, any EU Member State has the right to prohibit the cultivation of a genetically engineered variant within its territory, irrespective of the EC's decision (Mackelprang and Lemaux 2020; Anon n.d.-b). Since the inception of the EFSA in 2003, only one variety, an Amflora potato, has received approval for commercial cultivation in the EU. This potato was edited to generate additional starch for use in the paper industry, but it was on the market for only 2 years before being withdrawn. As of now, only the MON810 event has been approved for cultivation, an authorization that was granted prior to the formation of the EFSA (Smyth and Wesseler 2022).

The strictness of this legislation also permeates the regulatory framework for GE crops. In July 2018, the European Court of Justice (CJEU) declared that "...organisms obtained via mutagenesis techniques/methods are to be classified as genetically modified organisms..." This implies that all genome-edited organisms are to be designated as GMOs and thus subjected to significant regulatory obligations under the EU GMO Directive (Mackelprang and Lemaux 2020). Nonetheless, mutations induced by chemical or radiation methods are exempt from the directive. But according to the European Court of Justice (ECJ), genome editing techniques could not be excluded because they were developed after the enactment of the GMO directive (Friedrichs et al. 2019). The ruling from the European Court of Justice (ECJ) has led to confusion among scientists and has negatively affected agricultural innovation. As a consequence, merely 8% of agriculture-related CRISPR/Cas patents originate from Europe, in contrast to 26% from the US and 60% from China (Schmidt et al. 2020; Martin-Laffon et al. 2019).

Recently, the European Union has begun to reassess its regulatory structure, initiating a study to explore the application of EU legislation on genomic modification (Laaninen 2021). In February 2023, the Luxembourg-based Court of Justice of the European Union (CJEU) concluded that "organisms obtained by the in vitro application of a technique/method of mutagenesis which has conventionally been used in a number of in vivo applications and has a long safety record with regard to those applications are excluded from the scope of that directive..." This statement came after the Commission had gathered feedback and conducted a public consultation in 2022, with the possible adoption of new rules slated for 2023 (Reuters.Com n.d.).

7 Regulation of Different Genome Editing Techniques: How Regulations Differ for Techniques like CRISPR/Cas and Others

The CRISPR-based technology has impacted society, and its guidelines have been widely debated and implemented by various countries. Researchers are exploring the potential risks of genome-edited crops, such as the possibility of unintended gene editing in nontarget species, and are developing safety assessment methods to mitigate these risks. However, concerns about the safety of genome editing in crops remain, particularly given the potential for unintended consequences that may not be apparent until after the crops have been released into the environment. Even in countries that have adopted a more relaxed regulation of plant genome editing, there are still significant concerns related to the ethics and safety of this technology. One of the most important considerations in developing these guidelines is the need to balance the potential benefits of genome editing with the potential risks and ethical concerns (Lusser and Davies 2013; Rozas et al. 2022; Entine et al. 2021).

Overall, CRISPR's guidelines represent an important step forward in regulating plant editing technologies in a responsible and safe manner. However, there are still ongoing debates and challenges around how to implement these guidelines effectively and address the various concerns raised by different stakeholders. Governments around the world have set up regulatory frameworks to ensure that the use of the CRISPR/Cas system is safe and ethical. In the USA, the FDA is responsible for regulating the utilization of CRISPR/Cas systems. Any therapy that uses these systems must go through the FDA's rigorous approval process, which includes extensive testing to ensure the therapy is safe and effective.

8 Ethical and Safety Concerns Related to CRISPR

CRISPR-based systems are a revolutionary technology that has brought about a new era of plant genetic engineering. CRISPR/Cas9 and CRISPR-MAD7 systems raise ethical and safety concerns regarding the manipulation of the genetic engineering of living organisms (Teferra 2021). Researchers and society worldwide are currently discussing ethical issues related to genome editing in crops, such as the environmental impacts and the need for transparency and public engagement (Teferra 2021; Oliver 2014). Overall, the regulation, ethics, and safety concerns related to genome editing technologies in agriculture are complex and multifaceted (Wang et al. 2022). While these technologies offer tremendous potential for improving crop productivity, they also raise important questions about their impacts on human health, and social justice. To ensure responsible innovation and sustainable development, it is essential to evaluate and balance the potential benefits of genome editing with careful consideration of its risks and limitations.

Both CRISPR/Cas9 and CRISPR-MAD7 systems offer advantages in terms of precision and specificity of genomic modulation, reducing the risk of unintended consequences. Conversely, some critics argue that genome editing could have unintended consequences, such as off-target effects, genetic instability, and ecological disruption. One of the main concerns is the potential for off-target effects (Zhang et al. 2020; Eş et al. 2019; Rao et al. 2022). Although CRISPR/Cas9 and CRISPR-MAD7 allow for precise editing, there is always a risk that unintended mutations could occur (Anderson et al. 2016; Patil et al. 2018). This raises concerns about the safety of these crops and the potential impact they could have on the environment. This is a pertinent concern, but structural variations and rearrangements in plant chromosomes may also occur spontaneously with each generation since DNA is not a static element, which results in genetic variability and diversity over time (Anderson et al. 2016). Moreover, the partial or whole genome sequencing (WGS) of genome-edited plants shows that the frequency of off-target mutations is low in plants (Anderson et al. 2016; Schnell et al. 2015).

There are also concerns about the impact of genome editing on biodiversity and the potential for unintended effects on nontarget organisms (Abdul Aziz et al. 2022). For instance, the use of genome modification tools in agriculture could lead to the emergence of new pests and diseases, posing a risk to both crops and ecosystems. Moreover, there is a risk that genome-edited crops could crossbreed with wild plants, leading to unintended consequences, such as the spread of new invasive species (Abdul Aziz et al. 2022). There is also concern that genome-edited crops could have unintended environmental impacts, such as the creation of "superweeds" that are resistant to herbicides (Neve 2018; Collins 2018). Furthermore, there is a possibility that genetically edited crops could have unforeseen effects on human health, such as allergenicity or toxicity (Abdul Aziz et al. 2022; Steiner et al. 2013). CRISPR/Cas9 and CRISPR-MAD7 systems use proteins that could potentially trigger an immune response. This could lead to adverse reactions, which could be dangerous or even life-threatening. While it is theoretically possible that a new toxin, antinutrient, or allergen could be introduced into a plant, the likelihood of this occurrence is low. Furthermore, most genetically engineered crops are massively studied before being commercially released and are frequently monitored. In addition, the use of genome editing in agriculture raises concerns about social justice and the equitable distribution of benefits (Abdul Aziz et al. 2022; Shah et al. 2018). If genome-edited crops are expensive to produce or only available to large corporations, this could exacerbate food insecurity and limit access to healthy food for low-income populations (Shah et al. 2018). Criticizers argue that this technology could be used to create "designed crops," which would have desirable traits such as higher yields but could also result in increased social and economic inequality if only the wealthy can afford them (Abdul Aziz et al. 2022; Spök et al. 2022).

New technologies in agriculture can drive up the cost of food production as seed companies charge premiums for access to their proprietary technology. In other words, genetically modified crops could exacerbate existing inequalities by creating a technological divide between large-scale agricultural producers and small-scale farmers (Spök et al. 2022; Hassoun et al. 2022; Aman Mohammadi et al. 2023). There is a risk that the benefits of CRISPR/Cas9 and CRISPR-MAD7 will be concentrated in the hands of a few large seed companies, limiting the access of smaller farmers to these new technologies. This could create a situation where only wealthy farmers have access to the most advanced crop varieties, while smaller-scale farmers struggle to compete in the market. However, by editing the genome of crops such as soybeans and corn, it is possible to improve their resistance to insects and pathogens, which can increase yields and improve food security (Wang et al. 2022; Rao et al. 2022; Ahmad 2023; Bhatta and Malla 2020). This is particularly important for smallholder farmers in developing countries, who often lack access to expensive pesticides and other chemicals. Also, the CRISPR/Cas-based genome edition has the potential to improve the nutritional content of crops, such as increasing the levels of vitamins and minerals. This is especially important in regions where malnutrition is a significant problem, as it can help address the issue of nutrient deficiencies. Additionally, some people have ethical concerns about the use of genome editing in crops, arguing that it is unnatural and goes against the natural order of things (Spök et al. 2022; Kato-Nitta et al. 2021; Matsuo and Tachikawa 2022; Uddin et al. 2022). However, the potential benefits of using CRISPR/Cas9 and CRISPR-MAD7 to create new and improved crop varieties are enormous. CRISPR technology offers a powerful tool to develop crops that are more productive, nutritious, and sustainable since it will require fewer pesticides, reducing environmental pollution and improving the health of consumers.

This set of concerns of society and regulators is inherent to all emerging technologies that are currently being developed, and a careful balance between reliable sources of information and misinformation from scientific organizations and social media is essential (Araki and Ishii 2015; Teferra 2021; Ahmad 2023; Woźniak-Gientka et al. 2022; Ravikiran et al. 2023; Marette et al. 2021; Gonzalez-Avila et al. 2021). Therefore, as researchers continue to explore the potential benefits of genome editing in crops, it will be important to ensure that appropriate measures are in place to discuss these concerns and to ensure that the benefits of this technology can be realized without unintended consequences.

9 Perspectives

Considering the current scenario of basic and applied research in the development of products via NBTs, in the future, we can envision an increase in varieties being made available to producers and goods reaching consumers' shelves with different characteristics, opening a range of possibilities that range from better crop management and cost reduction to products with higher added value.

The use of CRISPR/Cas9 and CRISPR-MAD7 tools in crop genome editing has paved the way for advancing our understanding of plant biology and developing sustainable solutions for global challenges in agriculture and food production. Thus, CRISPR/Cas technology could provide modern solutions to the challenges of plant cultivation in the face of climate change by developing new varieties with improved nutritional value and by increasing crop yields and resilience to environmental stressors such as heat, cold, and salinity. Moreover, these technologies can also help reduce the use of pesticides and other harmful chemicals, thus minimizing the negative impact of agriculture on the environment. Lastly, scientists have been prospecting new CRISPR/Cas nucleases from metagenome databases, and as technology continues to evolve, it is likely that we will see even more exciting applications and breakthroughs in the field of crop genome editing in the years to come.

An important factor to be considered to ensure that these products do not encounter the resistance and rejection that GMOs faced in the past, albeit to a lesser extent today, is how the "message" will be delivered by scientists and how the media will convey this information to the public. It is crucial to provide a correct and clear understanding of the science behind these products, making them accessible and comprehensible without fear and taboos, while emphasizing the science, the positive aspects, and the advantages of these products, and always ensuring that the safety of people, animals, and the environment has been observed.

10 Final Considerations

In the context of regulatory discussions surrounding genome editing technologies, it is important to highlight the improvements that have been made. While there are still divergences in regulations worldwide, progress has been made in recognizing the potential benefits and opportunities offered by genome editing tools such as CRISPR/Cas.

One notable improvement is the growing acknowledgment of the democratizing effect of genome editing. These technologies have expanded the possibilities for start-ups, public research institutions, and universities to actively participate in the development of biotechnological products, previously dominated by large multinational companies. This shift has created a more inclusive and competitive landscape in the field of biotechnology.

Additionally, the release of edited products in various countries, such as the Japanese tomato and American soybean, signifies a significant advancement. These products have successfully navigated the regulatory pathways and are now available in the market. This demonstrates a tangible outcome of the ongoing regulatory discussions and provides evidence of the potential of genome editing technologies to deliver innovative and improved products to consumers.

Overall, the improvements lie in the increased recognition of the value and potential of genome editing, the broader participation in biotechnological advancements, and the successful commercialization of edited products. These advancements signify progress in regulatory frameworks, opening doors for further innovation and the realization of the full potential of genome editing technologies.

References

Abdul Aziz M, Brini F, Rouached H, Masmoudi K (2022) Genetically engineered crops for sustainably enhanced food production systems. Front Plant Sci 13:1027828

Academy of Science of South Africa (2016) a: the regulatory implications of new breeding techniques

Ahmad M (2023) Plant breeding advancements with "CRISPR/Cas" genome editing technologies will assist future food security. Front Plant Sci 14:1133036

Ahmad A, Munawar N, Khan Z, Qusmani AT, Khan SH, Jamil A, Ashraf S, Ghouri MZ, Aslam S, Mubarik MS et al (2021) An outlook on global regulatory landscape for genome-edited crops. Int J Mol Sci 22:11753

Aman Mohammadi M, Maximiano MR, Hosseini SM, Franco OL (2023) CRISPR/Cas engineering in food science and sustainable agriculture: recent advancements and applications. Bioprocess Biosyst Eng 46(4):483–497

Anderson JE, Michno JM, Kono TJ, Stec AO, Campbell BW, Curtin SJ, Stupar RM (2016) Genomic variation and DNA repair associated with soybean transgenesis: a comparison to cultivars and mutagenized plants. BMC Biotechnol 16(1):41

Anon (2018) Reject the "Gmo" fallacy, in terms of both safety concerns and socioeconomic issues. Geogr Rev 108(1):e1–e5

Anon (2019) Africa Biennial Biosciences Communication Symposium, 2019

Anon (2020) Biotechnology do: draft document on genome edited organisms: regulatory framework and guidelines for risk assessment

Anon (2022) Japan embraces CRISPR-edited fish. Nat Biotechnol 40(1):10–10

Anon (2023) Global gene editing regulation tracker and index

Anon (n.d.-a) This report contains assessments of commodity and trade issues made by usda staff and not necessarily statements of official u.

Anon (n.d.-b) Fact sheet: questions and answers on EU's policies on GMOs. https://ec.europa.eu/commission/presscorner/detail/en/MEMO_15_4778

Apps.Fas.Usda.Gov (n.d.) Agricultural Biotechnology Annual. https://apps.fas.usda.gov/newgainapi/api/report/downloadreportbyfilename?filename=Agricultural%20Biotechnology%20Annual_Buenos%20Aires_Argentina_2-15-2019.pdf

Araki M, Ishii T (2015) Towards social acceptance of plant breeding by genome editing. Trends Plant Sci 20(3):145–149

Aven T (2016) Risk assessment and risk management: Review of recent advances on their foundation. Eur J Oper Res 253:1–13

Bado S, Forster BP, Nielen S, Ali AM, Lagoda PJL, Till BJ, Laimer M (2015) Plant mutation breeding: current progress and future assessment. Plant Breed Rev 39:22–35

Beghin JC, Gustafson CR (2021) Consumer valuation of and attitudes towards novel foods Produced with new plant engineering techniques: a review. Sustainability 13(20):11348

Benítez Candia N, Fernández Ríos D, Vicién C (2020) Paraguay's path toward the simplification of procedures in the approval of GE crops. Front Bioeng Biotechnol 8:1023

Bhatta BP, Malla S (2020) Improving horticultural crops via CRISPR/Cas9: current successes and prospects. Plants (Basel, Switzerland) 9(10):1360

Bratlie S, Halvorsen K, Myskja BK, Mellegård H, Bjorvatn C, Frost P, Heiene G, Hofmann B, Holst-Jensen A, Holst-Larsen T et al (2019) A novel governance framework for GMO. EMBO Rep 20(5):1

Brookes G (2022) Genetically modified (GM) crop use 1996–2020: environmental impacts associated with pesticide use CHANGE. GM Crop Food 13(1):262–289

Brown J, Caligari PDS, Campos HA (2014) Plant breeding. Wiley Blackwell, Hoboken, NJ

Buchholzer M, Frommer WB (2023) An increasing number of countries regulate genome editing in crops. New Phytologist 237:12–15

Bullock DW, Wilson WW, Neadeau JF (2019) Genetic editing (GE) versus genetic modification (GM) in the research and development of new crop varieties: an economic comparison

Canadian Food Inspection Agency C (n.d.) Decision documents—determination of environmental and livestock feed safety Canadian Food Inspection Agency

Cibus (n.d.) Cibus Announces Approval of First Commercial Product SU Canola™ in Canada. https://www.cibus.com/press-release.php?date=031814

Collins JP (2018) Gene drives in our future: challenges of and opportunities for using a self-sustaining technology in pest and vector management. BMC Proc 12(Suppl 8):9

Correa HB, Letícia M, Vieira R, Volpi N, Guilherme S, Prado S, Hernandes J, Filho L (n.d.) CRISPR technology in plant genome editing biotechnology applied to agriculture technical editors

Crispr-Gene (n.d.) Crispr gene editing regs tracker. https://crispr-gene-editing-regs-tracker.geneticliteracyproject.org/united-states-crops-food/

Cummings C, Peters DJ (2022) Who trusts in gene-edited foods? Analysis of a representative survey study predicting willingness to eat-and purposeful avoidance of gene edited foods in the United States. Front Food Sci Technol 2:1

Davison J (2010) GM plants: science, politics and EC regulations. Plant Sci 178:94–98

Dederer HG, Hamburger D (2019) Regulation of genome editing in plant biotechnology: a comparative analysis of regulatory frameworks of selected countries and the EU. Springer International Publishing, Cham

Demorest ZL, Coffman A, Baltes NJ, Stoddard TJ, Clasen BM, Luo S, Retterath A, Yabandith A, Gamo ME, Bissen J et al (2016) Direct stacking of sequence-specific nuclease-induced mutations to produce high oleic and low linolenic soybean oil. BMC Plant Biol 16(1):225

Dima O, Heyvaert Y, Inzé D (2022) Interactive database of genome editing applications in crops and future policy making in the European Union. Trends Plant Sci 27(8):746–748

Doudna JA, Charpentier E (2014) Genome editing. The new frontier of genome engineering with CRISPR/Cas9. Science (New York, NY) 346(6213):1258096

Eckerstorfer MF, Engelhard M, Heissenberger A, Simon S, Teichmann H (2019) Plants developed by new genetic modification techniques-comparison of existing regulatory frameworks in the EU and Non-EU countries. Front Bioeng Biotechnol 7:1

Entine J, Felipe MSS, Groenewald JH, Kershen DL, Lema M, McHughen A, Nepomuceno AL, Ohsawa R, Ordonio RL, Parrott WA et al (2021) Regulatory approaches for genome edited agricultural plants in select countries and jurisdictions around the world. Transgenic Res 30(4):551–584

Eriksson D, Kershen D, Nepomuceno A, Pogson BJ, Prieto H, Purnhagen K, Smyth S, Wesseler J, Whelan A (2019) A comparison of the EU regulatory approach to directed mutagenesis with that of other jurisdictions, consequences for international trade and potential steps forward. In: New phytologist, vol 222. Blackwell Publishing Ltd, Oxford, pp 1673–1684

Eş I, Gavahian M, Marti-Quijal FJ, Lorenzo JM, Mousavi Khaneghah A, Tsatsanis C, Kampranis SC, Barba FJ (2019) The application of the CRISPR/Cas9 genome editing machinery in food and agricultural science: current status, future perspectives, and associated challenges. Biotechnol Adv 37(3):410–421

Europa.eu (n.d.) Genome-edited crops and 21st centrury food system challenges. https://www.europarl.europa.eu/thinktank/en/document/%20EPRS_IDA(2022)690194

European C, Joint Research C, Parisi C, Rodríguez-Cerezo E (2021) Current and future market applications of new genomic techniques. Publications Office of the European Union, Luxembourg

Eu-Sage.eu (n.d.) European sustainable agriculture through genome editing. https://www.eu-sage.eu

Executive Office of the USPEO (2019) Modernizing the regulatory framework for agricultural biotechnology products

Friedrichs S, Takasu Y, Kearns P, Dagallier B, Oshima R, Schofield J, Moreddu C (2019) An overview of regulatory approaches to genome editing in agriculture. Biotechnol Res Innov 3(2):208–220

Gaj T, Sirk SJ, Shui SL, Liu J (2016) Genome-editing technologies: principles and applications. Cold Spring Harb Perspect Biol 8(12):a023754

Gatica-Arias A (2020) The regulatory current status of plant breeding technologies in some Latin American and the Caribbean countries. In: Plant cell, tissue and organ culture, vol 141. Springer, Cham, pp 229–242

Gleim S, Lubieniechi S, Smyth SJ (2020) CRISPR/Cas9 Application in Canadian Public and private plant breeding. CRISPR J 3:44–51

Globus R, Qimron U (2018) A technological and regulatory outlook on CRISPR crop editing. J Cell Biochem 119(2):1291–1298

GM Waxy Maize (n.d.) The gene edited Trojan Horse is moving through the gates. https://grain.org/en/article/6640-gm-waxy-maize-the-gene-edited-trojan-horse-is-moving-through-the-gates

Gonzalez-Avila LU, Vega-López JM, Pelcastre-Rodríguez LI, Cabrero-Martínez OA, Hernández-Cortez C, Castro-Escarpulli G (2021) The challenge of CRISPR/Cas toward bioethics. Front Microbiol 12:657981

Gupta S, Kumar A, Patel R, Kumar V (2021) Genetically modified crop regulations: scope and opportunity using the CRISPR/Cas9 genome editing approach. Mol Biol Rep 48(5):4851–4863

Gupta D, Sharma G, Saraswat P, Ranjan R (2021) Synthetic biology in plants, a boon for coming decades. Mol Biotechnol 63(12):1138–1154

Hassoun A, Cropotova J, Trif M, Rusu AV, Bobiş O, Nayik GA, Jagdale YD, Saeed F, Afzaal M, Mostashari P et al (2022) Consumer acceptance of new food trends resulting from the fourth industrial revolution technologies: a narrative review of literature and future perspectives. Front Nutr 9:972154

Hoffman NE (2021) Revisions to USDA biotechnology regulations: the SECURE rule. Proc Natl Acad Sci U S A 118(22):e2004841118

Holdren JP, Shelanski H, Vetter D, Goldfuss C (2015) Memorandum for heads of food and drug administration, environmental protection agency, and department of agriculture from: Chief agricultural negotiator United States trade representative subject: modernizing the regulatory system for biotechnology products

Hua K, Zhang J, Botella JR, Ma C, Kong F, Liu B, Zhu JK (2019) Perspectives on the application of genome-editing Technologies in Crop Breeding. Mol Plant 12(8):1047–1059

Huesing JE, Andres D, Braverman MP, Burns A, Felsot AS, Harrigan GG, Hellmich RL, Reynolds A, Shelton AM, Van Rijssen WJ et al (2016) Global adoption of genetically modified (GM) crops: challenges for the public sector. J Agric Food Chem 64:394–402

ISAAA (2019) Global status of CommerCialized bioteCh/Gm CropS in 2019:biotech Crops drive Socio-economic development and Sustainable environment in the new frontier, vol 55. ISAAA Brief, Ithaca, NY

Jinek M, Chylinski K, Fonfara I, Hauer M, Doudna JA, Charpentier E (2012) A programmable dual-RNA-guided DNA endonuclease in adaptive bacterial immunity. Science (New York, NY) 337(6096):816–821

Jorasch P (2020) Potential, challenges, and threats for the application of new breeding techniques by the private plant breeding sector in the EU. Front Plant Sci 11:582011

Julkaisut (n.d.) Utilisation of New Genome Editing Techniques in Finland. https://julkaisut.valtioneuvosto.fi/bitstream/handle/10024/163143/VNTEAS_2021_39.pdf?%20sequence=1

Kanchiswamy CN, Malnoy M, Velasco R, Kim JS, Viola R (2015) Non-GMO genetically edited crop plants. In: Trends in biotechnology, vol 33. Elsevier, Amsterdam, pp 489–491

Kato-Nitta N, Inagaki Y, Maeda T, Tachikawa M (2021) Effects of information on consumer attitudes towards gene-edited foods: a comparison between livestock and vegetables. CABI Agric Biosci 2(1):14

Laaninen T (2021) New genomic techniques. European Commission study and first reactions. https://www.europarl.europa.eu/RegData/etudes/BRIE/2021/698760/EPRS_BRI(2021)698760_EN.pdf

Lassoued R, Macall DM, Hesseln H, Phillips PWB, Smyth SJ (2019) Benefits of genome-edited crops: expert opinion. Transgenic Res 28(2):247–256

Lassoued R, Phillips PWB, Smyth SJ, Hesseln H (2019) Estimating the cost of regulating genome edited crops: expert judgment and overconfidence. GM Crops Food 10(1):44–62

Li H, Yang Y, Hong W, Huang M, Wu M, Zhao X (2020) Applications of genome editing technology in the targeted therapy of human diseases: mechanisms, advances and prospects. Signal Transduct Target Ther 5(1):1

Lusser M, Davies HV (2013) Comparative regulatory approaches for groups of new plant breeding techniques. New Biotechnol 30:437–446

Lusser M, Parisi C, Rodriguez Cerezo E, Plan D (2011) New plant breeding techniques. State-of-the-art and prospects for commercial development. EUR 24760 EN. Luxembourg (Luxembourg): Publications Office of the European Union. JRC63971

Lusser M, Parisi C, Plan D, Rodríguez-Cerezo E (2012) Deployment of new biotechnologies in plant breeding. Nat Biotechnol 30:231

Mackelprang R, Lemaux PG (2020) Genetic engineering and editing of plants: an analysis of new and persisting questions. Annu Rev Plant Biol 71:659

Mackenzie R, Burhenne-Guilmin F, La Viña AGM, Werksman JD, Ascencio A, Kinderlerer J, Kummer K, Tapper R (2003) An explanatory guide to the Cartagena protocol on biosafety. IUCN Environmental Law Centre, Bonn, p 46

Mackenzie R, Burhenne-Guilmin F, La Viña AGM, Werksman JD, Ascencio A, Kinderlerer J, Kummer K, Tapper R (n.d.) The world conservation union the world conservation union IUCN environmental law Centre
Marette S, Disdier A-C, Beghin JC (2021) A comparison of EU and US consumers' willingness to pay for gene-edited food: evidence from apples. Appetite 159:105064
Martin-Laffon J, Kuntz M, Ricroch AE (2019) Worldwide CRISPR patent landscape shows strong geographical biases. Nat Biotechnol 37(6):613–620
Matsuo M, Tachikawa M (2022) Implications and lessons from the introduction of genome-edited food products in Japan. Front Genome Ed 4:899154
McDougall P (2011) The cost and time involved in the discovery, development and authorisation of a new plant biotechnology derived trait. Crop Life International, Brussels
McHughen A (2016) A critical assessment of regulatory triggers for products of biotechnology: Product vs. process. GM Crop Food 7:125–158
Medvedieva MO, Blume YB (2018) Legal regulation of plant genome editing with the CRISPR/Cas9 technology as an example. Cytol Genet 52(3):204–212
Menz J, Modrzejewski D, Hartung F, Wilhelm R, Sprink T (2020) Genome edited crops touch the market: a view on the global development and regulatory environment. Front Plant Sci 11:586027
Metje-Sprink J, Sprink T, Hartung F (2020) Genome-edited plants in the field. Curr Opin Biotechnol 61:1–6
Neve P (2018) Gene drive systems: do they have a place in agricultural weed management? Pest Manag Sci 74(12):2671–2679
Nonaka S, Arai C, Takayama M, Matsukura C, Ezura H (2017) Efficient increase of ɣ-aminobutyric acid (GABA) content in tomato fruits by targeted mutagenesis. Sci Rep 7(1):7057
Objective 2030 (n.d.) Deregulate most GMOs in Europe? https://www.infogm.org/7512-objective-2030-deregulate-most-gmos-in-europe?lang=fr#nb2-3
Oliver MJ (2014) Why we need GMO crops in agriculture. Mo Med 111(6):492–507
Ostp US (1986) Coordinated framework for regulation of biotechnology. Office of Science and Technology Policy, Washington, DC
Patil G, Vuong TD, Kale S, Valliyodan B, Deshmukh R, Zhu C, Wu X, Bai Y, Yungbluth D, Lu F et al (2018) Dissecting genomic hotspots underlying seed protein, oil, and sucrose content in an interspecific mapping population of soybean using high-density linkage mapping. Plant Biotechnol J 16(11):1939–1953
Podevin N, Davies HV, Hartung F, Nogué F, Casacuberta JM (2013) Site-directed nucleases: a paradigm shift in predictable, knowledge-based plant breeding. Trends Biotechnol 31:375
Produtos Avaliados (n.d.) Técnicas Inovadoras de Melhoramento de Precisão. http://ctnbio.mctic.gov.br/documents/566529/2304555/Tabela+TIMP/8c4a7218-f810-405b-94bf-a352d849f3dc?version=1.1
Ramessar K, Capell T, Twyman RM, Christou P (2008) Trace and traceability—a call for regulatory harmony. Nat Biotechnol 25:975–978
Rao Y, Yang X, Pan C, Wang C, Wang K (2022) Advance of clustered regularly interspaced short palindromic repeats-Cas9 system and its application in crop improvement. Front Plant Sci 13:839001
Ravikiran KT, Thribhuvan R, Sheoran S, Kumar S, Kushwaha AK, Vineeth TV, Saini M (2023) Tailoring crops with superior product quality through genome editing: an update. Planta 257(5):86
Realagriculture (n.d.) Canada moves forward on giving gene-editing the conventional plant breeding stamp of approval. https://www.realagriculture.com/2023/05/canada-moves-forward-on-giving-gene-editing-the-conventional-plant-breeding-stamp-of-approval/
Reuters.Com (n.d.) In-vitro plant gene editing technique excluded from GMO rules, EU court says. https://www.reuters.com/world/europe/in-vitro-plant-gene-editing-technique-excluded-gmo-rules-eu-court-says-2023-02-07/

Rozas P, Kessi-Pérez EI, Martínez C (2022) Genetically modified organisms: adapting regulatory frameworks for evolving genome editing technologies. Biol Res 55:1

Sánchez-León S, Gil-Humanes J, Ozuna CV, Giménez MJ, Sousa C, Voytas DF, Barro F (2018) Low-gluten, nontransgenic wheat engineered with CRISPR/Cas9. Plant Biotechnol J 16(4):902–910

Schmidt SM, Belisle M, Frommer WB (2020) The evolving landscape around genome editing in agriculture. EMBO Rep 21(6):e50680

Schnell J, Steele M, Bean J, Neuspiel M, Girard C, Dormann N, Pearson C, Savoie A, Bourbonnière L, Macdonald P (2015) A comparative analysis of insertional effects in genetically engineered plants: considerations for pre-market assessments. Transgenic Res 24(1):1–17

Shah T, Andleeb T, Lateef S, Noor MA (2018) Genome editing in plants: advancing crop transformation and overview of tools. Plant Physiol Biochem 131:12–21

Smyth SJ (2019) Regulation of genome editing in plant biotechnology: Canada. In: Dederer H-G, Hamburger D (eds) Regulation of genome editing in plant biotechnology: a comparative analysis of regulatory frameworks of selected countries and the EU. Springer International Publishing, Cham, pp 111–135

Smyth SJ (2022) Contributions of genome editing technologies towards improved nutrition, environmental sustainability and poverty reduction. Front Genome Ed 4:4

Smyth SJ, Gleim S, Lubieniechi S (2020) Regulatory barriers to innovative plant breeding in Canada. Front Genome Ed:2, 591592

Smyth S, McHughen A (2008) Regulating innovative crop technologies in Canada: the case of regulating genetically modified crops. Plant Biotechnol J 6:213–225

Smyth SJ, Phillips PWB (2014) Risk, regulation and biotechnology: the case of GM crops. GM Crop Food 5(3):170–177

Smyth SJ, Wesseler J (2022) The future of genome editing innovations in the EU. Trends Biotechnol 40(1):1–3

Sorek R, Lawrence CM, Wiedenheft B (2013) CRISPR-mediated adaptive immune systems in bacteria and archaea. Annu Rev Biochem 82:237–266

Spök A, Sprink T, Allan AC, Yamaguchi T, Dayé C (2022) Towards social acceptability of genome-edited plants in industrialised countries? Emerging evidence from Europe, United States, Canada, Australia, New Zealand, and Japan. Front Genome Ed 4:899331

Sprink T, Wilhelm R, Hartung F (2022) Genome editing around the globe: an update on policies and perceptions. Plant Physiol 190(3):1579–1587

Steiner HY, Halpin C, Jez JM, Kough J, Parrott W, Underhill L, Weber N, Hannah LC (2013) Editor's choice: evaluating the potential for adverse interactions within genetically engineered breeding stacks. Plant Physiol 161(4):1587–1594

Tagliabue G (2017) Product, not process! Explaining a basic concept in agricultural biotechnologies and food safety. Life Sci Soc Policy 13(1):3

Teferra TF (2021) Should we still worry about the safety of GMO foods? Why and why not? A review. Food Sci Nutr 9(9):5324–5331

Tripathi L, Dhugga KS, Ntui VO, Runo S, Syombua ED, Muiruri S, Wen Z, Tripathi JN (2022) Genome Editing for Sustainable Agriculture in Africa. Front Genome Ed 4:4

Turnbull C, Lillemo M, Hvoslef-Eide TAK (2021) Global regulation of genetically modified crops amid the gene edited crop boom—a review. Front Plant Sci 12:630396

Uddin A, Gallardo RK, Rickard B, Alston J, Sambucci O (2022) Consumer acceptance of new plant-breeding technologies: an application to the use of gene editing in fresh table grapes. PLoS One 17(12):e0270792

Us O (1986) Coordinated framework for regulation of biotechnology. office of Science and Technology Policy, Washington, DC

Us O (2017) Update to the coordinated framework for the regulation of biotechnology

Vieira LR, Freitas NC, Justen F, Miranda VDJ, Garcia BDO, Nepomuceno AL, Fuganti-Pagliarini R, Felipe MSS, Molinari HBC, Velini ED (2021) Regulatory framework of genome editing in Brazil and worldwide

Waltz E (2022) GABA-enriched tomato is first CRISPR-edited food to enter market. Nat Biotechnol 40(1):9–11

Wang Y, Zafar N, Ali Q, Manghwar H, Wang G, Yu L, Ding X, Ding F, Hong N, Wang G et al (2022) CRISPR/Cas genome editing Technologies for Plant Improvement against biotic and abiotic stresses: advances, limitations, and future perspectives. Cell 11(23)

Wasmer M (2019) Roads forward for European GMO policy-uncertainties in wake of ECJ judgment have to be mitigated by regulatory reform. Front Bioeng Biotechnol 7:7

Wiedenheft B, Sternberg SH, Doudna JA (2012) RNA-guided genetic silencing systems in bacteria and archaea. Nature 482(7385):331–338

Wolt JD, Wang K, Yang B (2016) The regulatory status of genome-edited crops. Plant Biotechnol J 14(2):510–518

Woźniak-Gientka E, Tyczewska A, Twardowski T (2022) Public opinion on biotechnology and genetic engineering in the European Union: polish consumer study. Biotechnologia 103(2):185–201

Zhang D, Hussain A, Manghwar H, Xie K, Xie S, Zhao S, Larkin RM, Qing P, Jin S, Ding F (2020) Genome editing with the CRISPR/Cas system: an art, ethics and global regulatory perspective. Plant Biotechnol J 18(8):1651–1669

Zhao J, Lai L, Ji W, Zhou Q (2019) Genome editing in large animals: current status and future prospects. Natl Sci Rev 6(3):402–420

Novel Genome-Editing Approaches for Developing Non-GM Crops for Sustainable Improvement and the Mitigation of Climate Changes

Naglaa A. Abdallah, Aladdin Hamwieh, and Michael Baum

Abstract The utilization of clustered regularly interspaced short palindromic repeat (CRISPR)/Cas-based genome-editing technologies holds significant promise in the realm of crop genome manipulation since it enables precise modifications and expedites the progress of crop breeding initiatives. Crop improvements need to be genetically stable and transgene-free to ensure sustainability, mitigate environmental stresses, and gain consumer and decision-maker acceptance. Edited plants with transgenic-based approaches can address many problems associated with transgenic plants. CRISPR/Cas genome editing allows the development of precise modification at the nucleotide level that is not different from that which occurred from natural recombination during conventional breeding. Various methods have accomplished genome editing without the incorporation of transgenes. These strategies involve the utilization of site-directed nucleases, specifically type 1 (SDN-1), as well as cisgenic editing employing SDN-2. A number of countries, including the United States, Japan, India, and Australia, classify genome-edited crops that lack transgenes or foreign deoxyribonucleic acid (DNA) as non-genetically modified (non-GM) and thereby exclude them from regulations governing genetically modified organisms (GMOs). *Agrobacterium*-mediated or biolistic transformations are often employed methods for introducing the CRISPR components into the plant genome. The first generation plants could be used to obtain transgenic-free plants through the segregation of heterozygous crops. However, the transformation process is expensive and time-consuming, and many species are recalcitrant to transformation. In addition, it will be impossible to get transgenic-free plants from plants that are propagated vegetatively. The target delivery of Cas-gRNA nucleoprotein using chemical

N. A. Abdallah (✉)
Department of Genetics Faculty of Agriculture, Cairo University, Cairo, Egypt

National Biotechnology Network of Expertise, ASRT, Cairo, Egypt
e-mail: naglaa.abdallah@agr.cu.edu.eg

A. Hamwieh · M. Baum
Department of Biotechnology, International Centre for Agricultural Research in the Dry Areas (ICARDA), Giza, Egypt

J.-T. Chen, S. Ahmar (eds.), *Plant Genome Editing Technologies*, Interdisciplinary Biotechnological Advances,
https://doi.org/10.1007/978-981-99-9338-3_3

or physical methods could be a promising tool for developing transgene-free edited plants. Also, viral vectors were used for the delivery of CRISPR/Cas components to obtain transgene-free edited plants. In a recent study, the technique of grafting was employed to introduce transgenic roots harboring tRNA-like sequences (TLS) that serve as molecular signals facilitating the transport of ribonucleic acids (RNAs) over long distances within plants. This approach was utilized to distribute the CRISPR/Cas components to both the shoots and seeds of the plants. This chapter presents a thorough examination of the many techniques employed in the acquisition of transgene-free plant genome editing, as well as the advancements made in comparison to other genetically modified (GM) plants and edited organisms.

Keywords CRISPR/Cas · Transgenic free · tRNA-like sequences (TLS) · Morphogenic regulators · CRISPR/Cas-Combo · Transgene killer CRISPR

1 Introduction

Global climate change has severely negative impacts on agricultural productivity and food security all around the world as it causes frequent floods, droughts, temperature variations, and higher soil salinity (Bing et al. 2020). It is expected that global temperatures are going to rise by 2 °C by the year 2050. Raising temperature will lead to rising sea, causing scarcity of arable land and increasing soil salinity. The implementation of these modifications will result in heightened abiotic pressures on plant life, the emergence of novel strains of plant pathogens and insect pests, intensified pathogen infections, diminished water resources and arable land, and significant obstacles to food security in light of the growing global population (IPCC Sixth Assessment Report 2022).

The deployment of genome-editing technologies could be one of the solutions for crop improvement and the mitigation of climate change. Genome editing is more acceptable than genetically modified (GM) crops as its precise genetic manipulation does not need exogenous deoxyribonucleic acid (DNA), and edited plants could reach the market quickly with affordable prices compared to GM crops (Gao 2021). The clustered regularly interspaced short palindromic repeat (CRISPR)/Cas system's simplicity and ease of manipulation have played a significant role in expediting advancements in functional genomics and crop trait enhancements. The CRISPR/Cas system, which has undergone significant advancements, has been extensively employed to address climate change by mitigating both biotic and abiotic pressures (Table 1).

Abdallah et al. (2021) have reported that there are three distinct gene-editing systems, namely SDN1, SDN2, and SDN3, which are associated with the repair mechanism following double-strand DNA (dsDNA) breaks. The repair of SDN1 is reliant on the process of nonhomologous end joining (NHEJ), which is associated with the potential for mutations and subsequent gene silencing. The SDN-2 method exhibits great fidelity due to its reliance on a tiny DNA donor that possesses

Table 1 Summary of gene-editing applications for climate change mitigation

Trait	Target	Species	Gene(s)	Edited method	Transformation method	References
Abiotic stress	Drought tolerance	Maize	*ZmLBD5*	SDN-1 CRISPR/Cas	*Agrobacterium tumefaciens*	Feng et al. (2022)
		Rapeseed	*BnaRGA*	SDN-1 CRISPR/Cas	*Agrobacterium tumefaciens*	Wu et al. (2020)
		Rice	*OsEBP89*	SDN-1 CRISPR/Cas	*Agrobacterium tumefaciens*	Zhang et al. (2020b)
		Soybean	*GmHsps_p23-like*	SDN-1 CRISPR/Cas	*Agrobacterium tumefaciens*	Zhang et al. (2022c)
		Tomato	*SlLBD40*	SDN-1 CRISPR/Cas	*Agrobacterium tumefaciens*	Liu et al. (2020b)
		Vitis vinifera	*VvEPFL9–1*	SDN-1 CRISPR/Cas	*Agrobacterium tumefaciens*	Clemens et al. (2022)
	Salinity tolerance	Rice	*OsRR22*	SDN-1 CRISPR/Cas	*Agrobacterium tumefaciens*	Han et al. (2022)
		Soybean	*GmAITR*	SDN-1 CRISPR/Cas	*Agrobacterium tumefaciens*	Wang et al. (2021a)
		Rice	*OsPIN9*	SDN-1 CRISPR/Cas	*Agrobacterium tumefaciens*	Xu et al. (2022)
		Tomato	*BAG9*	SDN-1 CRISPR/Cas	*Agrobacterium tumefaciens*	Huang et al. (2022)
	Aluminum resistance	Tomato	*slstop1*, *slszp1*	SDN-1 CRISPR/Cas	*Agrobacterium tumefaciens*	Zhang et al. (2022a)
	Cadmium detoxification & accumulation	Rice	*OsHIPP16*	SDN-1 CRISPR/Cas	*Agrobacterium tumefaciens*	Cao et al. (2022)

(continued)

Table 1 (continued)

Trait	Target	Species	Gene(s)	Edited method	Transformation method	References
Biotic stress	Virus resistance	Rice black-streaked dwarf virus	*eIF4G* gene	SDN1 CRISPR/Cas	*Agrobacterium tumefaciens*	Hoffie et al. (2023)
		Barley mild mosaic virus (BaMMV)	*PDIL5–1* gene	SDN1 CRISPR/Cas	*Agrobacterium tumefaciens*	Hoffie et al. (2023)
		Potato virus Y	PVY genome	RNA editing Cas13a	Agro- infiltration	Zhan et al. (2019)
		Grapevine leafroll-associated virus 3	GLRaV-3 genome	RNA editing LshCas13a	*Agrobacterium tumefaciens*	Jiao et al. (2022)
	Bacteria	Citrus *Xanthomonas citri*	*CsWRKY22* gene	SDN1 CRISPR/Cas	*Agrobacterium tumefaciens*	Long et al. (2021)
		Rice *Xanthomonas oryzae*	$EBE_{AvrXa23}$ Gene	SDN2 CRISPR/Cas	Particle bombardment	Wei et al. (2021)
	Fungi	Brassica *Sclerotinia sclerotiorum*	*BnMPK3 gene*	SDN2 CRISPR/Cas & transgenic	*Agrobacterium tumefaciens*	Zhang et al. (2022b)
		Soybean *Phytophthora sojae*	*BnMPK3*	SDN2 CRISPR/Cas & transgenic	*Agrobacterium tumefaciens*	Zhang et al. (2022b)
		Wheat powdery mildew resistance	*TaTMT3B*	SDN1 CRISPR/Cas	Particle bombardment	Li et al. (2022b)
		Wheat stripe rust	*TaCIPK14*	SDN1 CRISPR/Cas	*Agrobacterium tumefaciens*	He et al. (2022)
		Barley *Bipolaris* spot blotch and *fusarium* root rot	*Hv*MORC1 and *Hv*MORC6a	SDN1 CRISPR/Cas	*Agrobacterium tumefaciens*	Galli et al. (2022)
	Insect	Soybean leaf-chewing insects	*Glyma.07 g110300*	SDN1 CRISPR/Cas	*Agrobacterium tumefaciens*	Zhang et al. (2022d)
		Watermelon aphids	*vst1–1*	SDN1 CRISPR/Cas	*Agrobacterium tumefaciens*	Li et al. (2022a)

sequence homology to the target DNA, enabling an accurate repair of the DNA break. The repair process occurs through the utilization of the homologous direct repair (HDR) system, wherein a donor carrying the desired modifications and flanking sequences that are complimentary to both ends of the break is employed. The restoration of SDN3 necessitates the involvement of a foreign donor and exhibits a notable degree of accuracy and precision. The process of HDR-mediated repair involves the integration of the complete gene or genetic elements into the desired location, utilizing a donor sequence. This method is classified as transgenic.

In some countries, plants that are devoid of transgenes, resulting from small indels, base-pair alterations, and particular short sequence modifications, are not classified as genetically modified organisms (GMOs) and are thus granted exemption from GMO regulations. The generation of transgene-free edited plants poses a significant obstacle to the implementation of genome editing in crop breeding as legislative and regulatory frameworks play a crucial role in the approval and marketing of genetically modified crops. In order to achieve transgene-free edited plants and ensure the stability of the change, it is necessary to either segregate or cross both the Cas enzyme and the guide ribonucleic acid (gRNA) from transgenic plants or introduce them transiently (Liang et al. 2021). Conventional genome-editing systems deliver the DNA constructs to plants via *Agrobacterium*-mediated or particle-bombardment-mediated transformations, which require the insertion of foreign DNA. In order to limit the presence of incorporated foreign DNA, the practice of selfing or crossing with wild-type plants is employed to yield transgene-free edited plants. It is worth noting that these procedures are characterized by a significant investment of labor and time (Bhattacharjee et al. 2023). In this chapter, we will summarize the novel strategies used to develop transgene-free edited plants.

2 Base Editors and Prime Editors

Modified Cas enzymes were constructed to work as nickase or dead enzymes. Enzymes that have been altered or engineered have played a significant role in the advancement of novel editing methodologies, including base editing and prime editing. Base editing (SND-1) and prime editing (SDN-2) techniques, because of their independence from a DNA donor template, can be regarded as non-genetically modified (non-GM) products that do not necessitate adherence to biosafety rules. Precise base changes in the genome (Fig. 1a) could be achieved for the target editing of a single base pair through cytosine base editing (C:G to T:A) or adenine base editing (A:T to G:C) or dual base editing by editing C to T and A to G (Abdallah et al. 2021). Multiplexed orthogonal genome editing is performed using simultaneous and wide editing induced by a single system (SWISS) using dual-base editing for cytosine and adenine conversion, in addition to producing insertion/deletion (indel) mutations (Fig. 1b). The SWISS method necessitates the utilization of a nickase Cas (nCAS) enzyme, together with two sno-derived RNA (sdRNA) scaffolds that are recruited for the binding of cognate proteins. One of these sdRNAs is

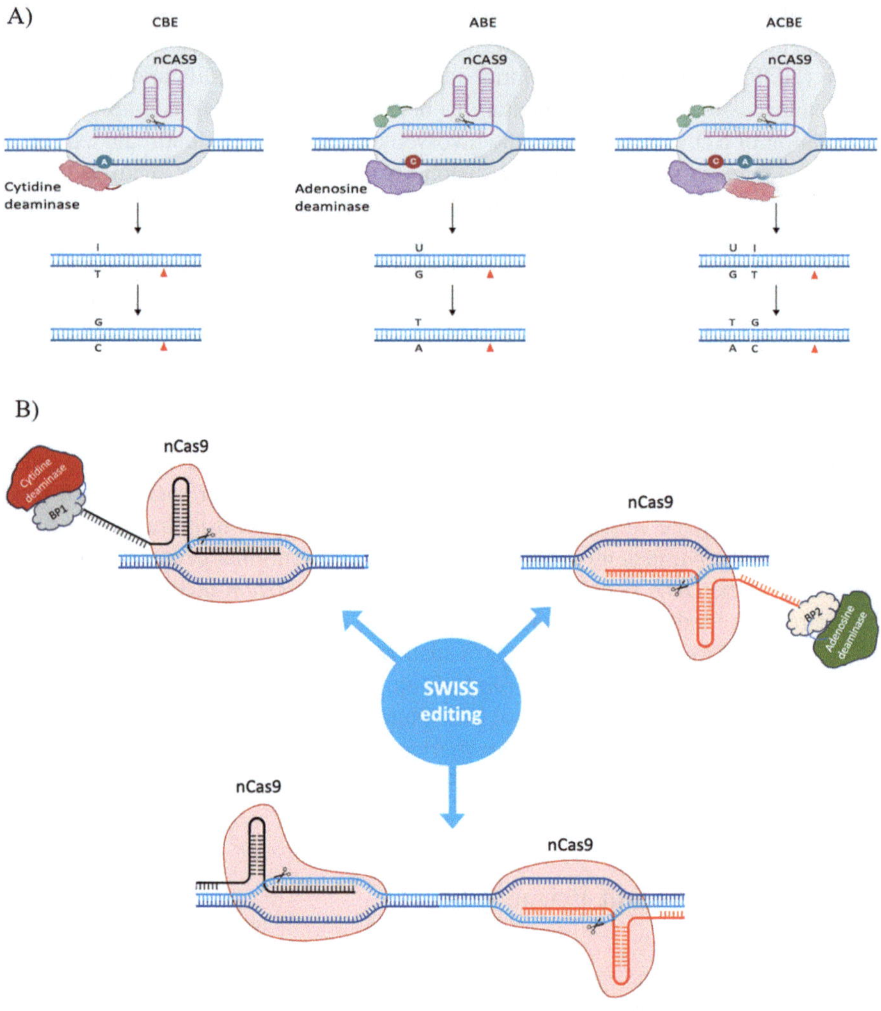

Fig. 1 (**a**) Base editing for CBE targeted C-to-T base editing, ABE targeted A-to-G base editing, and ACBE targeted both C-to-T and A-to-G base editing. (**b**) Swiss editing using multiple base editing and indel mutation via a single Cas9 nickase (nCas9) enzyme

coupled with cytidine deaminase, while the other is joined with adenosine deaminase. Furthermore, the generation of paired single-guide RNAs (sgRNAs) is achieved through the utilization of the nCAS enzyme, which serves to induce indel mutations. Li et al. (Li et al. 2020) demonstrated the successful application of the CRISPR-Cas9 system with the SWISS method in rice for performing base editing on OsALS-T2 (C:T substitution) and OsACC-T2 (A:G substitution), as well as generating OsBADH2-Indels-sgL and OsBADH2-Indels mutants.

In recent studies conducted by Anzalone et al. (2019) and Abdallah et al. (2021), it has been demonstrated that prime editing (PE) can facilitate accurate substitutions of tiny fragments or indels without requiring a donor DNA template. This process is achieved through the utilization of the nCas enzyme. The development of the system involved the fusion of a reverse transcriptase (RT) with a Cas9 nickase, as depicted in Fig. 2a. The guide RNA utilized in prime editing, known as pegRNA, incorporates a reverse transcription (RT) template positioned at its 3′ terminus. This RT template serves the purpose of initiating accurate modifications at the designated target site. The potential of correcting mutations can be achieved by utilizing programmable nucleases, such as prime editing (PE), which has the capability to insert tiny segments of up to 44 base pairs (Anzalone et al. 2019). The editing process known as PE was optimized to improve its editing capabilities and minimize the presence of nonedited sequences. This was achieved by introducing six specific mutations into the M-MLV-RT enzyme. In addition, a modified approach was employed wherein the unedited strand was directly cleaved using a specific sgRNA that induces nicking. To accomplish these modifications, engineered versions of the Moloney-murine leukemia virus reverse transcriptase (eM-MLV RT) were utilized, including the dual eM-MLV RT variant. These advancements were described in

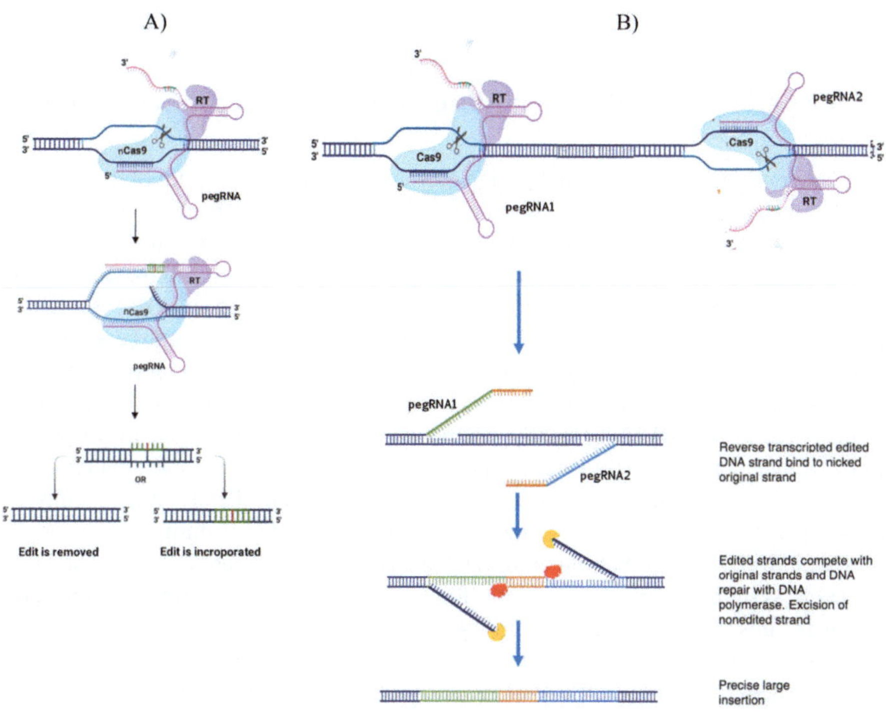

Fig. 2 Schematic illustrating the strategies for prime editing and GRAND editing to targeted insert DNA. (**a**) Prime editing system utilizes a Cas9 nickase fused with reverse transcriptase and apegRNA. (**b**) GRAND editing synthesizes ssDNA by RT using RTT. The edited strands complementary bind to each other at the 3' ends. The gap is filled, and the edited dsDNA is inserted into the target region of the genome

recent studies conducted by H. Li et al. (2022c), Nelson et al. (2022), and Zong et al. (2022). The inclusion of the 3′ flap is crucial in prime editing to ensure optimal efficiency in the process of insertion. GRAND editing was designed to enable the exact insertion of massive DNA pieces without the need for DNA donors. The technique employs a pair of pegRNAs, which consist of reverse transcription templates that are nonhomologous and complementary to one another. The GRAND editing technique employs a pair of pegRNAs, wherein reverse transcription templates (RTTs) exhibit partial complementarity to one another. Additionally, two Cas9 nickase-RT molecules are utilized, with one molecule inducing a nick on the target DNA strand that is opposite to the other (Fig. 2b). The corresponding binding site of pegRNA hybridized at the 3′ end of nicked DNA, followed by the initiation and extension of reverse transcription to form two new single-stranded DNAs complementary at the 3′ ends. Following a process of competition and equilibration between edited strands and original strands, it is observed that the original strands undergo degradation while the edited strands engage in hybridization at the 3′ complementary ends, thereby facilitating repair and gap filling. This strategy allows insertions of up to 1 kb but with low efficiency, but inserting a fragment less than 400 bp develops suitable efficiency (Wang et al. 2022).

3 Genetic Segregation for Eliminating Transgenic Sequences

The conventional methods of delivering editor genes to plant cells involve the use of DNA constructs containing traditional editor genes. These constructs are typically introduced into the cells through either *Agrobacterium tumefaciens*-mediated transformation or particle bombardment techniques (Altpeter et al. 2016). To identify edited plants, the cleaved amplified polymorphic sequence (CAPS) screening method is commonly employed. The incorporation of selection markers, such as genes conferring resistance to antibiotics or herbicides, in conjunction with the genome editing of the target gene, can facilitate the development of transgenic plants (Yin et al. 2017). Typically, T0 plants exhibit hemizygosity, meaning that only a single copy of the transgenes has been put into the genome at the specific site. The acquisition of transgene-free altered plants necessitates the process of selfing or crossing with wild-type plants, a labor-intensive and time-consuming endeavor.

He et al. (2018) and Yubing et al. (2019) have recently created a transgene killer CRISPR system in rice. This system allows for the self-elimination of transgenic plants in subsequent generations through the use of two marker selections: positive and negative markers. This approach depends on adding two suicide genes: *barnase* (toxic protein), under the control of the early embryo-specific promoter, and CMS2 (causes male sterility), under the control of the *35S* promoter. During transformation and regeneration, positive markers will be used to select transgenic plants and the CRISPR/Cas system will edit target genes, but when the plants undergo reproductive growth, suicide genes (negative markers) will be expressed and kill the pollen and embryos (He et al. 2018). In addition, visible selection markers, such as fluorescent-based screening, were used to shorten the time requirement for selecting transgene-free progenies in edited plants. The mCherry, under the control of the seed-specific promoter At2S3, along with the CRISPR/Cas vector, was used for transgene counterselection in *Arabidopsis (*Yu and Zhao 2019*)*. Also, as a seed-fluorescence-based marker, *Sl*-FAST2 was established for *Nicotiana benthamiana*. According to Stuttmann et al. (2021), it was observed that all plants derived from nonfluorescent seeds of genome-altered plants were devoid of transgenes. In the study conducted by Zhang et al. (2023), the CRISPR/Cas9-mediated genome-editing technique was employed in tobacco plants. Specifically, a dark red tag (Ros1) regulated by the AtUb10 promoter was utilized to facilitate the identification of mutants through visual screening. The plants exhibiting a dark red color were identified as containing the transgenes and were subsequently excluded from the selection process to acquire plants that were free of transgenes.

4 Mobile CRISPR Grafting Strategy

Prior research has demonstrated that short noncoding RNAs and big messenger RNA (mRNA) molecules can be transported between cells and over considerable distances through the plasmodesmata of the phloem in plants. The study conducted by Zhang et al. (2016) examines the impact of fusing the transcript of the disruption of meiotic control 1 with tRNA-like sequence (TLS) motifs in the rootstocks on the interference of male sporogenesis in grafted *Nicotiana* wild-type plants. In the study conducted by Zhang et al. (2016), genetically modified *Arabidopsis* plants were created by introducing Cas9 and gRNAs fused to TLS1 (tRNAMet sequence) or TLS2 (tRNAMet-ΔDT sequence, deleting the D and T loop), along with a brief poly-A tail at the 3′ end of each fusion. These transgenic plants were subsequently utilized as rootstocks for grafting wild-type seedlings. Transcripts were observed in the shoots of grafted transgene-free wild-type plants when utilizing rootstocks expressing Cas9-TLS1 and gNIA1-TLS1, or Cas9-TLS2 and gNIA1-TLS2 (Fig. 3). The absence of detectable transcripts of the selectable marker genes, specifically the hygromycin and kanamycin resistance genes, in the shoot samples suggests that there is no evidence of contamination resulting from the grafting process.

The researchers employed the mobile CRISPR grafting technique to generate transgenic *Arabidopsis thaliana* plants with targeted genetic modifications. Specifically, two guide RNAs (gRNAs) were developed to facilitate the deletion of a 1000-base-pair (bp) segment of the nitrate reductase1 (NIA1) gene. This gene encodes an enzyme involved in the conversion of nitrate (NO3) to ammonium (NH4) (Yang et al. 2023). The utilization of mutated NIA1 served as an evident phenotype, whereby the gradual manifestation of chlorosis in smaller plants was seen. In this study, genetically modified *Arabidopsis* plants were utilized as rootstocks for wild plants. These transgenic *Arabidopsis* plants were engineered to express gRNAs targeting the NIA1 gene, together with the Cas9 protein linked to tRNA-like sequence (TLS) motifs. The chlorotic phenotype, resulting from mutated

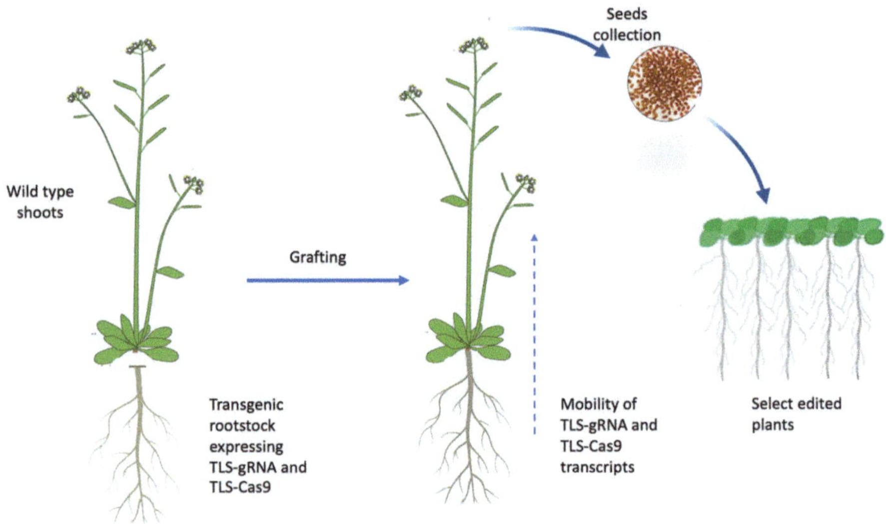

Fig. 3 Development of transgene-free genome editing using mobile CRISPR-Cas9 grafting. Transgenic rootstock expressing the Cas9 mRNA and gRNA fused to tRNA-like sequence (TLS) sequences is used to graft wild-type scion. Transcripts of the trans-genes mobile from the transgenic rootstock to grafted scion and perform genome editing

nia1, was observed in certain leaves of the recipient plant after 14 days of grafting, specifically when cultivated under NH4-deficient conditions. The training's efficacy suggests that there is a substantial presence of Cas9-TLS and gNIA1-TLS transcripts in grafted nontransgenic wild-type shoots. Also, the transferred transcript of Cas9-TLS could translate into a functional protein. The transgenic *Arabidopsis* was also used as rootstock for grafting *Brassica rapa* wild type and was able to achieve heritable gene editing. The offspring developed by the graft-mobile gene editing system was transgene-free. Although the frequency of edited plants using this system is low, it did not need to eliminate transgenes or the use of viral editing vectors. This system could be applied to plants that are graft-compatible with *Arabidopsis*, such as tomatoes.

5 Virus-Induced Genome Editing

According to Ellison et al. (2021), RNA viruses possess significant potential as a very effective and adaptable means of facilitating transgene-free genome editing. This is due to their ability to undergo RNA replication, which allows for multiplication, and their typical avoidance of reverse transcription into DNA during their life cycle. The RNA virus delivery method is classified as nontransgenic due to the absence of integration into host chromosomes during replication by RNA viruses. The utilization of negative-strand RNA-virus-based vectors is a viable option for delivering the whole CRISPR-Cas9 cassette to plant cells in a DNA-free manner. This method proves advantageous when contrasted to the limitations imposed by the size insertion and movement restrictions within the host constraints of positive-strand viral replicons. One of the primary constraints associated with the utilization of virus-mediated genome editing pertains to the host range of virus species, as well as the occurrence of off-target mutations (Zhou et al. 2019).

The utilization of the positive-strand RNA PVX as a means to transiently express Cas9 and gRNA in *N. benthamiana* resulted in impaired viral movement, which can be attributed to the inclusion of the sizeable Cas9 cassette. Plant agroinfiltration with the viral cDNA could result in the integration of T-DNA in the plant genome. The progeny will be free of PVX RNA and retain genomic edits, but any integrated T-DNA should be eliminated by segregation (Ariga et al. 2020). To eliminate the risk of T-DNA incorporation into the plant genome, agroinfiltration to develop infectious viral sources with the Cas9 and gRNA cassettes was performed. Infiltrated plants were used as a source for inoculating recipient plants and performing editing. However, indirect agroinfiltration demonstrated much less efficiency than direct agroinfiltration. The positive stranded foxtail mosaic virus (FoMV) has been used to develop transgene-free edited tobacco plants (Zhang et al. 2020a). The *N. benthamiana* plant leaves were subjected to agroinfiltration using two FoMV vectors. One vector had the Cas9 gene, while the other vector contained the gRNA responsible for inducing mutations in the phytoene desaturase (PDS) gene. Additionally, the viral RNA-silencing suppressor gene (p19) was included in order to enhance the efficacy of genome editing. In a study conducted by Uranga et al. (2021), two positive-stranded viruses, namely, PSV and Tobacco etch virus (TEV), were employed as agents for genome editing in tobacco plants. The Cas12a gene was incorporated into vectors designed to replace the *NIb* gene in both the tobacco etch virus (TEV) and the potato virus X (PVX). These vectors contained both the guide RNA (gRNA) and the NIb gene. The resulting constructs were then utilized for the coagroinfiltration of leaves from *Nicotiana benthamiana* plants. The effectiveness of gene editing in systemic leaves was observed to be 20%.

The Barley yellow striate mosaic virus (BYSMV) is classified as a negative-stranded virus, characterized by its significant cargo capacities and remarkable gene stability. The BYSMV vector containing the Cas9 and gRNA cassettes was used as an expression platform to edit *N. benthamiana* plants (Gao et al. 2019). Also, the negative-stranded RNA virus sonchus yellow net rhabdovirus (SYNV) was used to deliver Cas9 and sgRNA into tobacco leaves (Ma et al. 2020). The SYNV-based vector contains a tRNA SYNV–tgtRNA–Cas9 construct to target multiple sequences at a time. The agroinfiltration of the engineered SYNV into tobacco leaves showed mutagenesis efficiency ranging from 40 to 91%. The virus-free progenies of the regenerated mutants did not need selection as more than 90% were mutated. The mechanical transfer of the produced viruses to other plants is facilitated by the stability of the vector.

6 Transient Expression of CRISPR/Cas Ribonucleoproteins (RNPs)

Traditional transformation strategies usually develop transgenic plants and require several steps to eliminate the transgene. In addition, transformation systems are genotype-dependent; therefore, they require the adaptation of each genotype and may face low transformation efficiency. The utilization of various vehicle-based delivery systems for direct transfer to plant cells represents a recently developed approach for delivering genetic material directly to plant genomes. Protoplasts, zygotes, and pollen serve as vehicles for the direct administration of the CRISPR/Cas system, facilitating transgene-free genome editing (Fig. 4). Chemical or physical methods are used to deliver the assembled Cas enzyme with the target gRNA to avoid the need for gene transcription and translation and speed up the evaluation of edited plants (Kim et al. 2020). The assembled Cas-gRNA system is transiently expressed as it is exposed to degradation (Liang et al. 2017; Banakar et al. 2019).

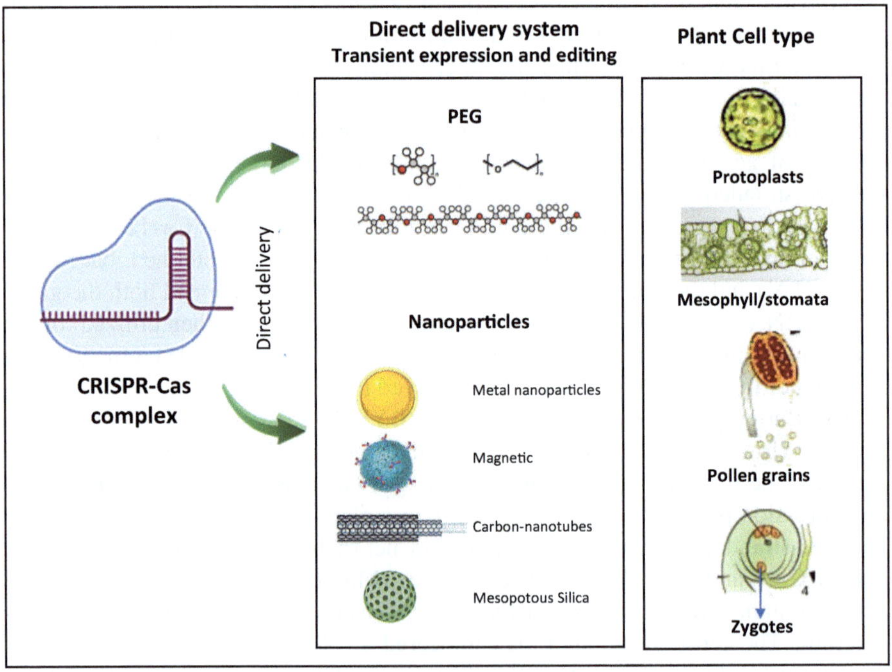

Fig. 4 Advancement of transgene-free genome editing techniques in plants, specifically through the utilization of in vitro/chemically generated assembled ribonucleoproteins (RNPs) consisting of Cas/gRNA. The utilization of polyethylene glycol (PEG) as a means of delivering preassembled Cas/gRNA ribonucleoproteins (RNPs) into plant protoplasts. The prevailing categories of nanoparticles are presently employed in the field of biotechnology. Nanoparticles, including metal, magnetic, and silica nanoparticles and carbon nanotubes, exhibit promising potential as effective carriers for delivering nucleoproteins into plant cells via stomata pores, pollen grains, or zygotes

This system develops non-GM crops and avoids regulatory oversight (Wolter and Puchta 2017).

The transport of constructed CRISPR-Cas ribonucleoproteins to plant protoplasts was achieved using various methods, including polyethylene glycol-calcium (PEG-Ca2+)-mediated cell transfection, electroporation, and lipofection (Yue et al. 2021; Baek et al. 2016; Liu et al. 2020a). The Cas-gRNA complex, once built and accompanied by PEG, is transported into the nucleus, where it initiates the process of genome editing. Nevertheless, the process of regenerating plants from protoplasts is characterized by inefficiency in numerous plant species, and it is worth noting that protoplast regeneration might lead to genome instability, as indicated by studies conducted by Yue et al. (2021) and Fossi et al. (2019). In addition, transformation with PEG may limit reproducibility due to its cytotoxic effects. The lipofection of tobacco BY2 protoplasts and the electro-transfection of cabbage protoplasts were implemented for editing using the assembled CRISPR/Cas9 RNP but had low efficiency (Liu et al. 2020a; Lee et al. 2020) (Fig. 4).

Nanoparticles, such as inorganic, carbon-based, silicon-based, and polymeric nanoparticles, could deliver any biomolecules (DNA, RNA, protein, or RNP) to plants (Fig. 4). Nanoparticle-mediated delivery is a promising protocol for improvement in cargo delivery as it is species-independent in a nonintegrating manner and is a good candidate for improving the gene editing efficiency of desired plants. The utilization of nanovehicle-based approaches for the direct delivery of DNA to plant genomes is a potential solution to address the challenge of genetic manipulation in some recalcitrant plant genotypes of significance. Furthermore, it serves the purpose of safeguarding the transferred DNA from the enzymatic degradation caused by nucleases that are found within the cellular cytoplasm. The diminutive dimensions of nanoparticles facilitate their passage across the cellular membrane, while the chemical composition of these nanoparticles enables the binding of CRISPR RNA-protein complexes and shields them from degradation. Silica-based nanoporous materials have distinct characteristics that render them highly favorable for the delivery of substances to targeted plant species. The potential attributes of porosity structure, biocompatibility, biodegradability, and surface chemistry render them viable options for controlled and targeted delivery systems. The use of enzymatic or light-mediated cleavage release methods with nanovehicle/RNP delivery method is a promising method for improving gene editing efficiency (Ahmar et al. 2021; Demirer et al. 2021; Nadakuduti and Enciso-Rodríguez 2021; Wang et al. 2021b). Nanoparticles with a diameter of approximately 10 nm have the capability to enter the mesophyll tissue of plants via stomatal holes. However, particles larger than 10 nm require chemical or physical techniques in order to be successfully transported into plants. The delivery of constructed Cas/gRNA RNPs using particle bombardment has been demonstrated to be effective for transfection into pollen or zygote, as reported by Gong et al. (2021). The study conducted by Toda et al. (2019) demonstrated the use of polyethylene glycol (PEG) as a means of delivering assembled Cas9/gRNA ribonucleoprotein (RNP) to the early stages of gamete development in rice. Furthermore, it has been observed that pollen grains can be effectively utilized as carriers for the transport of constructed ribonucleoprotein complexes

(RNPs) by the application of magnetic nanotechnological methods (Zhao et al. 2017). The pollen magnetofection technology is characterized by its simplicity, equipment-free nature, independence from culture conditions and genotype, as well as ability to transport numerous genes. Gold-based nanoparticles have been found to be a superior choice for delivering CRISPR-Cas9 editing compared to other types of nanoparticles, such as lipid NP, AuNP, and AuNC. This is due to their remarkable physicochemical stability and exceptional biocompatibility, as demonstrated in studies conducted by Chen et al. (2019) and Vats et al. (2022). The main obstacles to using nanoparticles as delivery systems are optimization of the correct dose, biosafety knowledge of the nanoparticles, binding affinity of cargos and nanoparticles, and mechanism of delivery. Successfully edited wheat was developed using foliar application of plasmid-coated carbon dots (Doyle et al. 2019). The plasmid carried the Cas9 gene and gRNAs to make a deletion in the SPO11 genes. The delivery system using foliar application (spraying on) was also successful in maize and could be extended to other plants.

7 Morphogenic Transcription Factor Mediated to Accelerating Genome Edited

Delivering CRISPR reagents (Cas9 and sgRNAs) for developing edited plants requires an effective and genotype-flexible transformation system. Plant regeneration has been improved by the ectopic overexpression of plant transcription factors known as morphogenic regulators (MRs), such as BABY BOOM (BBM), WUSCHEL2 (WUS2), shoot meristemless (STM), and growth-regulating factor (GRF)/GRF-interacting factor (GRF-GIF) (Debernardi et al. 2020; Lowe et al. 2016). These reprograming genes promote somatic embryogenesis or the regeneration of shoots, improving the efficiency of plant transformation. This approach has been successfully used to transform different crops, such as cotton, rice, soybean, wheat, and maize, including genotypes otherwise recalcitrant to *Agrobacterium*-mediated transformation (Debernardi et al. 2020; Lowe et al. 2016, 2018 Aesaert et al. 2022; Hoerster et al. 2020; Che et al. 2022; Masters et al. 2020; Zhang et al. 2019;).

The overexpression of the MR genes Bbm and Wus2 in maize has been found to facilitate direct somatic embryogenesis and significantly improve the efficiency of *Agrobacterium*-mediated transformation in many crops, including maize, sorghum, sugarcane, rice, and wheat (Lowe et al. 2016; Gordon-Kamm et al. 2019).

Bbm/Wus2 gene expression cassettes were used to transform maize, enhancing the percentage of transformed embryogenic calli to 45% (Lowe et al. 2016). However, the overexpression of Bbm/Wus2 inhibits plant regeneration and fertility. To avoid the harmful effect of Bbm/Wus2, scientists used an inducible promoter-driven cre/loxP system to remove the Bbm/Wus2 cassettes. Two loxP sites flank the Bbm, Wus2, and cre genes, and activating the cre gene will remove these genes

(Lowe et al. 2016; Masters et al. 2020; Mookkan et al. 2017). Also, inducible or tissue-specific expression promoters were used to control the expression of Bbm/Wus2 cassettes (Aregawi et al. 2022; Beyene et al. 2022). In addition, the nonintegrating Wus2 strategy was also used to avoid the deleterious effects of Wus2 overexpression. The success of the process relies on the utilization of the cotransformation method involving two distinct strains of *Agrobacterium*. One strain carries the Wus2 expression cassette, while the other strain carries selectable and visual marker cassettes. The efficacy of the technique relies on the optimization of the ratio between the two strains of *Agrobacterium*, aiming to achieve an adequate level of Wus2 expression while minimizing excessive integration into the plant genome.

The overexpression of growth-regulating factors (GRFs) also enhances plant growth and development (Debernardi et al. 2020). The miR396 is a key element in regulating the expression of several GRFs and affects plant growth and development. In wheat, the overexpression of wheat GRF4 and its cofactor GRF-interactive factor1 (GIF1) improved transformation frequencies 7.8-fold as well as regeneration efficiency (Debernardi et al. 2020). According to Qiu et al. (2022), it was observed that the negative regulation of GRF4 by miR396 was enhanced by point mutation at the miR396 target location, resulting in a significant improvement in regeneration and base editing efficiency, with fold increases ranging from two to nine. The overexpression of the STM gene from *Brassica oleracea* in *Arabidopsis* resulted in a twofold increase in somatic embryogenesis. Similarly, the overexpression of the maize STM gene KN1 in tobacco led to a threefold increase in shoot organogenesis.

8 CRISPR-Combo

The CRISPR-Combination (CRISPR-Combo) platform was created with the purpose of assigning two separate capabilities to Cas9 and sgRNA in plants. These functionalities include genome editing, which encompasses targeted mutagenesis or base editing, as well as gene activation. Notably, this platform achieves these functionalities with a single Cas9 protein. The activation of genes was successfully accomplished by the utilization of the CRISPR-Act3.0 system, which involves the recruitment of transcriptional activators by the sgRNA2.0 scaffold. This scaffold is equipped with two MS2 RNA aptamers, enabling the binding of transcriptional activators (Pan et al. 2021). The MS2 adaptor demonstrates an affinity for the complex formed by the MS2 bacteriophage coat protein and the SunTag activator, as depicted in Fig. 5. Each SunTag peptide has the ability to recruit a total of ten activator copies. The study conducted by Pan et al. (2022) demonstrates the efficacy of CRISPR-Combo in facilitating orthogonal genome editing, specifically by double-strand break (DSB)-mediated genome editing or base editing, as well as transcriptional activation in several plant species, including *Arabidopsis*, rice, tomato, and poplar. This methodology exhibits a notable degree of editing efficiency and offers

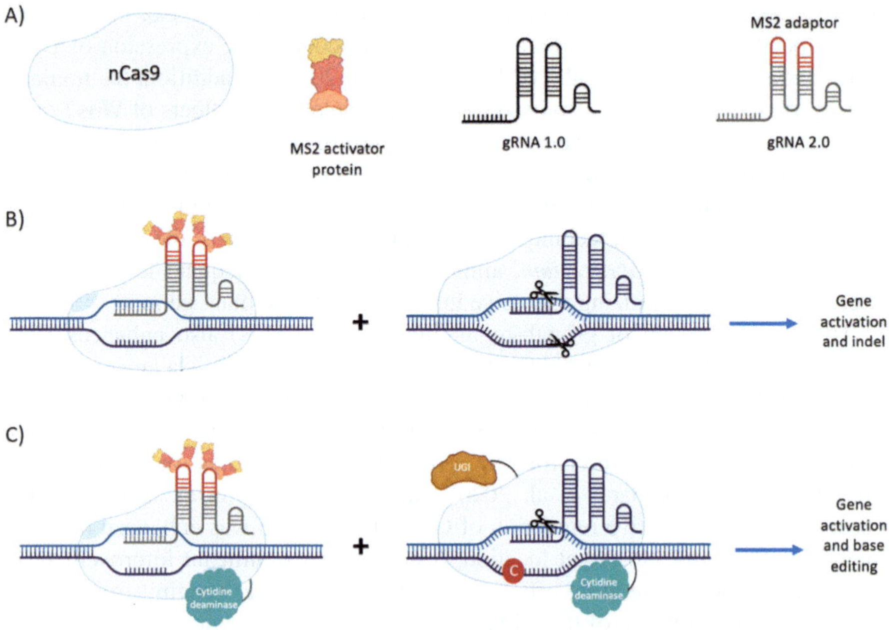

Fig. 5 A schematic representation of the process of genome editing and transcriptional activation facilitated by the Cas9-Act3.0 system. The Cas9-Act3.0 system necessitates the presence of Cas9 nuclease, a complex of MS2 activators, and two distinct sgRNA scaffolds, namely, gR1.0 and gR2.0. The gR2.0 construct incorporates two MS2 RNA-binding sequences, which facilitate the interaction of two MS2 activator complexes. Cas9-Act3.0-mediated double-strand breaks: a review of genome editing and transcriptional activation. The technique employed in this study involves the use of Cas9-Act3.0 to facilitate C-to-T base editing and transcriptional activation

significant time and cost savings for the molecular identification of transgene-free plants that have undergone genome editing.

CRISPR-Combo-mediated speed breeding was achieved by activating the florigen gene (FT) by inducing early flowering in *Arabidopsis*. With the use of flora dip transformation, the flowering time of edited plants is reduced, and the edited plants become transgene-free and do not require a growth chamber for speed breeding (Pan et al. 2022). In addition, CRISPR-Combo systems are designed to selectively activate endogenous morphogenic genes through the utilization of targeted sgRNAs. The application of CRISPR-Combo in poplar was employed to speed up plant regeneration in rice through the augmentation of endogenous morphogenic genes, namely, PtWUS and PtWOX11. The application of genome-editing techniques in poplar plants resulted in an increase in the regenerative capacity of callus and the development of root organs from petiole and stem cuttings, leading to improved production of shoot and root biomass. The acceleration of genome-edited plant regeneration was accomplished through the activation of the endogenous morphogenic gene OsBBM1, which bears a resemblance to auxin, in rice. Nevertheless, the utilization of a robust promoter, such as DEX, for the overexpression of the OsBBM1 gene hinders shoot regeneration and necessitates the application of exogenous cytokinin (Khanday et al. 2020). The task necessitated the substitution of the native promoter with the desired target promoter. The issue could be circumvented through the utilization of CRISPR-Combo, which enables the overexpression of an endogenous morphogenic gene by incorporating transcriptional regulation, hence obviating the necessity of introducing exogenous cytokinin. According to Pan et al. (2022), the expression of OsBBM1 in leaves is subject to tissue-specific control, which serves to inhibit the excessive production of the hormone and mitigate its adverse consequences.

9 Conclusion

Novel plant breeding strategies, such as genome editing, have the potential to provide solutions to deal with future challenges in global climate change and food security. During the last 10 years, genome editing, particularly the CRISPR/Cas9 system, has stimulated the ongoing revolution in crop improvement. This system has made considerable progress in developing plants tolerant to biotic and abiotic stresses and has improved nutrient content and yield. However, regulatory approval and consumer acceptance play a vital role in commercializing the existing genome-edited crops. Transgenic-free edited plants could have good prospects of commercialization as they may overcome ethical and safety considerations. Transgene-free edited plants can be achieved through the introduction of the Cas9 enzyme and guide RNA (gRNA), either through transitory means or by outcrossing from transgenic edited offspring. The process of genetically modifying plants through the utilization of *Agrobacterium* or particle-bombardment-mediated DNA delivery methods is characterized by its labor-intensive nature and time-consuming requirements. Furthermore, it is important to note that not all crop varieties are amenable to transformation or capable of successful regeneration posttransformation. RNA and RNP delivery of CRISPR/Cas9 reagents is used to avoid the integration of DNA into the plant genome and to decrease off-target mutations. These systems realize that on the host cells, using embryonic cells has low efficiency, with low percentages of edited plants. When protoplasts are used, efficiency increases but still has challenges in isolation, culturing, and plant regeneration. The primary limitation pertaining to the utilization of RNA viruses for the delivery of CRISPR/Cas9 transcripts and the subsequent development of transgenic, genetically modified plants is the host range that is inherently linked to particular viruses. The transference of mobile Cas9-TLS and gRNA-TLS fusions from transgenic rootstock to wild-type shoots via grafting facilitated genome editing, resulting in the production of seeds with heritably modified genomes in the flowers. The main limitation of using this method is the development of transgenic stock for each target, and monocot grafting is not established. Although the efficiency of this protocol is low, it is easy to implement for obtaining transgenic-free plants and could graft any compatible plants together, such as *Arabidopsis* with *Brassica* and tomato. Novel techniques have been developed to accelerate and improve genome editing, such as CRISPR-Combo to achieve the dual functionality of a single Cas9, morphogenic transcription factor mediated for the easy screening of edited plants, and GRAND editing for precisely inserting large DNA fragments without DNA donors. Some countries do not regulate edited crops with no foreign gene integration. But now is the time, more than ever, to develop transgenic-free edited plants to improve varieties of various crops to mitigate global climate change, improve crop productivity, and overcome food security.

References

Abdallah NA, Hamwieh A, Radwan K, Fouad N, Prakash C (2021) Genome editing techniques in plants: a comprehensive review and future prospects toward zero hunger. GM Crop Food 12(2):601–615

Aesaert S, Impens L, Coussens G, Van Lerberge E, Vanderhaeghen R, Desmet L et al (2022) Optimized transformation and gene editing of the B104 public maize inbred by improved tissue culture and use of morphogenic regulators. Front Plant Sci 13:883847. https://doi.org/10.3389/fpls.2022.883847

Ahmar S, Mahmood T, Fiaz S, Poblet F, Shafique MS, Chattha MS et al (2021) Advantage of nanotechnology-based genome editing system and its application in crop improvement. Front Plant Sci 12:663849

Altpeter F, Springer NM, Bartley LE, Blechl AE, Brutnell TP, Citovsky V et al (2016) Advancing crop transformation in the era of genome editing. Plant Cell 28(7):1510–1520. https://doi.org/10.1105/tpc.16.00196

Anzalone AV, Randolph PB, Davis JR, Sousa AA, Koblan LW, Levy JM et al (2019) Search-and-replace genome editing without double-strand breaks or donor DNA. Nature 576(7785):149–157

Aregawi K, Shen J, Pierroz G, Sharma MK, Dahlberg J, Owiti J, Lemaux PG (2022) Morphogene-assisted transformation of Sorghum bicolor allows more efficient genome editing. Plant Biotechnol J 20(4):748–760

Ariga H, Toki S, Ishibashi K (2020) Potato virus X vector-mediated DNA-free genome editing in plants. Plant Cel Physiol 61(11):1946–1953. https://doi.org/10.1093/pcp/pcaa123

Baek K, Kim DH, Jeong J, Sim SJ, Melis A, Kim J-S et al (2016) DNA-free two-gene knockout in Chlamydomonas Reinhardtii via CRISPR-Cas9 ribonucleoproteins. Sci Rep 6(1):30620. https://doi.org/10.1038/srep30620

Banakar R, Eggenberger AL, Lee K, Wright DA, Murugan K, Zarecor S et al (2019) High-frequency random DNA insertions upon co-delivery of CRISPR-Cas9 ribonucleoprotein and selectable marker plasmid in rice. Sci Rep 9(1):19902. https://doi.org/10.1038/s41598-019-55681-y

Beyene G, Chauhan RD, Villmer J, Husic N, Wang N, Gebre E et al (2022) CRISPR/Cas9-mediated tetra-allelic mutation of the 'green Revolution' SEMIDWARF-1 (SD-1) gene confers lodging resistance in tef (Eragrostis tef). Plant Biotechnol J 20(9):1716–1729

Bhattacharjee S, Bhowmick R, Kant L, Paul K (2023) Strategic transgene-free approaches of CRISPR-based genome editing in plants. Mol Gen Genomics 298(3):507–520. https://doi.org/10.1007/s00438-023-01998-3. Epub 2023 Feb 25. PMID: 36840794; PMCID: PMC9958309

Bing L, Ma M, Liu L, Wang J, Niu L, Xi F (2020) An investigation of the global uptake of CO_2 by lime from 1930 to. Earth Syst Sci Data 15(2431–2444):2023. https://doi.org/10.5194/essd-15-2431-2023

Cao HW, Zhao YN, Liu XS, Rono JK, Yang ZM (2022) A metal chaperone OsHIPP16 detoxifies cadmium by repressing its accumulation in rice crops. Environ Pollut 311:120058. https://doi.org/10.1016/j.envpol.2022.120058

Che P, Wu E, Simon MK, Anand A, Lowe K, Gao H et al (2022) Wuschel2 enables highly efficient CRISPR/Cas-targeted genome editing during rapid de novo shoot regeneration in sorghum. Commun Biol 5(1):344. https://doi.org/10.1038/s42003-022-03308-w

Chen R, Huang H, Liu H, Xi J, Ning J, Zeng W et al (2019) Friend or foe? Evidence indicates endogenous exosomes can deliver functional gRNA and Cas9 protein. Small 15(38):1902686. https://doi.org/10.1002/smll.201902686

Clemens M, Faralli M, Lagreze J, Bontempo L, Piazza S, Varotto C et al (2022) VvEPFL9-1 knock-out via CRISPR/Cas9 reduces stomatal density in grapevine. Front Plant Sci 13:878001. https://doi.org/10.3389/fpls.2022.878001

Debernardi JM, Tricoli DM, Ercoli MF, Hayta S, Ronald P, Palatnik JF, Dubcovsky J (2020) A GRF–GIF chimeric protein improves the regeneration efficiency of transgenic plants. Nat Biotechnol 38(11):1274–1279

Demirer GS, Silva TN, Jackson CT, Thomas JB, Ehrhardt D, Rhee SY et al (2021) Nanotechnology to advance CRISPR–Cas genetic engineering of plants. Nat Nanotechnol 16:243–250. https://doi.org/10.1038/s41565-021-00854-y

Doyle C, Higginbottom K, Swift TA, Winfield M, Bellas C, Benito-Alifonso D et al (2019) A simple method for spray-on gene editing in planta. *bioRxiv*. https://doi.org/10.1101/805036

Ellison EE, Chamness JC, Voytas DF (2021) Viruses as vectors for the delivery of gene-editing reagents. In: Genome editing for precision crop breeding. Burleigh Dodds Science Publishing, Cambridge, pp 97–122. https://doi.org/10.1201/9781003048237-5

Feng X, Xiong J, Zhang W, Guan H, Zheng D, Xiong H et al (2022) ZmLBD5, a class-II LBD gene, negatively regulates drought tolerance by impairing abscisic acid synthesis. Plant J 112(6):1364–1376

Fossi M, Amundson K, Kuppu S, Britt A, Comai L (2019) Regeneration of *Solanum tuberosum* plants from protoplasts induces widespread genome instability. Plant Physiol 180(1):78–86. https://doi.org/10.1104/pp.18.00906

Galli M, Hochstein S, Iqbal D, Claar M, Imani J, Kogel KH (2022) CRISPR/Sp Cas9-mediated KO of epigenetically active MORC proteins increases barley resistance to bipolaris spot blotch and fusarium root rot. J Plant Dis Prot 129(4):1005–1011. https://doi.org/10.1007/s41348-022-00574-y

Gao C (2021) Genome engineering for crop improvement and future agriculture. Cell 184(6):1621–1635. https://doi.org/10.1016/j.cell.2021.01.005

Gao Q, Xu WY, Yan T, Fang XD, Cao Q, Zhang ZJ et al (2019) Rescue of a plant cytorhabdovirus as versatile expression platforms for planthopper and cereal genomic studies. New Phytol 223(4):2120–2133. https://doi.org/10.1111/nph.15889

Gong Z, Cheng M, Botella JR (2021) Non-GM genome editing approaches in crops. Front Genome Ed 3:817279. https://doi.org/10.3389/fgeed.2021.817279

Gordon-Kamm B, Sardesai N, Arling M, Lowe K, Hoerster G, Betts S, Jones T (2019) Using morphogenic genes to improve recovery and regeneration of transgenic plants. Plan Theory 8(2):38. https://doi.org/10.3390/plants8020038. PMID: 30754699; PMCID: PMC6409764

Han X, Chen Z, Li P, Xu H, Liu K, Zha W et al (2022) Development of novel rice germplasm for salt-tolerance at seedling stage using CRISPR-Cas9. Sustainability 14(5):2621. https://doi.org/10.3390/su14052621

He Y, Zhu M, Wang L, Wu J, Wang Q, Wang R, Zhao Y (2018) Programmed self-elimination of the CRISPR/Cas9 construct greatly accelerates the isolation of edited and transgene-free rice plants. Mol Plant 11(9):1210–1213. https://doi.org/10.1016/j.molp.2018.05.005

He F, Wang C, Sun H, Tian S, Zhao G, Liu C, Wan C, Guo J, Huang X, Zhan G (2022) Simultaneous editing of three homoeologs of TaCIPK14 confers broad-spectrum resistance to stripe rust in wheat. Plant Biotechnol J 21(2):354–368

Hoerster G, Wang N, Ryan L, Wu E, Anand A, McBride K et al (2020) Use of non-integrating Zm-Wus2 vectors to enhance maize transformation: non-integrating WUS2 enhances transformation. In Vitro Cell Dev Biol Plant 56(3):265–279. https://doi.org/10.1007/s11627-019-10042-2

Hoffie RE, Perovic D, Habekuß A, Ordon F, Kumlehn J (2023) Novel resistance to the Bymovirus BaMMV established by targeted mutagenesis of the PDIL5-1 susceptibility gene in barley. Plant Biotechnol J 21(2):331–341

Huang H, Liu C, Yang C, Kanwar MK, Shao S, Qi Z, Zhou J (2022) BAG9 confers Thermotolerance by regulating cellular redox homeostasis and the stability of heat shock proteins in *Solanum lycopersicum*. Antioxidants 11(8):1467. https://doi.org/10.3390/antiox11081467

IPCC Sixth Assessment Report (2022) Climate change 2022: impacts, adaptation and vulnerability. https://www.ipcc.ch/report/ar6/wg2/

Jiao B, Hao X, Liu Z, Liu M, Wang J, Liu L, Liu N, Song R, Zhang J, Fang Y (2022) Engineering CRISPR immune systems conferring GLRaV-3 resistance in grapevine. Hortic Res 9:uhab023

Khanday I, Santos-Medellín C, Sundaresan V (2020) Rice embryogenic trigger BABY BOOM1 promotes somatic embryogenesis by upregulation of auxin biosynthesis genes. BioRxiv:2020–2008. https://doi.org/10.1101/2020.08.24.265025

Kim YJ, Lee H, Cha H, Park JH (2020) Non-viral gene disruption by CRISPR/Cas9 delivery using cell-permeable and protein-stabilizing 30Kc19 protein. Biotechnol Bioprocess Eng 25(5):724–733. https://doi.org/10.1007/s12257-020-0068-8

Lee MH, Lee J, Choi SA, Kim Y-S, Koo O, Choi SH et al (2020) Efficient genome editing using CRISPR-Cas9 RNP delivery into cabbage protoplasts via electro-transfection. Plant Biotechnol Rep 14(6):695–702. https://doi.org/10.1007/s11816-020-00645-2

Li C, Zong Y, Jin S, Zhu H, Lin D, Li S et al (2020) SWISS: multiplexed orthogonal genome editing in plants with a Cas9 nickase and engineered CRISPR RNA scaffolds. Genome Biol 21:1–15. https://doi.org/10.1186/s13059-020-02051-x

Li M, Guo S, Zhang J, Sun H, Tian S, Wang J, Zuo Y, Yu Y, Gong G, Zhang H (2022a) Sugar transporter VST1 knockout reduced aphid damage in watermelon. Plant Cell Rep 41(1):277–279

Li S, Lin D, Zhang Y, Deng M, Chen Y, Lv B, Li B, Lei Y, Wang Y, Zhao L (2022b) Genome-edited powdery mildew resistance in wheat without growth penalties. Nature 602(7897):455–460

Li H, Zhu Z, Li S, Li J, Yan L, Zhang C, Ma Y, Xia L (2022c) Multiplex precision gene editing by a surrogate prime editor in rice. Mol Plant 15(7):1077–1080

Liang Z, Chen K, Li T, Zhang Y, Wang Y, Zhao Q et al (2017) Efficient DNA-free genome editing of bread wheat using CRISPR/Cas9 ribonucleoprotein complexes. Nat Commun 8(1):14261. https://doi.org/10.1038/ncomms14261

Liang Z, Wu Y, Guo Y, Liu Y, Ma L, Wu Y (2021) Bi-functional selection markers assist segregation of transgene-free, genome-edited mutants. Sci China Life Sci 64:1567–1570. https://doi.org/10.1007/s11427-020-1820-9. Epub 2021 Jan 27. PMID: 33521855

Liu W, Rudis MR, Cheplick MH, Millwood RJ, Yang J-P, Ondzighi-Assoume CA et al (2020a) Lipofection-mediated genome editing using DNA-free delivery of the Cas9/gRNA ribonucleoprotein into plant cells. Plant Cel Rep 39(2):245–257. https://doi.org/10.1007/s00299-019-

Liu L, Zhang J, Xu J, Li Y, Guo L, Wang Z et al (2020b) CRISPR/Cas9 targeted mutagenesis of SlLBD40, a lateral organ boundaries domain transcription factor, enhances drought tolerance in tomato. Plant Sci 301:110683. https://doi.org/10.1016/j.plantsci.2020.110683

Long Q, Du M, Long J, Xie Y, Zhang J, Xu L, He Y, Li Q, Chen S, Zou X (2021) Transcription factor WRKY22 regulates canker susceptibility in sweet orange (Citrus sinensis Osbeck) by enhancing cell enlargement and CsLOB1 expression. Hortic Res 8:50

Lowe K, Wu E, Wang N, Hoerster G, Hastings C, Cho MJ et al (2016) Morphogenic regulators baby boom and Wuschel improve monocot transformation. Plant Cell 28(9):1998–2015

Lowe K, La Rota M, Hoerster G, Hastings C, Wang N, Chamberlin M et al (2018) Rapid genotype "independent" Zea mays L.(maize) transformation via direct somatic embryogenesis. In Vitro Cell Dev Biol Plant 54:240–252

Ma X, Zhang X, Liu H, Li Z (2020) Highly efficient DNA-free plant genome editing using virally delivered CRISPR-Cas9. Nat Plants 6:773–779. https://doi.org/10.1038/s41477-020-0704-5

Masters A, Kang M, McCaw M, Zobrist JD, Gordon-Kamm W, Jones T, Wang K (2020) Agrobacterium-mediated immature embryo transformation of recalcitrant maize inbred lines using morphogenic genes. J Vis Exp 156:e60782. https://doi.org/10.3791/60782. PMID: 32116304

Mookkan M, Nelson-Vasilchik K, Hague J, Zhang ZJ, Kausch AP (2017) Selectable marker independent transformation of recalcitrant maize inbred B73 and sorghum P898012 mediated by morphogenic regulators BABY BOOM and WUSCHEL2. Plant Cell Rep 36:1477–1491. https://doi.org/10.1007/s00299-017-2169-1. Epub 2017 Jul 5. PMID: 28681159; PMCID: PMC5565672.

Nadakuduti SS, Enciso-Rodríguez F (2021) Advances in genome editing with CRISPR systems and transformation technologies for plant DNA manipulation. Front Plant Sci 11:637159. https://doi.org/10.3389/fpls.2020.637159

Nelson JW, Randolph PB, Shen SP, Everette KA, Chen PJ, Anzalone AV et al (2022) Engineered pegRNAs improve prime editing efficiency. Nat Biotechnol 40(3):402–410. https://doi.org/10.1038/s41587-021-01039-7

Pan C, Wu X, Markel K, Malzahn AA, Kundagrami N, Sretenovic S et al (2021) CRISPR–Act3. 0 for highly efficient multiplexed gene activation in plants. Nat plants 7(7):942–953

Pan C, Li G, Malzahn AA, Cheng Y, Leyson B, Sretenovic S et al (2022) Boosting plant genome editing with a versatile CRISPR-combo system. Nat Plants 8(5):513–525

Qiu F, Xing S, Xue C, Liu J, Chen K, Chai T, Gao C (2022) Transient expression of a TaGRF4-TaGIF1 complex stimulates wheat regeneration and improves genome editing. Sci China Life Sci 65(4):731–738

Stuttmann J, Barthel K, Martin P, Ordon J, Erickson JL, Herr R, Ferik F, Kretschmer C, Berner T, Keilwagen J, Marillonnet S, Bonas U (2021) Highly efficient multiplex editing: one-shot generation of 8× *Nicotiana benthamiana* and 12× Arabidopsis mutants. Plant J 106:8–22. https://doi.org/10.1111/tpj.15197

Toda E, Koiso N, Takebayashi A, Ichikawa M, Kiba T, Osakabe K et al (2019) An efficient DNA- and selectable-marker-free genome-editing system using zygotes in rice. Nat Plants 5(4):363–368. https://doi.org/10.1038/s41477-019-0386-z

Uranga M, Vazquez-Vilar M, Orzáez D, Daròs J-A (2021) CRISPRCas12a genome editing at the whole-plant level using two compatible RNA virus vectors. CRISPR J 4(5):761–769. https://doi.org/10.1089/crispr.2021.0049

Vats S, Kumawat S, Brar J, Kaur S, Yadav K, Magar SG et al (2022) Opportunity and challenges for nanotechnology application for genome editing in plants. Plant Nano Biol 1:100001

Wang JW, Cunningham FJ, Goh NS, Boozarpour NN, Pham M, Landry MP (2021a) Nanoparticles for protein delivery in planta. Curr Opin Plant Biol 60:102052. https://doi.org/10.1016/j.pbi.2021.102052

Wang T, Xun H, Wang W, Ding X, Tian H, Hussain S et al (2021b) Mutation of GmAITR genes by CRISPR/Cas9 genome editing results in enhanced salinity stress tolerance in soybean. Front Plant Sci 12:779598. https://doi.org/10.3389/fpls.2021.779598. PMID: 34899806

Wang J, He Z, Wang G, Zhang R, Duan J, Gao P et al (2022) Efficient targeted insertion of large DNA fragments without DNA donors. Nat Methods 19(3):331–340. PMID: 35228726

Wei Z, Abdelrahman M, Gao Y, Ji Z, Mishra R, Sun H, Sui Y, Wu C, Wang C, Zhao K (2021) Engineering broad-spectrum resistance to bacterial blight by CRISPR-Cas9-mediated precise homology directed repair in rice. Mol Plant 14(8):1215–1218

Wolter F, Puchta H (2017) Genome engineering mit CRISPR/Cas—Revolution in der Pflanzenzüchtung. BIOspektrum 23:159–161. https://doi.org/10.1007/s12268-017-0782-8

Wu J, Yan G, Duan Z, Wang Z, Kang C, Guo L, Liu K et al (2020) Roles of the *Brassica napus* DELLA protein BnaA6.RGA, in modulating drought tolerance by interacting with the ABA signaling component BnaA10.ABF2. Front Plant Sci 11:577

Xu H, Yang X, Zhang Y, Wang H, Wu S, Zhang Z et al (2022) CRISPR/Cas9-mediated mutation in auxin efflux carrier OsPIN9 confers chilling tolerance by modulating reactive oxygen species homeostasis in rice. Front Plant Sci 13:967031. https://doi.org/10.3389/fpls.2022.967031

Yang L, Machin F, Wang S, Saplaoura E, Kragler F (2023) Heritable transgene-free genome editing in plants by grafting of wild-type shoots to transgenic donor rootstocks. Nat Biotechnol 1–10:958. https://doi.org/10.1038/s41587-022-01585-8

Yin K, Gao C, Qiu JL (2017) Progress and prospects in plant genome editing. Nat Plants 3:17107

Yu H, Zhao Y (2019) Fluorescence marker-assisted isolation of Cas9-free and CRISPR-edited *Arabidopsis* plants. In: Qi Y (ed) Plant genome editing with CRISPR systems. Methods in molecular biology, vol 1917. Humana, New York, NY. https://doi.org/10.1007/978-1-4939-8991-1_11

Yubing HE, Min ZHU, Lihao W, Junhua WU, Qiaoyan W, Rongchen W, Yunde Z (2019) Improvements of TKC technology accelerate isolation of transgene-free CRISPR/Cas9-edited rice plants. Rice Sci 26(2):109–117. https://doi.org/10.1016/j.rsci.2018.11.001

Yue JJ, Yuan JL, Wu FH, Yuan YH, Cheng QW, Hsu CT, Lin CS (2021) Protoplasts: from isolation to CRISPR/Cas genome editing application. Front Genome Ed 3:717017. https://doi.org/10.3389/fgeed.2021.717017

Zhan X, Zhang F, Zhong Z, Chen R, Wang Y, Chang L et al (2019) Generation of virus-resistant potato plants by RNA genome targeting. Plant Biotechnol J 17(9):1814–1822

Zhang W, Thieme CJ, Kollwig G, Apelt F, Yang L, Winter N et al (2016) tRNA-related sequences trigger systemic mRNA transport in plants. Plant Cell 28(6):1237–1249

Zhang Q, Zhang Y, Lu MH, Chai YP, Jiang YY, Zhou Y et al (2019) A novel ternary vector system united with morphogenic genes enhances CRISPR/Cas delivery in maize. Plant Physiol 181(4):1441–1448. https://doi.org/10.1104/pp.19.00767

Zhang X, Kang L, Zhang Q, Meng Q, Pan Y, Yu Z et al (2020a) An RNAi suppressor activates in planta virus-mediated gene editing. Funct Integr Genomics 20(4):471–477. https://doi.org/10.1007/s10142-019-00730-y

Zhang Y, Li J, Chen S, Ma X, Wei H, Chen C et al (2020b) An APETALA2/ethylene responsive factor, OsEBP89 knockout enhances adaptation to direct-seeding on wet land and tolerance to drought stress in rice. Mol Gen Genomics 295:941–956. https://doi.org/10.1007/s00438-020-01669-7

Zhang L, Dong D, Wang J, Wang Z, Zhang J, Bai RY et al (2022a) A zinc finger protein SlSZP1 protects SlSTOP1 from SlRAE1-mediated degradation to modulate aluminum resistance. New Phytol 236(1):165–181. https://doi.org/10.1111/nph.18336

Zhang K, Zhuo C, Wang Z, Liu F, Wen J, Yi B, Shen J, Ma C, Fu T, Tu J (2022b) BnaA03. MKK5-BnaA06. MPK3/BnaC03. MPK3 module positively contributes to *Sclerotinia sclerotiorum* resistance in *Brassica napus*. Plants 11(5):609

Zhang Y, Guo W, Chen L, Shen X, Yang H, Fang Y et al (2022c) CRISPR/Cas9-mediated targeted mutagenesis of GmUGT enhanced soybean resistance against leaf-chewing insects through flavonoids biosynthesis. Front Plant Sci 13:802716. https://doi.org/10.3389/fpls.2022.802716

Zhang YZ, Li GL, Hu SW, Liu JF, Jiang YS, Liu SY et al (2022d) Cloning and drought resistance analysis of soybean *GmHsps_p23*-like gene. Phyton Internat J Exp Bot 91:1183–1198. https://doi.org/10.32604/phyton.2022.018853

Zhang J, Xing J, Mi Q, Yang W, Xiang H, Xu L et al (2023) Highly efficient transgene-free genome editing in tobacco using an optimized CRISPR/Cas9 system, pOREU3TR. Plant Sci 326:111523. https://doi.org/10.21203/rs.3.rs-1637574/v1

Zhao X, Meng Z, Wang Y, Chen W, Sun C, Cui B et al (2017) Pollen magnetofection for genetic modification with magnetic nanoparticles as gene carriers. Nat Plants 3(12):956–964. https://doi.org/10.1038/s41477-017-0063-z

Zhou X, Sun K, Zhou X, Jackson AO, Li Z (2019) The matrix protein of a plant rhabdovirus mediates superinfection exclusion by inhibiting viral transcription. J Virol 93(20):10–1128. https://doi.org/10.1007/s00438-023-01998-3

Zong Y, Liu Y, Xue C, Li B, Li X, Wang Y et al (2022) An engineered prime editor with enhanced editing efficiency in plants. Nat Biotechnol 40(9):1394–1402. https://doi.org/10.1038/s41587-022-01254-w

[illegible]

Genome-Editing Technologies in Crop Improvement

Richa Das, Pradeep Kumar, Shreni Agrawal, Kajal Singh, Nancy Singh, Sakshi Singh, Amit Kumar Singh, Vishnu D. Rajput, Praveen Kumar Shukla, Tatiana Minkina, Indrani Bhattacharya, Sunil Kumar Mishra, and Kavindra Nath Tiwari

Abstract Global climate change, together with harmful biotic and abiotic factors, is limiting agricultural yields, rendering it challenging to meet the rising need for food supply on a global scale. The development of hybrid cultivars with greater agricultural output was facilitated by the use of applied genetics for improving crops. However, it becomes a challenge for crop breeders to develop new varieties using current gene pools. The time needed to develop novel varieties of crops with the necessary agronomic features limits the utilization of traditional breeding methods. This is dependent not only on the amount of time required for growing in a particular season and the time needed to reach maturity but also on the usage of numerous cycles and phases of breeding processes, which involve crossing, selection, and testing. Furthermore, conventional physical and chemical mutagenesis approaches do not permit targeted genomic effects. Genome-editing approaches have evolved as significant tools for accurately altering genomes at specific

R. Das (✉) · S. Agrawal · I. Bhattacharya
Department of Biotechnology, Parul Institute of Applied Science, Parul University, Vadodara, Gujarat, India

P. Kumar · P. K. Shukla · K. N. Tiwari
Department of Botany, MMV, Banaras Hindu University, Varanasi, UP, India

K. Singh · N. Singh
Department of Biosciences, Galgotias University, Greater Noida, Gautam Buddh Nagar, UP, India

S. Singh
Department of Bioscience and Biotechnology, Banasthali Vidhyapith, Niwai, Tonk, Rajasthan, India

A. K. Singh · S. K. Mishra
Department of Pharmaceutical Engineering and Technology, Indian Institute of Technology, Banaras Hindu University, Varanasi, UP, India

V. D. Rajput (✉) · T. Minkina
Academy of Biology and Biotechnology, Southern Federal University, Rostov-on-Don, Russia

J.-T. Chen, S. Ahmar (eds.), *Plant Genome Editing Technologies*, Interdisciplinary Biotechnological Advances,
https://doi.org/10.1007/978-981-99-9338-3_4

locations. Genome-editing tools cause breaks in double-strand breaks (DSBs) at a target site and are repaired either through homologous recombination (HDR) or nonhomologous end joining (NHEJ) pathways. The amendment of the targeted genome at a precise location and the exclusion of foreign deoxyribonucleic acid (DNA) in crops that are genomically edited are some of the noteworthy properties of genome-editing tools, leading to their wide implementation in crop improvement. This chapter aims to highlight different genome-editing techniques, like zinc-finger nucleases (ZFNs), clustered regularly interspaced palindromic repeat (CRISPR)/CRISPR-associated (Cas) protein systems, and transcription activator-like effector nucleases (TALENs), followed by issues and challenges, and their applications in crop improvement.

Keywords Zinc-finger nucleases (ZFNs) · Clustered regularly interspaced palindromic repeat (CRISPR)/CRISPR-associated (Cas) protein · Transcription activator-like effector nucleases (TALENs) · Crop improvement

1 Introduction

Nowadays, environmental challenges and expanding human populations make it difficult to maintain the sustainable production of agricultural products. Approximately 720 to 811 million individuals still experience hunger. According to reports, over two billion people currently fall under the category of "food insecurity" (Herman et al. 2019). By 2050, 9.7 billion people will be on earth, which would require a 70% increase in agricultural output while reducing the environmental impact (FAO; IFAD; UNICEF; WFP; WHO. The State of Food Security and Nutrition in the World 2021; Lau et al. 2022). Along with the rapid population growth, climate change has a substantial impression on the production of food, creating considerable hazards to the availability of food due to extreme temperature or weather fluctuations, floods, droughts, and elevated soil salinity. Numerous studies show that changes in climatic patterns hugely affect agricultural production, which is more severe in developing nations than in developed nations, mostly because these nations are situated in tropical latitudes, where climate change is more sensitive. Due to climate change, abiotic and biotic forces on crops are likewise becoming more intense. Given that conventional agricultural methods cannot produce enough food to fulfill both present and future demands, novel techniques need to be developed to improve plant tolerance as the severity of climate change on crops increases (Yim 2017; Hamdan et al. 2022).

Several breeding approaches, such as mutation and crossbreeding, have been employed to improve the performance of crops even during harsh climatic changes. However, breeding techniques may be time-consuming and labor-intensive, taking up to 6–15 years or 8–10 years (Gao 2021; Lyzenga et al. 2021)

to develop a plant variety for agricultural output that is modified genetically. Moreover, the results of the breeding procedures are also uncertain. Therefore, a large population of mutagenized plants must undergo time-consuming screening and selection processes to find the desired traits (Ma et al. 2021). In addition to these breeding methods, a number of transgenic technologies employing recombinant deoxyribonucleic acid (DNA) technology have also been developed for crop improvement. This method primarily involves the movement of transgenes or other components of genes, whose functions are known, within plants in order to produce genetically modified (GM) plant varieties having advantageous properties (Zhang et al. 2018). This technique can provide higher quality and more nutritionally dense food than any other conventional method. However, this method has significant drawbacks due to limited public support for GM crops and, in turn, complicated and stringent safety regulatory processes (Herman et al. 2019). Therefore, genome-editing techniques can be one of the greatest approaches established in order to increase agricultural productivity and food security (Abdallah et al. 2015).

Genome editing has received a lot of consideration, especially among those who work in the agricultural sector. This is mostly attributable to the fact that it is simple to apply, accurate, and has the ability to develop new plant types by the exact insertion of specific characteristics or elimination of undesired genes. There are numerous methods for editing the genome that either use site-specific recombinase (SSR) or a site-specific nuclease (SSN) strategy (Abdallah et al. 2015). The three primary genome-editing methods are zinc-finger nucleases (ZFNs), clustered regularly interspaced palindromic repeat (CRISPR)/CRISPR-associated (Cas) protein systems, and transcription activator-like effector nucleases (TALENs). All these approaches enable the exact targeting and modification of particular target sequences through the following steps: (1) the target DNA sequence is recognized by an exogenous nuclease, which is engineered and composed of a nuclease domain and recognition module, (2) which then binds to the sequence and causes double-strand breaks (DSBs) at or near the target site. (3) After that, the double-strand breaks (DSBs) undergo repair through either homologous recombination (HR) or nonhomologous end joining (NHEJ). Insertions and deletions (indels) are frequently caused by NHEJ, which is prone to errors, whereas DSBs are precisely repaired by HR (Wada et al. 2020).

According to several investigations, these novel methods for modifying genomes produce more precise results than traditional approaches to crop improvement or conventional genetic engineering techniques. Thus, these technologies are very effective tools to apply to secure food supply globally. Genome editing promises that new crops will be generated more quickly with a very minimal chance of side effects. The main benefit of these approaches is that they can be used on any kind of crop, even those with complicated genomes that are challenging to breed with conventional methods and can be performed in any lab (Abdallah et al. 2015). In utilizing these methods, remarkable genetic alterations have been made to plants, e.g., improving their level of tolerance to biotic stresses (pathogenic fungi, bacteria, or viruses) or abiotic stresses (cold, salt, drought), boosting metabolic pathways, improving nutritional quality, increasing grain yield and quality, producing haploid seeds, and resisting herbicides (El-Mounadi et al. 2020).

The agricultural applications of this genome-editing method are already very promising, but they are still constrained by problems such as poor HR efficiency, off-target effects, and restricted protospacer adjacent motif (PAM) sequences. To overcome these restrictions, discoveries are fortunately constantly being added to genome-editing technology (Zhang et al. 2018).

2 Genome-Editing Technologies

Presently, there are three major genome-editing techniques categorized according to their mode of action (Table 1). The production of DSBs by SSNs at a targeted site is most often used in plant genomics (Mohanta et al. 2017; Jansing et al. 2019). For the recovery of these DSBs, cells can use their endogenous repair mechanisms. They can also be recovered through homologous recombination (HDR) or nonhomologous end joining (NHEJ) (Kamburova et al. 2021). Such an event results in generating single-base mutations arising through frameshift or indels of nucleotides (Mishra and Zhao 2018; Chen and Gao 2014). The generation of DSB can be accomplished by employing nucleases like CRISPR/Cas9, TALENs, and ZFNs. These nucleases or genome-editing tools have different modes of action and structural variations, which makes them different from each other based on their effectiveness, target choice, and specificity (Kamburova et al. 2021).

2.1 Zinc-Finger Nucleases (ZFNs)

ZFNs are among the earliest nucleases that are constructed artificially and intended for use in genome editing (Gaj et al. 2013). Cleavage and a DNA-binding domain are combined to construct ZFNs. The former is not sequence specific; it is a restriction endonuclease (type II) called FokI. The latter, on the other hand, is sequence specific and comprises Cys_2His_2 zinc fingers (ZFs) (Kim et al. 1996; Li et al. 2020a), where a single ZF has 30 amino acids, consisting of two β sheets, which are positioned in an antiparallel fashion opposite an α-helix (Beerli and Barbas 2002). Upon its binding with the target DNA, a single ZF binds specifically with three nucleotides (Fig. 1). This is achieved through the α-helix interaction of ZF with the DNA's major groove (Buck-Koehntop et al. 2012; Fairall et al. 1993). FokI identifies 5′-GGATG-3′:5'-CATCC-3′ (pentadeoxyribonucleotide), which is nonpalindromic, and cleaves nine to 13 nucleotides away from the site of recognition (Kim et al. 1994). It should also be emphasized that one zinc finger does not have enough specificity to bind to the target DNA. However, ZFNs constructed artificially often contain three or four zinc fingers, which allow them to bind to the site containing 18–24 oligonucleotides, followed by the dimerization of FokI, which is required for

Table 1 Gene-editing methods

Property	ZFN	TALEN	CRISPR/Cas9	References
Protein responsible for the recognition of the target site	ZF protein	TALE (RVD repeats)	sgRNA	Abdallah et al. (2015) and Li et al. (2020a)
Type of endonuclease involved	FokI enzyme	FokI enzyme	Cas9 and its different types	
Size of target sequence	9 to18 nucleotides per ZFN monomer, 18 to 36 nucleotides for a ZFN pair	14 to 20 nucleotides per TALEN monomer, 28 to 40 nucleotides for a TALEN pair	20 nucleotides guide sequence + PAM	
Construction of DNA-binding domain	3 to 4 ZF domains	8 to 31 TALE	sgRNA	
Action mechanism	Introduces DSBs at the target site	Introduces DSBs at the target site	Introduces DSBs at the target site by wild-type Cas9 or nicks at single strands by Cas9 nickases	
Repair mechanism	NHEJ	HDR	NHEJ	
Dimerization required	Yes	Yes	No	
Cleavage efficacy	Low	Moderate	High	
Affordability	Requires lots of resources and consumes more time	Affordable but consumes more time	Extremely affordable and consumes less time	
Level of difficulty while delivering	Comparatively simple because a wide range of viral vectors can use ZFN expression components because of their small size	Difficult since the functional elements are large	Moderately difficult because large Cas9 proteins may cause packaging issues for AAV (viral vector); however, small orthologs of the protein exist	

the effective cleavage of DNA (Gaj et al. 2013). In the course of dimerization, individual ZFN binds with the forward and reverse strands of DNA. Both ZFNs bound to the opposite strands need to be separated from each other by a spacer sequence, which is five to seven nucleotides in length (Zhang et al. 2017). ZFN functions as a dimer, generating DSBs with cohesive 5′ overhangs. These DSBs are repaired by HDR or NHEJ, giving rise to indels in the target DNA (Kamburova et al. 2021). The three factors that determine the specificity and target site recognition of ZFNs are

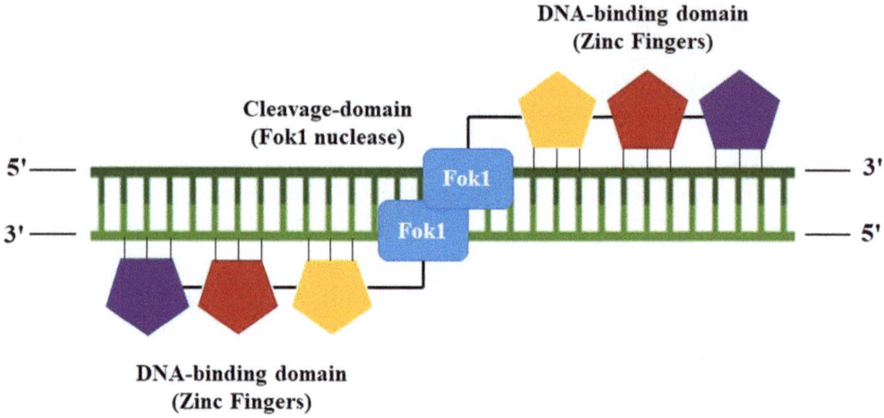

Fig. 1 Diagrammatic presentation of zing-finger nucleases. A cleavage domain (Fok1 nuclease) and a DNA-binding domain (zinc fingers) are combined to construct ZFNs. Fok1 is not sequence specific; it helps in inducing double-strand breaks (DSBs). The zinc fingers are sequence specific, and individual zinc finger binds specifically with three nucleotides

(a) the sequence of the amino acids of ZFs, (b) the number of ZFs, and (c) cleavage-domain interaction with the target site. ZFNs offer a number of benefits that make them a popular technique for gene editing. Both domains can be optimized individually, therefore giving freedom to scientists to develop novel assemblies having better specificity and affinity (Li et al. 2020a). ZFNs integrate rapidly into the locus. Mutations arising from ZFNs are permanent and can be inherited by the progeny (Ghosh et al. 2021). ZFNs have been used effectively in plant gene modification and were initially employed for correcting preintegrated selectable marker genes. In one study using protoplasts of tobacco, a defective fusion between genes encoding the GUS reporter and neomycin phosphotransferase (NPTII) selectable marker proteins was corrected with the aid of co-delivering of a gene encoding a ZFN, which cleaved in the junction between the two genes, and a donor template DNA comprising sequences homologous to those flanking the ZFN cleavage site and the missing GUS/NPTII gene sequences (Wright et al. 2005). ZFNs are also used for endogenous cleavage to integrate transgenes like herbicide-resistance genes (Petolino 2015). In another study a gene encoding a ZFN designed to cleave the *ENDOCHITINASE-50* (*CHN50*) gene of tobacco was co-delivered via *Agrobacterium* along with a PAT herbicide-resistance gene flanked on each side by 750 bp of CHN50 gene sequence (Cai et al. 2009).

2.2 *Transcription Activator-Like Effector Nucleases (TALENs)*

They are the second type of genome-editing tools, displaying better efficiency and specificity when compared with ZFNs. TALENs also comprise a DNA-binding and a cleavage domain for the generation of DSBs. Like ZFN, TALEN is sequence specific and customizable, comprising repeated sequences that are conserved, called transcription activator-like effectors (TALEs). The latter is not sequence specific. TALE is a protein isolated from the phytopathogenic bacteria called *Xanthomonas*. TALE binds with the host DNA and alters the transcription process (Boch et al. 2009; Bogdanove et al. 2010). Individual TALE repeat comprises of 33 to 35 amino acids that bind to a single nucleotide of the target (Fig. 2). All of the TALE repeats share similar amino acid sequences, except at positions 12 and 13. They are called repeat variable diresidues (RVDs) (Fig. 2) (Bogdanove and Voytas 2011), which determine binding specificity (Boch et al. 2009). NN, NG, NI, and HD are the four common RVDs, which associate with G, T, A, and C, respectively. NK is the least common type, showing specificity for G higher than NN (Morbitzer et al. 2010; Miller et al. 2011). FokI is combined with the DNA-binding domain for the generation of a DSB site specifically, which stimulates the recombination of DNA to achieve target gene modification induced by TALEN. For the cleavage of the target strands, the dimerization of the cleavage domain, FokI, is necessary. To achieve this, TALEN is designed in pairs, which bind with the two opposite strands of the target (Fig. 2), with a space of 12–30 bp between the binding sites (Li et al. 2011). Unlike in ZFNs, where a single zinc finger binds with three nucleotides, in TALENs, each TALE binds with a single nucleotide (Gaj et al. 2013; Jankele and Svoboda 2014). Though TALE domains possess high specificity, they allow nucleotide mismatches (Mohanta et al. 2017; Jansing et al. 2019; Gaj et al. 2013). The first reported crop

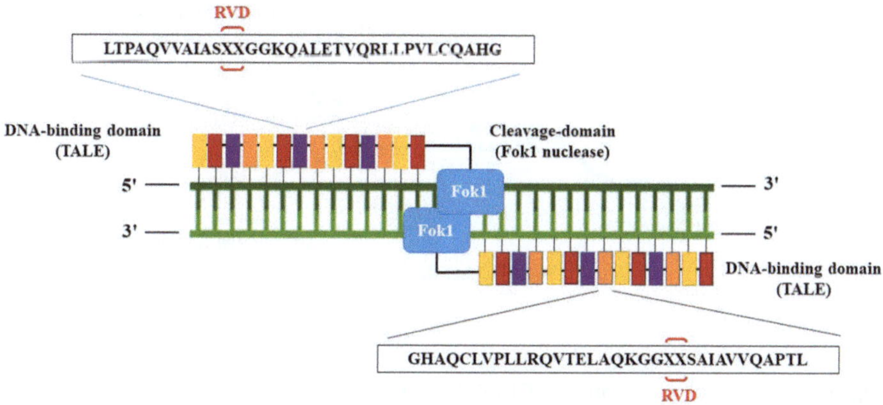

Fig. 2 Diagrammatic presentation of TALENs. A cleavage domain (Fok1 nuclease) and a DNA-binding domain (TALE proteins) are combined to construct TALENs. The TALE repeats are sequence specific and share similar amino acid sequences, except at positions 12 and 13. They are called repeat variable diresidues (RVDs)

that was improved using TALENs is rice. The bacterial protein of *Xanthomonas oryzae* binds to the OsSWEET 14 gene at its promoter site, which encodes the sucrose transporter. The binding of proteins activates genes responsible for causing diseases. A pair of TALENs was constructed to target the promoter region. This led to the mutation of the effector binding site within the promoter, resulting in the silencing of the gene, therefore providing resistance to *Xanthomonas oryzae* (Li et al. 2012).

2.3 Clustered Regularly Interspaced Palindromic Repeat (CRISPR)/CRISPR-associated (Cas9)

CRISPRs were initially identified in *Escherichia coli* (Ishino et al. 1987). They provide immunity to the bacteria against invading DNA by inducing DNA cleavage guided by ribonucleic acid (RNA) (Li et al. 2020a). Based on the structure and procedure of the Cas genes, CRISPR-Cas can be categorized into two groups (Jinek et al. 2012). Class 1 requires multiprotein effector proteins to function, whereas for class 2 functioning, only a single protein is needed. Class 1 is further segregated into types I, III, and IV. Class 2 is segregated into types II, V, and VI (Makarova et al. 2011, 2015). Type II is mostly employed. Only a single protein, called Cas9, is used by the system. The protein is obtained from *Streptococcus pyogenes* (SpCas9), which targets specific sequences of DNA, making it an impressive genomic-editing tool (Jiang et al. 2013). This genomic technique entails a CRISPR-associated protein 9 (Cas9) endonuclease, transactivating crRNA (tracrRNA), CRISPR RNA (crRNA), and ribonuclease III (RNase III) (El-Mounadi et al. 2020; Mohanta et al.

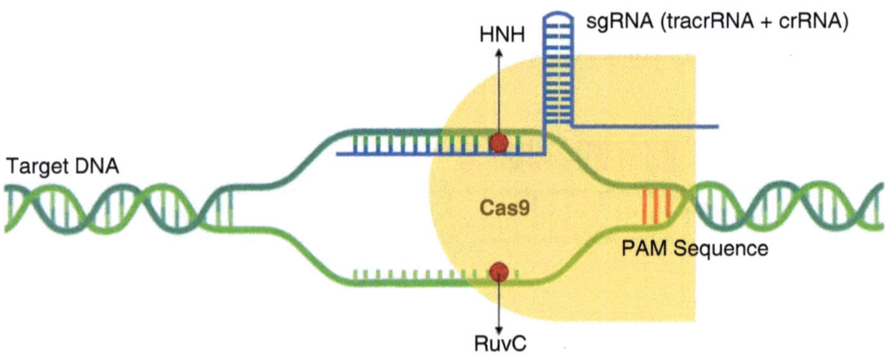

Fig. 3 Diagrammatic presentation of CRISPR/Cas9. This genomic technique entails a CRISPR-associated protein 9 (Cas9) endonuclease, transactivating crRNA (tracrRNA) and CRISPR RNA (crRNA). The tracrRNA and crRNA assemble to form single-stranded guide RNA (sgRNA). Cas9 is an endonuclease that is responsible for inducing cuts at double-stranded DNA. His-Asn-His (HNH) and RuvC-like domains are the two Cas9 domains. The domain HNH cuts the complementary strand of crRNA. The opposing strand of dsDNA is cleaved by the RuvC-like domain

2017; Jansing et al. 2019). The tracrRNA and crRNA assemble to form single-stranded guide RNA (sgRNA) (Kumar et al. 2019). Cas9 is an endonuclease that is responsible for cutting double-stranded DNA (dsDNA) (Kamburova et al. 2021). His-Asn-His (HNH) and RuvC-like domains are the two Cas9 domains that are involved in producing cleavage in the dsDNA upstream of the protospacer adjacent motif (PAM). The PAM sequence is 5′ NGG or 5′-NAG (Jiang and Doudna 2017; Hille and Charpentier 2016; Manghwar et al. 2019). The domain HNH cuts the complementary strand of crRNA. The opposing strand of dsDNA is cleaved by the RuvC-like domain (Fig. 3). As a result, DSBs are produced and corrected by HDR or NHEJ (Kumar et al. 2019; Jiang and Doudna 2017; Hille and Charpentier 2016; Manghwar et al. 2019; Hille et al. 2018; Liu et al. 2017). The sgRNA is a 100-mer RNA that comprises tracrRNA and crRNA. Its 5′-end contains a guide sequence that is 20 nucleotides long, which helps in the identification of the target site. This sequence is followed by PAM (Liu et al. 2017). Its 3′-end has a loop-like structure, which helps the guide sequence to bind properly with the target sequence. The sgRNA and Cas9 together form the ribonucleoprotein (RNP) complex to generate cleavage (Manghwar et al. 2019; Liu et al. 2017). The RNP complex efficiently cleaves the target DNA. The crRNA was identified as the target site, which also aids in directing the RNP to a specific site, where it binds to the target site by forming an R-loop-like structure (Manghwar et al. 2019). The formation of a loop structure activates both domains of the Cas9 endonuclease, which then leads to the cleaving of the dsDNA and the generation of blunt ends (Hille and Charpentier 2016). CRISPR-Cas9 has generated impressive results, such as improvement in nutritional quality (Li et al. 2018a), generation of male sterility in maize (Li et al. 2017) and wheat (Okada et al. 2019), and generation of disease-resistant plants (Zhang et al. 2017b) and herbicide-resistant plants (Sun et al. 2016). On September 15, 2021, the world's first CRISPR-Cas9-based genome-edited tomato, which possesses the highest amounts of ɣ-aminobutyric acid (GABA), was launched by Japan (Ezura 2022).

3 Applications

Applications have been attributed in recent years to the development of important plant trait improvements, gene regulatory adjustments for the control of biotic and abiotic stresses, the regulation of biosynthetic pathways, and crop quality improvement (Fig. 4). A few applications are listed in Table 2.

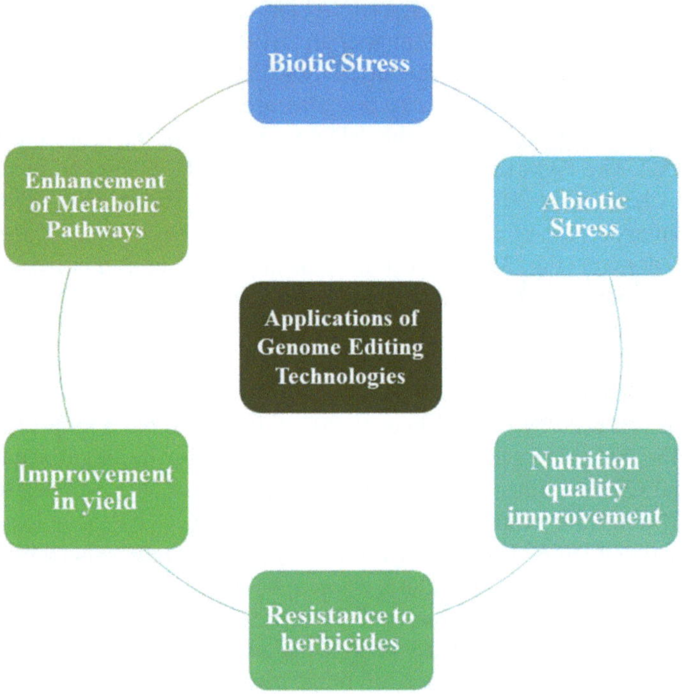

Fig. 4 Applications of genome-editing technologies in crop improvement. These techniques have been widely employed for nutrition quality improvement, resistance to herbicides, improvement in yield, the enhancement of metabolic pathways, and combating abiotic and biotic stresses

3.1 Abiotic Stress

Change in climatic patterns is the most important cause of abiotic stress, which is mostly brought on by toxicity of ions, salinity, drought, heat, floods, and radiation and results in an annual crop loss of more than 50%. In light of this, genome-editing technology broadened the spectrum of remedies for the issues brought on by abiotic stress in plants (Pandey et al. 2017). The most common abiotic stress that causes substantial harm to the plant community due to global warming is drought (Ghosh and Dey 2022). Several studies showed that the drought tolerance of several crops has been improved through genome editing. The H^+-ATPase is encoded by a gene called *OST2* (*OPEN STOMATA 2*). In the cells of the plants, H^+-ATPase is responsible for the formation of proton gradients. Drought stress resistance has been reported to be conferred through a precise alteration of this gene using CRISPR/Cas9. This modification affects stomatal closing in response to water shortage situations (Osakabe et al. 2016). In soils, the buildup of salts as a result of seawater drift, transpiration, or evaporation impairs the plant's capacity to take in nutrients through its roots, stunting its growth. The salinity tolerance of plants increases when the *OsRR22* gene transcription decreases (Takagi et al. 2015). Using CRISPR/Cas9,

Table 2 Applications

Sr. no.	Crop	Gene target	Technique employed	Application	Reference
1	Maize	*ZmIPK*	TALENs/CRISPR/ Cas 9	Reduction of the nutritionally harmful substance-phytic acid	Liang et al. (2014)
2	Tomato	*SlWUS*	CRISPR/Cas 9	Increasing the size of the fruit	Rodríguez-Leal et al. (2017)
3	Wheat	*TaEDR1*	CRISPR/Cas 9	Resistance against the disease powdery mildew	Zhang et al. (2017b)
4	Rice	*Os09g29100*	TALENs	Increased resistance against bacterial leaf spotting	Cai et al. (2017)
5	*Arabidopsis thaliana*	*elF(iso)4E*	CRISPR/Cas 9	Resistance to potyvirus	Pyott et al. (2016)
6	Tobacco	*MEL1*	*ZFN*	Resistance to herbicides	Cai et al. (2009)
7	Cucumber	*elF4E*^{1}	CRISPR/Cas 9	Resistance to yellow vein virus in cucumber, yellow mosaic virus in zucchini, and ring spot mosaic virus in papaya	Chandrasekaran et al. (2016)
8	Potato	*VInv*	TALENs	Lowering sugar accumulation	Clasen et al. (2016)
9	Soybean	*FAD2-1A* and *FAD2-1B*		Increase in the quality of the oil	Haun et al. (2014)
10	Barley	*HvPM19*	CRISPR/Cas 9	Positive control of grain dormancy	Lawrenson et al. (2015)

the gene *OsRR22* was altered, and two homologous T2 generations demonstrated improved salt tolerance with a slight difference between the modified and wild-type strains (Zhang et al. 2019). Moreover, plants also have a preferred temperature, and any change in that temperature can significantly impair the development and production of those plants (Fahad et al. 2017; Kazemi-Shahandashti and Maali-Amiri 2018). By employing genome editing, scientists have sought to make plants more resilient to temperature stress. It has been noted that lettuce seedlings generated by CRISPR/Cas9 exhibit improved thermal tolerance at the germination stage when the 9-cis-EPOXYCAROTENOID DIOXYGENASE4 (NCED4) gene is knocked out (Bertier et al. 2018). According to Kazemi et al. (Kazemi-Shahandashti and Maali-Amiri 2018), in plants, there are two forms of cold stress: freezing stress (0 °C or lower) and chilling stress (0–20 °C), both of which have an impact on plant development. One study suggested that wild-type tomato plants were less damaged by freezing when CRISPR/Cas9 was employed to knock out C-repeat binding factors (SlCBFs) (Li et al. 2018). The molecular control of different genes that are taking part in response to abiotic stress has also been studied via genome editing.

3.2 *Biotic Stress*

Globally, 20–40% of the losses in agricultural productivity are attributable to biotic stressors such as bacterial, fungal, and viral diseases (Walker 1984). As an alternative, several kinds of herbicides, insecticides, and fungicides are utilized, which result in damage to the environment, both directly and indirectly. Therefore, by using genome-editing techniques, researchers developed plants extremely resistant to fungal, bacterial, viral, and insect-borne diseases with no or fewer side effects (Chen et al. 2019). Herbicide resistance is a major cause of biotic stress in crops, which can be effectively induced by using genome editing. For plants to develop herbicide resistance, several genes have undergone modification using the genome techniques that are discussed in previous sections, among which EPSPS (5-enolpyruvylshikimate-3-phosphate) and ALS (acetolactate synthase) are the most significant ones. Numerous herbicides have been shown to block the action of ALS in plants, including pyrimidinylthio (or oxy)-benzoates, imidazolinones, sulfonylamino-carbonyl-triazolinones, triazolopyrimidines, and sulfonylureas (Mazur et al. 1987; Zhou et al. 2007). In 2009, this gene was accurately altered with the use of ZFN. As a result, tobacco plants acquired sulfonylurea herbicide resistance (Townsend et al. 2009). A few years later, CRISPR/Cas9 and TALENs were employed to modify the gene ALS in watermelon, soybean, potato, and maize (Sun et al. 2016; Butler et al. 2015; Svitashev et al. 2015; Tian et al. 2018). Moreover, numerous bacterial species may infect plants and cause diseases, frequently showing various symptoms. Because it is difficult to identify disease before it is visible and because there are few effective insecticides, plant pathogenic organisms are challenging to manage. Plant resistance to microbial infections has been discovered to be enhanced through the CRISPR/Cas9 editing of plant genomes (Schloss and Handelsman 2004). Besides this, the agricultural productivity ratio also drastically decreases due to viral infections in plants, which are inevitable parasites (Das et al. 2019; Shahriar et al. 2021). These viruses depend on plant system machinery for reproduction and proliferation. As a result, host vulnerability can be altered or pathogenicity-related genes can be targeted to increase viral resistance (Tyagi et al. 2021). For this purpose, several scientists have made use of the CRISPR-Cas method, which directly protects plants from DNA and RNA viruses by focusing on and dividing viral genes (Mushtaq and Molla 2021; Rath et al. 2015). Fungus is one type of detrimental organism that damages crops by causing destructive diseases such as powdery mildew, which has a negative impact on agricultural output. In relation to this, scientists have removed three TaMLO alleles from wheat using CRISPR/Cas9, thereby producing plants resistant to the disease (Wang et al. 2014).

3.3 CRISPR/Cas9 System for Crop Improvement

In order to improve crop nutritional quality, such as proteins, vitamins, carbohydrates, oils, colors, extended shelf life, and many other factors, genome-editing techniques are currently being employed more often to develop modified crops. Several investigations have demonstrated that crop quality is significantly impacted by the quantity and portion of protein content. Therefore, to regulate the nutritional value of seeds, genome-editing tools may be employed to alter the storage protein composition of seeds (Yang et al. 2022). Recently, researchers have worked hard to boost the quality of oil using the CRISPR/Cas9 technology, mostly concentrating on a few oil crops, including camelina, soybean, rapeseed, etc. Okuzaki and associates focused on BnaA.FAD2.a (FAD2_Aa) in *Brassica napus* to raise the concentration of oleic acid. Additionally, these genetically modified oil crops can be useful for decreasing cholesterol and systolic blood pressure in humankind (Okuzaki et al. 2018). Furthermore, white rice is an essential diet in developing countries that are deficient in provitamin A (mostly β-carotene). Thus, Dong et al. effectively created marker-free rice plants with a high concentration of carotenoid in seeds by inserting a 5.2-kilo base pair of carotenoid biosynthetic sequences made up of protein-coding sequences of ZmPsy and SSU-crtI at two genomic safe harbors in rice (Dong et al. 2020). It is well recognized that a number of heavy metals, including cadmium (Cd) and cesium (Cs), are extremely hazardous and are readily absorbed by plants and transferred to their edible sections. For this purpose, Nieves-Cordones and associates employed CRISPR-Cas9 to knock out the gene OsHAK1, leading to a significant decrease in the concentration of radioactive cesium in mutant plants cultivated on 137Cs + polluted Fukushima soil (Nieves-Cordones et al. 2017). Generally, fruits mature rapidly due to ethylene, a hormone naturally produced by plants, which leads to their short shelf span. The short shelf life of fruits severely restricts agricultural transportation, marketing, and preservation, resulting in massive post-harvest losses. For this reason, Hu et al. (2021) have inhibited the activity of the O 1 (Ma07_t19730.1) gene, which showed decreased ethylene production and prolonged shelf life under natural ripening conditions (Hu et al. 2021).

3.4 *Improvement of a Biosynthetic Pathway Through Genome Editing*

Humans and industries both consume and process plants that contain a lot of nutrients and metabolites. Numerous novel metabolic engineering technologies are employed to increase both the quantity and quality of crops (Schaart et al. 2016; Alagoz et al. 2016). Interestingly, metabolic pathways frequently comprise several important genes, but single gene knockout would result in less accumulation of fascinating metabolites. It is reported that by using genome-editing technologies, a single gene present in the metabolic pathway can be altered to improve the metabolic activity of plants. For example, researchers have generated opium poppies with high levels of benzylisoquinoline alkaloids (Alagoz et al. 2016). The quantity of a certain metabolite can be increased by enhancing a biosynthetic pathway and/or suppressing a catabolic process. The most common use of metabolic engineering with genome-editing technologies is the inactivation of a target enzyme. Generally, it is more difficult to improve biosynthetic enzymes since gain-of-function mutations must be accurately inserted into the target enzyme (Sukegawa et al. 2022). Besides the methods listed above, altering the expression level of a gene of interest can be a different approach to creating mutants having a variety of characteristics. For example, Rodrguez-Leal and associates used CRISPR-Cas9 in tomato to construct many changes in the promoter of a target gene, leading to a modification of features like fruit size, plant architecture, and inflorescence branching (Rodríguez-Leal et al. 2017).

4 Challenges

Advancements in genome-editing technologies are constantly developing. Although there has been progress, limitations remain.

Despite having ideal binding specificity, ZFNs are prone to errors since they are subjected to nucleotide mismatch (Gaj et al. 2013). ZFNs can bind at off-target sites and cause unwanted cleavage and mutations (Ates et al. 2020), leading to toxicity, like apoptosis (Porteus and Carroll 2005; Szczepek et al. 2007). The unwanted binding of ZFN dimers at the off-target sites can be avoided by employing obligate heterodimers of ZFNs bound to the engineered cleavage domain, FokI, to cause heterodimerization. However, obligate heterodimers are less efficient in dimerization when compared to the wild-type FokI domains, therefore highly affecting the cleavage process (Miller et al. 2007). The proper assembly of zinc-finger domains is very important for it to bind with the target sequence with high affinity. This has made it tedious for nonprofessionals to design ZFNs regularly. To address this issue, an academic group has created a library containing components of zinc fingers as well as protocols for screening the best zinc-finger domains that bind with their target sequence with high affinity. However, obtaining optimized ZFNs can still take several months for nonprofessionals (Ramirez et al. 2008; Maeder et al. 2008, 2009).

One of the major limitations of TALENs is their size. The cDNA encoding a TALEN is 3 kb, whereas the one encoding a ZFN is only 1 kb, therefore making it difficult for TALENs to get expressed in cells. The size also limits their use in therapeutic purposes, where they need to be packed in viral vectors. Moreover, the repeating DNA-binding domains may impede the capacity to be packed and transported by viral vector vehicles (Holkers et al. 2013; Yang et al. 2013). Another major limitation is the assembly of many identical repetitive sequences required for the construction of a TALEN array, which is technically difficult. This process is laborious and not cost-effective (Bhardwaj and Nain 2021; Yamagata 2023). This problem can be addressed by incorporating strategies that will help fasten custom TALE assembly. A few of them include Golden Gate molecular cloning, high-throughput solid-phase assembly, and connection-independent cloning techniques (Cermak et al. 2011; Schmid-Burgk et al. 2013; Reyon et al. 2012; Briggs et al. 2012).

Like ZFN, the CRISPR-Cas system also binds at off-target sites, causing unwanted gene editing at these sites (Cradick et al. 2013; Pattanayak et al. 2013). Such unwanted gene editing can disrupt tumor-suppressing genes, cause chromosomal translocations and deletions, and also activate oncogenes (Kosicki et al. 2018; Shin et al. 2017). The mismatches between the sgRNA and DNA, and bulges in nucleic acids downstream of PAM, are responsible for such unwanted cleavages at the off-target sites. Therefore, while designing sgRNAs, thorough screening is necessary before they can be used for different therapeutic purposes (Lin et al. 2014). Off-targets can be predicted using in silico tools. CRISPRitz is one of the recent tools that account for mismatches and bulges while discriminating between the noncoding and coding sites of off-targets (Cancellieri et al. 2019). Nevertheless, for the further validation of off-targets, there will be a need to proceed with deep-sequencing methods to precisely quantify the specificity of Cas9. Targeting both strands with fCas9 and Cas9 nickases (Cho et al. 2014; Guilinger et al. 2014), which consist of catalytically inactive dCas9 adhered to FokI for the formation of DSB in a staggered fashion, showed improved specificity of the CRISPR-Cas system. This does not affect the cleavage at the on-target site. To reduce gene editing at off-target sites, mutant versions of Cas9 have been developed to allow us to continue the usage of Cas9. Such mutated Cas9 types have been able to reduce gene editing at the off-target sites. These mutants do not affect the activity at the on-target sites. A few examples are HypaCas9 (Chen et al. 2017), Sniper-Cas9 (Lee et al. 2018), and evoCas9 (Casini et al. 2018). In a study, the Cas9 mutant HiFi Cas9 was able to reduce off-target binding when compared with the wild type. The mutant also maintained good gene editing at the on-target site, $CD34^+$ HSPCs (Vakulskas et al. 2018). Another challenge with the CRISPR-Cas9 system is its size, which is approximately 4.3 kb. This hugely affects its application. To overcome this challenge, other homologs of Cas9 were developed from different bacterial species. *Staphylococcus aureus* is one such bacteria that was employed to develop the novel CRISPR-Cas9, SaCas9. In this case, sgRNA binds to the sites upstream of a 5′-NNGRRT PAM sequence. SaCas9 has proved to be successful (Li et al. 2020b; Jarrett et al. 2018). Like other homologs of Cas9, the reagents of SaCas9 are available commercially to enable their delivery directly to the site as performed RNPs (sgRNA-Cas9). The SaCas9 has a compact size of approximately 3.3 kb. The rate of release of the substrate is also fast, which avoids further nuclease activity post cleavage of the target DNA (Yourik et al. 2019). However, SaCas9 is also associated with a limitation. Due to the requirement of a long sequence of PAM, the targeting range of SaCas9 is limited. Luna et al. successfully developed the mutants of SaCas9, SaCas9-RL, and SaCas9-NR, having a broader range and better activity when compared with the wild-type SaCas9. These mutants of SaCas9 can be promising for therapeutic purposes (Luan et al. 2019).

5 Conclusion

Crop improvement is critical to address the increasing needs of a changing world (for example, climate change, growing population, and limited availability of land). Over the years, different technologies were employed for crop improvement, plant breeding being one of the techniques. Plant breeding can improve quality, yield, and resistance to pests and diseases. However, breeding techniques may be time-consuming and labor-intensive, taking up to 6–15 years or 8–10 years, as well as developing a plant that is modified genetically. Moreover, the results of breeding procedures are also uncertain, therefore necessitating the need for novel and efficient techniques. Presently, there are three major genome-editing techniques categorized according to their mode of action. The generation of DSBs by SSNs at a targeted site is most often used in plant genomics. The generation of DSB can be achieved by employing nucleases like TALENs, ZFNs, and CRISPR/Cas9. The first artificial endonucleases are ZFNs. Cleavage and a DNA-binding domain are combined to construct ZFNs. The former is not sequence specific, whereas the latter is sequence specific. The cleavage domain is called FokI, which causes DSBs. The DNA-binding domain comprises Cys_2His_2 ZFs. The number of amino acids that a single ZF comprises is 30. The ZFs bind specifically with three nucleotides of the DNA. TALENs also comprise two domains for the generation of DSBs. Like ZFN, the former is not sequence specific, whereas the latter is sequence specific and customizable. The latter comprises repeated sequences called TALE proteins. The number of amino acids that each TALE comprises is 33–35. Each TALE binds to one nucleotide. FokI is fused with TALEs to produce DSBs. This stimulates the recombination of DNA to achieve target gene modification induced by TALEN. CRISPR was initially identified in *E. coli*. It consists of Cas9 endonuclease, crRNA, tracrRNA, and RNase III. Cas9 cleaves a dsDNA. As a result, DSBs are produced and corrected by HDR or NHEJ. These genome-editing techniques are employed for crop improvement, ranging from combating abiotic and biotic stress, improving crop nutritional quality and biosynthetic pathways. These techniques prove to be efficacious in crops like maize, rice, wheat, tomato, cucumber, potato, and many others. Although there has been progress in recent years, limitations remain. These challenges need to be tackled to declare them as promising genome-editing tools.

Acknowlegements Vishnu D. Rajput and Tatiana Minkina acknowlege the financial support of the Ministry of Science and Higher Education of the Russian Federation (agreement no. 075-15-2022-1122) and the Strategic Academic Leadership Program of the Southern Federal University "Priority 2030".

References

Abdallah NA, Prakash CS, McHughen AG (2015) Genome editing for crop improvement: challenges and opportunities. GM Crops Food 6(4):183–205. PMID: 26930114; PMCID: PMC5033222. https://doi.org/10.1080/21645698.2015.1129937

Alagoz Y, Gurkok T, Zhang B, Unver T (2016) Manipulating the biosynthesis of bioactive compound alkaloids for next-generation metabolic engineering in opium poppy using CRISPR-Cas 9 genome editing technology. Sci Rep 6:30910

Ates I, Rathbone T, Stuart C, Bridges PH, Cottle RN (2020) Delivery approaches for therapeutic genome editing and challenges. Gene 11(10):1113. https://doi.org/10.3390/genes11101113

Beerli RR, Barbas CF (2002) Engineering polydactyl zinc-finger transcription factors. Nat Biotechnol 20:135–141

Bertier LD, Ron M, Huo H, Bradford KJ, Britt AB, Michelmore RW (2018) High-resolution analysis of the efficiency, heritability, and editing outcomes of CRISPR/Cas9-induced modifications of NCED4 in lettuce (*Lactuca sativa*). G3: Genes Genomes Genetics 8(5):1513–1521

Bhardwaj A, Nain V (2021) TALENs-an indispensable tool in the era of CRISPR: a mini review. J Genet Eng Biotechnol 19:125

Boch J et al (2009) Breaking the code of DNA binding specificity of TAL-type III effectors. Science 326:1509–1512

Bogdanove AJ, Voytas DF (2011) TAL effectors: customizable proteins for DNA targeting. Science 333:1843–1846

Bogdanove AJ, Schornack S, Lahaye T (2010) TAL effectors: finding plant genes for disease and defense. Curr Opin Plant Biol 13:394–401

Briggs AW et al (2012) Iterative capped assembly: rapid and scalable synthesis of repeat-module DNA such as TAL effectors from individual monomers. Nucleic Acids Res 40:e117

Buck-Koehntop BA et al (2012) Molecular basis for recognition of methylated and specific DNA sequences by the zinc finger protein kaiso. Proc Natl Acad Sci U S A 109:15229–15234

Butler NM, Atkins PA, Voytas DF, Douches DS (2015) Generation and inheritance of targeted mutations in potato (*Solanum tuberosum* L.) using the CRISPR/Cas system. PLoS One 10(12):e0144591

Cai CQ, Doyon Y, Ainley WM, Miller JC, Dekelver RC, Moehle EA, Rock JM, Lee YL, Garrison R, Schulenberg L, Blue R, Worden A, Baker L, Faraji F, Zhang L, Holmes MC, Rebar EJ, Collingwood TN, Rubin-Wilson B, Gregory PD, Urnov FD, Petolino JF (2009) Targeted transgene integration in plant cells using designed zinc finger nucleases. Plant Mol Biol 69:699–709. https://doi.org/10.1007/s11103-008-9449-7

Cai L, Cao Y, Xu Z, Ma W, Zakria M, Zou L, Cheng Z, Chen G (2017) A transcription activator-like effector Tal7 of *Xanthomonas oryzae* pv. Oryzicola activates rice gene Os09g29100 to suppress rice immunity. Sci Rep 7(5089):1–13

Cancellieri S, Canver MC, Bombieri N, Giugno R, Pinello L (2019) CRISPRitz: rapid, high-throughput and variant-aware in silico off-target site identification for CRISPR genome editing. Bioinformatics 36:2001–2008

Casini A, Olivieri M, Petris G, Montagna C, Reginato G, Maule G, Lorenzin F, Prandi D, Romanel A, Demichelis F et al (2018) A highly specific SpCas9 variant is identified by in vivo screening in yeast. Nat Biotechnol 36:265–271

Cermak T et al (2011) Efficient design and assembly of custom TALEN and other TAL effector-based constructs for DNA targeting. Nucleic Acids Res 39:e82

Chandrasekaran J, Brumin M, Wolf D, Leibman D, Klap C, Pearlsman M, Gal-On A (2016) Development of broad virus resistance in non-transgenic cucumber using CRISPR/Cas9 technology. Mol Plant Pathol 17(7):1140–1153

Chen K, Gao C (2014) Targeted genome modification technologies and their applications in crop improvements. Plant Cell Rep 33:575–583. https://doi.org/10.1007/s00299-013-1539-6

Chen JS, Dagdas YS, Kleinstiver BP, Welch MM, Sousa AA, Harrington LB, Sternberg SH, Joung JK, Yildiz A, Doudna JA (2017) Enhanced proofreading governs CRISPR-Cas9 targeting accuracy. Nature 550:407–410

Chen K, Wang Y, Zhang R, Zhang H, Gao C (2019) CRISPR/Cas genome editing and precision plant breeding in agriculture. Annu Rev Plant Biol 70:667–697. https://doi.org/10.1146/annurev-arplant-050718-100049

Cho SW, Kim S, Kim Y, Kweon J, Kim HS, Bae S, Kim JS (2014) Analysis of off-target effects of CRISPR/Cas-derived RNA-guided endonucleases and nickases. Genome Res 24:132–141

Clasen BM, Stoddard TJ, Luo S, Demorest ZL, Li J, Cedrone F, Tibebu R, Davison S, Ray EE, Daulhac A, Coffman A, Yabandith A, Retterath A, Haun W, Baltes NJ, Mathis L, Voytas DF, Zhang F (2016) Improving cold storage and processing traits in potato through targeted gene knockout. Plant Biotechnol J 14:169–176

Cradick TJ, Fine EJ, Antico CJ, Bao G (2013) CRISPR/Cas9 systems targeting β-globin and CCR5 genes have substantial off-target activity. Nucleic Acids Res 41:9584–9592

Das A, Sharma N, Prasad M (2019) CRISPR/Cas9: a novel weapon in the arsenal to combat plant diseases. Front Plant Sci 9:2008

Dong OX, Yu S, Jain R, Zhang N, Duong PQ, Butler C et al (2020) Marker-free carotenoid-enriched rice generated through targeted gene insertion using CRISPR-Cas9. Nat Commun 11:1178. https://doi.org/10.1038/s41467-020-14981-y

El-Mounadi K, Morales-Floriano ML, Garcia-Ruiz H (2020) Principles, applications, and biosafety of plant genome editing using CRISPR-Cas9. Front Plant Sci 11:56

Ezura H (2022) Letter to the editor: the World's first CRISPR tomato launched to a Japanese market: the social-economic impact of its implementation on crop genome editing. Plant Cell Physiol 63(6):731–733. https://doi.org/10.1093/pcp/pcac048

Fahad S, Bajwa AA, Nazir U et al (2017) Crop production under drought and heat stress: plant responses and management options. Front Plant Sci 8:1147

Fairall L, Schwabe JW, Chapman L, Finch JT, Rhodes D (1993) The crystal structure of a two zinc-finger peptide reveals an extension to the rules for zinc-finger/DNA recognition. Nature 366:483–487

FAO; IFAD; UNICEF; WFP; WHO. The State of Food Security and Nutrition in the World (2021) Transforming food Systems for Food Security, improved nutrition and affordable healthy diets for all. World Health Organization, Rome, pp 1–240

Gaj T, Gersbach CA, Barbas CF (2013) ZFN, TALEN and CRISPR/Cas-based methods for genome engineering. Trends Biotechnol 31(7):397–405. https://doi.org/10.1016/j.tibtech.2013.04.004

Gao C (2021) Genome engineering for crop improvement and future agriculture. Cell 184:1621–1635

Ghosh S, Dey G (2022) Biotic and abiotic stress tolerance through CRISPR-Cas mediated genome editing. J Plant Biochem Biotechnol 31:1–12

Ghosh D, Kumar A, Sinha N (2021) Targeted genome editing. In: Advances in animal genomics. Elsevier, Amsterdam. https://doi.org/10.1016/B978-0-12-820595-2.00006-0

Guilinger JP, Thompson DB, Liu DR (2014) Fusion of catalytically inactive Cas9 to FokI nuclease improves the specificity of genome modification. Nat Biotechnol 32:577–582

Hamdan MF, Karlson CKS, Teoh EY, Lau S-E, Tan BC (2022) Genome editing for sustainable crop improvement and mitigation of biotic and abiotic stresses. Plan Theory 11:2625. https://doi.org/10.3390/plants11192625

Haun W, Coffman A, Clasen BM, Demorest ZL, Lowy A, Ray E, Retterath A, Stoddard T, Juillerat A, Cedrone F, Mathis L, Voytas DF, Zhang F (2014) Improved soybean oil quality by targeted mutagenesis of the fatty acid desaturase 2 gene family. Plant Biotechnol J 12(7):934–940

Herman RA, Fedorova M, Storer NP (2019) Will following the regulatory script for GMOs promote public acceptance of gene-edited crops? Trends Biotechnol 37:1272–1273

Hille F, Charpentier E (2016) CRISPR-Cas: biology, mechanisms and relevance. Philos Trans R Soc B Biol Sci 371:20150496. https://doi.org/10.1098/rstb.2015.0496

Hille F, Richter H, Wong SP, Bratovič M, Ressel S, Charpentier E (2018) The biology of CRISPR-Cas: backward and forward. Cell 172(6):1239–1259. https://doi.org/10.1016/j.cell.2017.11.032

Holkers M et al (2013) Differential integrity of TALE nuclease genes following adenoviral and lentiviral vector gene transfer into human cells. Nucleic Acids Res 41(5):e63. https://doi.org/10.1093/nar/gks1446

Hu C, Sheng O, Deng G, He W, Dong T, Yang Q et al (2021) CRISPR/Cas9-mediated genome editing of MaACO1 (Aminocyclopropane-1-carboxylate oxidase 1) promotes the shelf life of Banana fruit. Plant Biotechnol J 19:654–656. https://doi.org/10.1111/pbi.13534

Ishino Y, Shinagawa H, Makino K, Amemura M, Nakata A (1987) Nucleotide sequence of the iap gene, responsible for alkaline phosphatase isozyme conversion in *Escherichia coli*, and identification of the gene product. J Bacteriol 169:5429–5433

Jankele R, Svoboda P (2014) TAL effectors: tools for DNA targeting. Brief Funct Genomics 13:409–419. https://doi.org/10.1093/bfgp/elu013

Jansing J, Schiermeyer A, Schillberg S, Fischer R, Bortesi L (2019) Genome editing in agriculture: technical and practical considerations. Int J Mol Sci 20:2888. https://doi.org/10.3390/ijms20122888

Jarrett KE, Lee C, De Giorgi M, Hurley A, Gillard BK, Doerfler AM, Li A, Pownall HJ, Bao G, Lagor WR (2018) Somatic editing of Ldlr with adeno-associated viral-CRISPR is an efficient tool for atherosclerosis research. Arterioscler Thromb Vasc Biol 38:1997–2006

Jiang F, Doudna JA (2017) CRISPR-Cas9 structures and mechanisms. Annu Rev Biophys 46:505–529. https://doi.org/10.1146/annurev-biophys-062215-010822

Jiang W, Bikard D, Cox D, Zhang F, Marraffini LA (2013) RNA-guided editing of bacterial genomes using CRISPR-Cas systems. Nat Biotechnol 31:233–239

Jinek M et al (2012) A programmable dual-RNA-guided DNA endonuclease in adaptive bacterial immunity. Science 337:816–821

Kamburova VS, Salakhutdinov IB, Shermatov SE, Abdurakhmonov IY (2021) Using of genome editing methods in plant breeding. IntechOpen, London. https://doi.org/10.5772/intechopen.96431

Kazemi-Shahandashti S-S, Maali-Amiri R (2018) Global insights of protein responses to cold stress in plants: signaling, defence, and degradation. J Plant Physiol 226:123–135

Kim YG, Li L, Chandrasegaran S (1994) Insertion and deletion mutants of FokI restriction endonuclease. J Biol Chem 269(50):31978–31982

Kim YG, Cha J, Chandrasegaran S (1996) Hybrid restriction enzymes: zinc finger fusions to Fok I cleavage domain. Proc Natl Acad Sci U S A 93:1156–1160

Kosicki M, Tomberg K, Bradley A (2018) Repair of double-strand breaks induced by CRISPR-Cas9 leads to large deletions and complex rearrangements. Nat Biotechnol 36:765–771

Kumar R, Kaur A, Pandey A, Mamrutha HM, Singh GP (2019) CRISPR-based genome editing in wheat: a comprehensive review and future prospects. Mol Biol Rep 46:3557–3569. https://doi.org/10.1007/s11033-019-04761-3

Lau S-E, Teo WFA, Teoh EY, Tan BC (2022) Microbiome engineering and plant biostimulants for sustainable crop improvement and mitigation of biotic and abiotic stresses. Discov Food 2:9

Lawrenson T, Shorinola O, Stacey N, Li C, ØstergaardL PN, Uauy C, Harwood W (2015) Induction of targeted, heritable mutations in barley and *Brassica oleracea* using RNA-guided Cas9 nuclease. Genome Biol 16:1–13

Lee JK, Jeong E, Lee J, Jung M, Shin E, Kim YH, Lee K, Jung I, Kim D, Kim S et al (2018) Directed evolution of CRISPR-Cas9 to increase its specificity. Nat Commun 9:3048

Li T et al (2011) TAL nucleases (TALNs): hybrid proteins composed of TAL effectors and FokI DNA-cleavage domain. Nucleic Acids Res 39:359–372

Li T, Liu B, Spalding MH, Weeks DP et al (2012) High-efficiency TALEN-based gene editing produces disease resistant rice. Nat Biotechnol 30:390–392

Li J, Zhang H, Si X, Tian Y, Chen K, Liu J et al (2017) Generation of thermosensitive male-sterile maize by targeted knockout of the ZmTMS5 gene. J Genet Genomics 44:465–468. https://doi.org/10.1016/j.jgg.2017.02.002

Li R, Zhang L, Wang L, Chen L, Zhao R, Sheng J, Shen L (2018) Reduction of tomato-plant chilling tolerance by CRISPR–Cas9-mediated SlCBF1 mutagenesis. J Agric Food Chem 66:9042–9051

Li A, Jia S, Yobi A, Ge Z, Sato SJ, Zhang C et al (2018a) Editing of an alpha-Kafirin gene family increases, digestibility and protein quality in sorghum. Plant Physiol 177:1425–1438. https://doi.org/10.1104/pp.18.00200

Li H, Yang Y, Hong W et al (2020a) Applications of genome editing technology in the targeted therapy of human diseases: mechanisms, advances and prospects. Sig Transduct Target Ther 5:1. https://doi.org/10.1038/s41392-019-0089-y

Li A, Tanner MR, Lee CM, Hurley AE, De Giorgi M, Jarrett KE, Davis TH, Doerfler AM, Bao G, Beeton C et al (2020b) AAV-CRISPR gene editing is negated by pre-existing immunity to Cas9. Mol Ther 28:1432–1441

Liang Z, Zhang K, Chen K, Gao C (2014) Targeted mutagenesis in Zea mays using 939 TALENs and the CRISPR/Cas system. J Genet Genomics 41:63–68

Lin Y, Cradick TJ, Brown MT, Deshmukh H, Ranjan P, Sarode N, Wile BM, Vertino PM, Stewart FJ, Bao G (2014) CRISPR/Cas9 systems have off-target activity with insertions or deletions between target DNA and guide RNA sequences. Nucleic Acids Res 42:7473–7485

Liu X, Wu S, Xu J, Sui C, Wei J (2017) Application of CRISPR/Cas9 in plant biology. Acta Pharm Sin B 7(3):292–302. https://doi.org/10.1016/j.apsb.2017.01.002

Luan B, Xu G, Feng M, Cong L, Zhou R (2019) Combined computational–experimental approach to explore the molecular mechanism of SaCas9 with a broadened DNA targeting range. J Am Chem Soc 141:6545–6552

Lyzenga WJ, Pozniak CJ, Kagale S (2021) Advanced domestication: harnessing the precision of gene editing in crop breeding. Plant Biotechnol J 19:660–670

Ma L, Kong F, Sun K, Wang T, Guo T (2021) From classical radiation to modern radiation: past, present, and future of radiation mutation breeding. Front Public Health 9:11

Maeder ML et al (2008) Rapid "open-source" engineering of customized zinc-finger nucleases for highly efficient gene modification. Mol Cell 31(2):294–301. https://doi.org/10.1016/j.molcel.2008.06.016

Maeder ML, Thibodeau-Beganny S, Sander JD, Voytas DF, Joung JK (2009) Oligomerized pool engineering (OPEN): an 'open-source' protocol for making customized zinc-finger arrays. Nat Protoc 4(10):1471–1501. https://doi.org/10.1038/nprot.2009.98

Makarova KS et al (2011) Evolution and classification of the CRISPR-Cas systems. Nat Rev Microbiol 9:467–477

Makarova KS et al (2015) An updated evolutionary classification of CRISPR-Cas systems. Nat Rev Microbiol 13:722–736

Manghwar H, Lindsey K, Zhang X, Jin S (2019) CRISPR/Cas system: recent advances and future prospects for genome editing. Trends Plant Sci 24(12):1102–1125. https://doi.org/10.1016/j.tplants.2019.09.006

Mazur BJ, Chui C-F, Smith JK (1987) Isolation and characterization of plant genes coding for acetolactate synthase, the target enzyme for two classes of herbicides. Plant Physiol 85(4):1110–1117

Miller JC, Holmes MC, Wang J, Guschin DY, Lee Y-L, Rupniewski I, Beausejour CM, Waite AJ, Wang NS, Kim KA et al (2007) An improved zinc-finger nuclease architecture for highly specific genome editing. Nat Biotechnol 25:778–785

Miller JC, Tan S, Qiao G, Barlow KA, Wang J, Xia DF, Meng X, Paschon DE, Leung E, Hinkley SJ et al (2011) A TALE nuclease architecture for efficient genome editing. Nat Biotechnol 29:143–148

Mishra R, Zhao K (2018) Genome editing technologies and their applications in crop improvement. Plant Biotechnology Reports 12:57–68. https://doi.org/10.1007/s11816-018-0472-0

Mohanta TK, Bashir T, Hashem A, Abdallah EF, Bae H (2017) Genome editing tools in plants. Gene 8:399. https://doi.org/10.3390/genes8120399

Morbitzer R, Romer P, Boch J, Lahaye T (2010) Regulation of selected genome loci using de novo-engineered transcription activator-like effector (TALE)-type transcription factors. Proc Natl Acad Sci U S A 1078:21617–21622

Mushtaq M, Molla KA (2021) CRISPR technologies for plant biotechnology innovation. In: Ricroch A, Chopra S, Kuntz M (eds) Plant biotechnology. Springer, Cham

Nieves-Cordones M, Mohamed S, Tanoi K, Kobayashi NI, Takagi K, Vernet A et al (2017) Production of low-Cs + rice plants by inactivation of the K + transporter Os HAK 1 with the CRISPR -Cas system. Plant J 92:43–56. https://doi.org/10.1111/tpj.13632

Okada A, Arndell T, Borisjuk N, Sharma N, Watson-Haigh NS, Tucker EJ et al (2019) CRISPR/Cas9-mediated knockout of Ms1 enables the rapid generation of male-sterile hexaploid wheat lines for use in hybrid seed production. Plant Biotechnol J 17:1905–1913. https://doi.org/10.1111/pbi.13106

Okuzaki A, Ogawa T, Koizuka C, Kaneko K, Inaba M, Imamura J et al (2018) CRISPR/Cas9-mediated genome editing of the fatty acid desaturase 2 gene in *Brassica napus*. Plant Physiol Biochem 131:63–69. https://doi.org/10.1016/j.plaphy.2018.04.025

Osakabe Y, Watanabe T, Sugano SS et al (2016) Optimization of CRISPR/Cas9 genome editing to modify abiotic stress responses in plants. Sci Rep 6(1):26685

Pandey P, Irulappan V, Bagavathiannan M, Senthil-Kumar M (2017) Impact of combined abiotic and biotic stresses on plant growth and avenues for crop improvement by exploiting physio-morphological traits. Front Plant Sci 8:537

Pattanayak V, Lin S, Guilinger JP, Ma E, Doudna JA, Liu DR (2013) High-throughput profiling of off-target DNA cleavage reveals RNA-programmed Cas9 nuclease specificity. Nat Biotechnol 31:839–843

Petolino JF (2015) Genome editing in plants via designed zinc finger nucleases. In Vitro Cell Dev Biol Plant 51(1):1–8. https://doi.org/10.1007/s11627-015-9663-3

Porteus MH, Carroll D (2005) Gene targeting using zinc finger nucleases. Nat Biotechnol 23:967–973

Pyott DE, Sheehan E, Molnar A (2016) Engineering of CRISPR/Cas9-mediated potyvirus resistance in transgene-free Arabidopsis plants. Mol Plant Pathol 17(8):1276–1288

Ramirez CL et al (2008) Unexpected failure rates for modular assembly of engineered zinc fingers. Nat Methods 5(5):374–375. https://doi.org/10.1038/nmeth0508-374

Rath D, Amlinger L, Rath A, Lundgren M (2015) The CRISPR-Cas immune system: biology, mechanisms and applications. Biochimie 117:119–128

Reyon D et al (2012) FLASH assembly of TALENs for high-throughput genome editing. Nat Biotechnol 30:460–465

Rodríguez-Leal D, Lemmon ZH, Man J, Bartlett ME, Lippman ZB (2017) Engineering quantitative trait variation for crop improvement by genome editing. Cell 171:470–480.e478

Schaart JG, van de Wiel CC, Lotz LA, Smulders MJ (2016) Opportunities for products of new plant breeding techniques. Trends Plant Sci 21:438–449

Schloss PD, Handelsman J (2004) Status of the microbial census. Microbiol Mol Biol Rev 68(4):686–691. https://doi.org/10.1128/mmbr.68.4.686-691.2004

Schmid-Burgk JL, Schmidt T, Kaiser V, Höning K, Hornung V (2013) A ligation-independent cloning technique for high-throughput assembly of transcription activator–like effector genes. Nat Biotechnol 31:76–81

Shahriar SA, Islam MN, CNW C, Rahim MA, Paul NC, Uddain J, Siddiquee S (2021) Control of plant viral diseases by CRISPR/Cas9: resistance mechanisms, strategies and challenges in foodcrops. Plan Theory 10:1264

Shin HY, Wang C, Lee HK, Yoo KH, Zeng X, Kuhns T, Yang CM, Mohr T, Liu C, Hennighausen L (2017) CRISPR/Cas9 targeting events cause complex deletions and insertions at 17 sites in the mouse genome. Nat Commun 8:15464

Sukegawa S, Toki S, Saika H (2022) Genome editing technology and its application to metabolic engineering in rice. Rice (N Y) 15(1):21. PMID: 35366102; PMCID: PMC8976860. https://doi.org/10.1186/s12284-022-00566-4

Sun Y, Zhang X, Wu C, He Y, Ma Y, Hou H et al (2016) Engineering herbicide-resistant rice plants through CRISPR/Cas9-mediated homologous recombination of Acetolactate synthase. Mol Plant 9:628–631. https://doi.org/10.1016/j.molp.2016.01.001

Svitashev S, Young JK, Schwartz C, Huirong Gao S, Falco C, Mark Cigan A (2015) Targeted mutagenesis, precise gene editing, and site-specific gene insertion in maize using Cas9 and guide RNA. Plant Physiol 169(2):931–945

Szczepek M, Brondani V, Büchel J, Serrano L, Segal DJ, Cathomen T (2007) Structure-based redesign of the dimerization interface reduces the toxicity of zinc-finger nucleases. Nat Biotechnol 25:786–793

Takagi H, Tamiru M, Abe A, Yoshida K, Uemura A, Yaegashi H, Obara T, Oikawa K, Utsushi H, Kanzaki E, Mitsuoka C, Natsume S, Kosugi S, Kanzaki H, Matsumura H, Urasaki N, Kamoun S, Terauchi R (2015) MutMap accelerates breeding of a salt-tolerant rice cultivar. Nat Biotechnol 33:445–449

Tian S, Jiang L, Cui X et al (2018) Engineering herbicide-resistant watermelon variety through CRISPR/Cas9-mediated base-editing. Plant Cell Rep 37(9):1353–1356

Townsend JA, Wright DA, Winfrey RJ et al (2009) High-frequency modification of plant genes using engineered zinc-finger nucleases. Nature 459(7245):442–445

Tyagi S, Kumar R, Kumar V, Won S, Shukla P (2021) Engineering disease resistant plants through CRISPR-Cas9 technology. GM Crops Food 12:125–144

Vakulskas CA, Dever DP, Rettig GR, Turk R, Jacobi AM, Collingwood MA, Bode NM, McNeill MS, Yan S, Camarena J et al (2018) A high-fidelity Cas9 mutant delivered as a ribonucleoprotein complex enables efficient gene editing in human hematopoietic stem and progenitor cells. Nat Med 24:1216–1224

Wada N, Ueta R, Osakabe Y, Osakabe K (2020) Precision genome editing in plants: state-of-the-art in CRISPR/Cas9-based genome engineering. BMC Plant Biol 20:234

Walker PT (1984) Quantification and economic assessment of crop losses due to pests, diseases and weeds. In: Advancing agricultural production in Africa: cabs first scientific conference, Arusha, Tanzania, 12–18 February

Wang Y, Cheng X, Shan Q, Zhang Y, Liu J, Gao C et al (2014) Simultaneous editing of three homoeoalleles in hexaploid bread wheat confers heritable resistance to powdery mildew. Nat Biotechnol 32(9):947–951. https://doi.org/10.1038/nbt.2969

Wright DA, Townsend JA, Winfrey RJ Jr, Irwin PA, Rajagopal J, Lonosky PM, Hall BD, Jondle MD, Voytas DF (2005) High-frequency homologous recombination in plants mediated by zinc-finger nucleases. Plant J 44:693–705. https://doi.org/10.1111/j.1365-313X.2005.02551.x

Yamagata M (2023) Programmable proteins: target specificity, programmability and future directions. SynBio 1(1):65–76. https://doi.org/10.3390/synbio1010005

Yang L et al (2013) Optimization of scarless human stem cell genome editing. Nucleic Acids Res 41(19):9049–9061. https://doi.org/10.1093/nar/gkt555

Yang Y, Xu C, Shen Z, Yan C (2022) Crop quality improvement through genome editing strategy. Front Genom Edit 3:44

Yim H (2017) Incorporating climate change adaptation into sustainable development. J Int Dev Coop 2017:139–171

Yourik P, Fuchs RT, Mabuchi M, Curcuru JL, Robb GB (2019) Staphylococcus aureus Cas9 is a multiple-turnover enzyme. RNA 25:35–44

Zhang H, Zhang J, Lang Z, Botella JR, Zhu J-K (2017) Genome editing—principles and applications for functional genomics research and crop improvement. Crit Rev Plant Sci 36:1–19. https://doi.org/10.1080/07352689.2017.1402989

Zhang Y, Bai Y, Wu G, Zou S, Chen Y, Gao C et al (2017b) Simultaneous modification of three homoeologs of TaEDR1 by genome editing enhances powdery mildew resistance in wheat. Plant J 91:714–724. https://doi.org/10.1111/tpj.13599

Zhang Y, Massel K, Godwin ID et al (2018) Applications and potential of genome editing in crop improvement. Genome Biol 19:210

Zhang A, Liu Y, Wang F, Li T, Chen Z, Kong D, Bi J, Zhang F, Luo X, Wang J, Tang J, Yu X, Liu G, Luo L (2019) Enhanced rice salinity tolerance via CRISPR/Cas9-targeted mutagenesis of theOsRR22 gene. Mol Breed 39:47

Zhou Q, Liu W, Zhang Y, Liu KK (2007) Action mechanisms of acetolactate synthase-inhibiting herbicides. Pestic Biochem Physiol 89(2):89–96

Plant Breeding Becomes Smarter with Genome Editing

Lakshay Goyal, Meghna Mandal, Dharminder Bhatia, and Kutubuddin Ali Molla

Abstract The conventional concept of plant breeding as art and science is shifting toward less art and more science due to the rapid development in genomics, transcriptomics, and the availability of advanced molecular tools. Conventional plant breeding methods were made rapid and more precise with the help of molecular markers. However, these methods lack ways to reduce or avoid unwanted variation incorporated during the transfer of any desirable trait. Clustered regularly interspaced short palindromic repeats-CRISPR-associated protein 9 (CRISPR-Cas9), a genome-editing tool, has revolutionized plant sciences and made genetic studies more accessible and easier. This tool can supplement breeding methods in terms of creating precise and targeted variation for a trait and that too with negligible background variation. It can also be used to create new variations through chromosome engineering and targeted recombination, aiding in breaking undesirable linkages that otherwise cannot be broken by conventional plant breeding methods. Besides this, genome editing is also assisting in rapid de novo domestication of crop wild relatives, generating haploid-inducer lines, and developing apomictic hybrids. Genome editing is poised to remove many major bottlenecks of plant breeding.

Keywords CRISPR/Cas · Crop breeding · De-novo domestication · Genome-edited crops

---

L. Goyal · M. Mandal · D. Bhatia
Department of Plant Breeding and Genetics, Punjab Agricultural University, Ludhiana, Punjab, India

K. A. Molla (✉)
ICAR-National Rice Research Institute, Cuttack, Odisha, India
e-mail: kutubuddin.molla@icar.gov.in

J.-T. Chen, S. Ahmar (eds.), *Plant Genome Editing Technologies*, Interdisciplinary Biotechnological Advances,
https://doi.org/10.1007/978-981-99-9338-3_5

1 Introduction

Plant breeding deals with developing superior plant varieties to satisfy human needs. In ancient agriculture, several varieties have been developed through the intentional and unintentional selection of different genes and their genetic variation. Such types of selections and varietal development were based on the phenotype and physical appearance of plants. A lack of understanding of science and genetics renders plant breeding to be considered an art. But with the advancement of science and technology, varietal development programs were designed considering the genetic basis and inheritance of traits. Knowingly or unknowingly, natural genetic variations have been selected over time to develop modern improved crop plants (Pattnaik et al. 2023). Breeding through mutagenesis, induced by chemical or physical agents, is a modern technology to generate artificial genetic variation rapidly. However, mutagenesis involves the generation of random and uncontrolled genetic variations throughout the genome, many of which may or may not be useful for breeding superior varieties. Genome editing (GE), the newest technology, allows us to generate novel genetic variations at target loci with high accuracy. GE has revolutionized the field of plant breeding by providing precise and highly efficient tools to make targeted changes in genomes.

Genome editing (GE) by CRISPR/Cas9 deals with the manipulation of DNA sequences at targeted sites either by insertions, deletions, or substitutions. CRISPR/Cas9 has become a widely used tool to carry out GE. With the rapid developments in the CRISPR/Cas9 field, several cutting-edge techniques and tools are available at our disposal for plant breeding. CRISPR-Cas9 is fundamentally based on the identification of genomic target sites by the complementarity of single guide RNAs (sgRNAs) and the creation of double-strand breaks (DSBs) by the Cas9 enzyme. Later on, the fate of these DSBs is decided by two pathways. One is nonhomologous end joining (NHEJ), and the other is homology-directed repair (HDR). NHEJ is an error-prone repair pathway, which in the course of repairing DSBs generates different insertions or deletions (indels) at the target sites. These indels often disrupt the open reading frame of the gene, leading to the knockout of the gene function. In the case of noncoding DNA sequences (for example, promoter), the same sequence of events generates novel genetic variation.

Furthermore, HDR involves repairing the DSB based on the availability of a template. An additional supply of donor DNA templates is required for HDR. The donor template contains the desired edit that can be incorporated precisely at the target site via HDR. Among these two pathways, NHEJ has been widely used in GE. The use of HDR is generally limited by innate low efficiency and difficulty in the delivery of donor templates in plant cells (Molla et al. 2022). Such limitations can be overcome by using advanced versions of CRISPR/Cas9, such as base editors (Komor et al. 2016; Gaudelli et al. 2017) and prime editors (Anzalone et al. 2019). These tools are being used to induce mutations that are hitherto not possible by NHEJ. They have expanded the uses of GE from altering a single base pair to replacing a region of nucleotides even when donor DNA templates are not supplied (Molla and Yang 2019; Molla et al. 2021). These tools are benefitting agricultural sciences by shaping the genome of plants without any further genetic erosion.

Amid rising threats of climate change, developing climate-resilient varieties resistant to biotic and abiotic stresses is the need of the hour. Identifying novel sources of stress resistance would be the key to generating resistant varieties (Goyal et al. 2022). In conventional practice, varieties are developed by introgression breeding, which involves the transfer of resistant alleles from wild relatives and other germplasm. The use of conventional practices is limited by the need for the identification of resistant donors, linkage drag, and the time required to develop a variety. Later on, varietal development programs were supplemented with molecular markers to hasten the speed of breeding programs and selecting complex quantitative traits. With the advent of omics technologies, plant breeders have been blessed with genomics-assisted breeding technologies. Genomics-assisted selection could predict breeding values, reduce cycle time, and improve selection accuracy.

Using these tools, several mega varieties have been released in different regions across the world. However, plant breeding still suffers from many bottlenecks such as dependency on natural variations, prolonged selection cycles, linkage drag, and the need for repeated backcross to reconstitute the recipient genome. GE can assist in removing many bottlenecks of conventional plant breeding in many unprecedented ways. By creating novel variations, it can satisfy the demand for new breeding programs. It is very difficult to get recombinant between two closely linked loci through the cross-breeding method. However, GE can generate recombinants between two closely linked loci by creating DSB. Crossing overs are limited in the telomeric region of a chromosome; as a result, breeders cannot properly utilize genetic diversity from that region (Karmakar et al. 2022). GE can assist in inducing crossovers at the telomeric regions to generate many unexplored recombinations. In this chapter, we have discussed different ways in which GE can be used to supplement different plant breeding strategies. GE has extended the breeder's toolbox by providing different techniques, such as synthetic apomixis, HI-IMGE (Haploid Inducer-Mediated Genome Editing), de novo domestication, and many more. Combining GE with conventional methods can open several unforeseen ways and strategies for crop improvement to ensure global food security.

2 Relevance of Genome Editing in Plant Breeding for Removing Bottlenecks and Increasing Preciseness

Plant breeding aims to develop superior crop varieties by harnessing favorable alleles from the available gene pool. The basic idea is to intercross parents with superior qualities, raising the progenies and selecting the best-performing lines from the progenies having a favorable complementation of traits (Lyzenga et al. 2021). Developing a new variety this way can take 5 to 15 years, depending upon the species, its growth habits, parental lines, and the choice of breeding program

(Acquaah 2012). The efficiency of these breeding methodologies has been reasonably increased by the use of molecular markers. However, when a desirable locus is present so close to an undesirable locus that the occurrence of meiotic recombination is highly unlikely between them, the two loci are then essentially coinherited (Choi 2017). Breaking genetic linkages between two loci will depend upon the distance between the two loci, species, population size, and genomic location (since the rate of recombination may vary along the length of a chromosome). This is the biggest bottleneck of gene introgression by conventional methods. Also, varietal development using conventional breeding methods requires several cycles of repeated selection. This involves culling some individuals in every cycle, for which some genetic variation, even if it is unrelated to the gene of interest, is lost. Genome editing can come to the rescue over here. By using gene editing, strong effect improvement alleles can be fixed in the early generations. As a result, culling will be minimized, thereby preventing genetic erosion. The expanded resource base in terms of genetic diversity can then be suitably utilized to select and improve other complex quantitative traits (Lyzenga et al. 2021).

Simple arithmetic can illustrate the challenges that exist with the use of backcross breeding. Let us consider the case of rice, which has a relatively simple, small genome of 430 Mb. The number of genes is predicted to be around 32,000 to 50,000 (Goff et al. 2002). One percent of the rice genome would span a physical region of 4.3 Mb, containing approximately 320 to 500 genes (assuming a linear, regular distribution of the genes along the chromosomes). By similar logic, 0.1% of the genome would span 430 kb and may contain 32 to 50 genes. With conventional backcrossing, the recurrent parent genome is recovered by 50% following each cycle of backcrossing, which alternatively means that 50% of the donor parent genome is lost every cycle. So after six backcross generations, 0.8% of the donor parent genome remains (Allard 1999). This translates into 3.44 Mb of the genomic region and about 256 to 400 genes from the donor parent being retained in the backcross progeny. However, the target in most cases is to retain or transfer a single gene and complete the exclusion of all other genes. The problem is compounded by genomic realities that genes are mostly clustered in gene-rich regions, meaning that the actual number of genes in a 3.44 Mb region can be much higher than 256–400. Through oversimplification, this simple sum highlights the relative inefficiency and impreciseness of backcross breeding. The complexities compound in the case of polyploidy species like wheat, with a genome size of 15.5 Gb (Aury et al. 2022), and the number of genes is estimated to be 110,000 (International Wheat Genome Sequencing Consortium (IWGSC) 2018). If a wild donor accession and the recurrent parent differ from each other at 30% of the loci, one can expect that more than a thousand genes from the donor parent would continue to be present in the converted variety, even after four rounds of backcrossing (Dhugga 2022).

Conventional breeding relies on the effect of homologous recombination in areas proximal to the gene of interest. However, homologous recombination is not absolutely random; some genomic regions are more predisposed to recombination than others, such as centromeres and telomeres, which show sparse recombination. In such cases, genome editing could alternatively be employed to promote homologous recombination in those areas (Filler Hayut et al. 2017; Sarno et al. 2017). Experiments done in yeast targeted an entire chromosome arm using 95 guide RNAs to induce mitotic recombination (Sadhu et al. 2016). Such a strategy could greatly facilitate functional genomic studies and introgressing genes with unexplored beneficial effects. Additionally, the same strategy could be used to reshuffle alleles in low-recombination regions along the length of a chromosome, thereby creating novel genetic variation (Lyzenga et al. 2021).

Backcross breeding has been used since the advent of strategy-driven and institutionalized plant breeding. While all other plant breeding methods have amply created permutations and combinations of useful genetic variation found in cultivated crops, backcross breeding was placed on a higher pedestal owing to its specific ability to correct specific lacunae in the cultivar background, without altering much of the background. Backcross breeding met its intended objectives for long and was further modernized with the advent of molecular markers into marker-assisted backcross breeding. Molecular markers made it easy to keep track of the inheritance of the desired trait at the genotypic levels over generations. Additionally, the use of genome-wide markers increased the speed of recovery of recurrent parent genomes. Classic examples include the introgression of the *Sub1* gene in the mega rice variety Swarna (MTU-7029), where recovery of >95% of the recurrent parent genome could be achieved in just two cycles of backcrossing (Neeraja et al. 2007). Though backcross breeding is indeed effective, more so with the use of molecular markers, linkage drag cannot be overcome fully. The reduction of linkage drag to tolerable levels requires extensive cycles of crossing, adding to the time and resource budget of the breeding program.

In the course of the development of inbred varieties by conventional means, a time duration of 8–12 years is required. While a variety is being bred for one or a few traits for the span of a decade, unanticipated changes may occur at the field level, for which the variety developed may actually miss its goal. These “near-miss” varieties can be saved by using genome editing (Lyzenga et al. 2021). For instance, in wheat, fusarium head blight (FHB) susceptibility is linked to the *Rht-D1b* semi-dwarfing allele, possibly by tight physical linkage. So breeding for FHB resistance will inadvertently cause the wheat plants to grow taller and be susceptible to lodging since they would lose the superior *Rht-D1b* allele (He et al. 2014; Srinivasachary et al. 2008; Hilton et al. 1999). Since the *Rht-D1b* semi-dwarfing allele is associated with a single SNP, it represents an easy target for gene editing. This would ensure that FHB resistance is achieved without compromising height, and the near-miss variety would not be lost (Lyzenga et al. 2021).

3 CRISPR/Cas-Mediated GE and Its Comparison with Conventional Mutagenesis Techniques

Genetic variability is one of the key requisites for the improvement of plant varieties using any plant breeding intervention. As the taxonomic distance between accessions increases, the exploitable genetic variation also increases. However, this is concurrently accompanied by increased barriers to crossability, making the utilization of this naturally existing variation difficult. To break this barrier, various means were used to generate novel, heritable genetic variation using physical or chemical mutagens. Chemical mutagens (ethyl methanesulfonate (EMS), colchicine) and physical mutagens (mostly gamma irradiation) have been used in the past to generate novel genetic variation. However, the use of such mutagens is plagued by several problems, such as high levels of nonspecificity, large-scale structural modification of chromosomes, tricky protocols for polyploidy crops, and the carcinogenic nature of chemical and physical mutagens (Mao et al. 2019). CRISPR-Cas is a "targeted" mutagenesis approach. With appropriate designing of guide RNAs, vector selection, efficient plant transformation, and regeneration protocol, absolute target specificity can be achieved. This is akin to a dart game: while a conventional mutagenesis approach can make the dart land anywhere on the dartboard, the CRISPR-Cas strategy is sure to hit the bull's eye each and every time with a negligible off-target. The extremely high target specificity of the CRISPR-Cas method can justify the resources invested in this technique over any other conventional mutagenesis approach. The two approaches, conventional mutagenesis and CRISPR-Cas, differ in many aspects (Table 1). Conventional mutagenesis creates novel genetic variation; this actually becomes the starting point for further selection and hybridization. On the contrary, genome editing can introduce very specific edits into the targeted loci of the genome of an organism. So the need for subsequent selection and hybridization is eliminated or is required in very small amounts.

4 Genome Editing for Enhancing Yield

Seed and fruit characteristics are important yield determinants, influencing crop economics via consumer preference and amenability to mechanical processing. In the case of food grains, grain size, seed weight, and grain number per unit of reproductive structure are the primary yield determinants. In rice, grain size and grain number are regulated by the *GS3* (Fan et al. 2006) and *Gn1a* (Ashikari et al. 2005) genes, respectively. It has been observed that CRISPR/Cas9-based knockout of *GS3* and *Gn1a* resulted in increased grain length in five different rice backgrounds. Interestingly, the combined knockout of both genes produced a larger number of grains as compared to *gs3* alone (Shen et al. 2018). Moreover, increased grain weight has been reported by the editing of the *GW2* gene in wheat. Both *GW2-B1* and *GW2-D1* affected thousand-grain weight by affecting grain width and length.

Table 1 Comparison of genome editing with conventional mutagenesis and backcross breeding

S. no.	Factor	Backcross breeding	Conventional mutagenesis	Genome editing
1	Target specificity	Less specificity due to linkage drag	Random mutagenesis, nonspecific	Highly specific with no linkage drag
2	Pleiotropy reduction	Remotely possible	Remotely possible	Can reduce the pleiotropic effect of genes via the editing of regulatory elements. For example, *OsIPA1* promoter editing reduces pleiotropy
3	Genetic variation	Explore only existing variations	Generate novel variations	Generate novel variations
4	Genetic recombination	Incorporate new variations completely based on genetic recombination	Variations are not based on recombination	Variations are not based on recombination
5	Telomeric recombination	The rare occurrence of telomeric recombination hinders the exploitation of variation from chromosome ends	Not possible	Can induce telomeric recombination, allowing to generate many unexplored variations
6	Crossability barriers	Affected by crossability barrier between species	Not affected	Not affected
7	Tissue culture dependency	Not dependent	Not dependent	Dependent; in planta transformation abrogates the need for tissue culture
8	Regulatory concerns	No regulatory concerns	No regulatory concerns	Varies from country to country if the final product is transgene-free

The thousand-grain weight increased progressively in the single, double, and triple mutants compared to the wild type (Wang et al. 2018b). These illustrations show that GE can be successfully employed as a tool to identify the effect of single gene even in crops with complex genome architecture. In the quest to develop high-yielding varieties, some cultivars with high-quality attributes have perished due to their yield limits. To avoid the extinction of such cultivars, GE can be employed to boost their yield output. The genes *GA20ox2* (Han et al. 2019) and *GA20ox3* (Zhang et al. 2020) were edited in rice and maize, respectively, to confer semi-dwarf phenotypes in otherwise high-yielding but lodging-susceptible varieties. Plant architecture is the three-dimensional organization of the entire plant—including both below- and aboveground parts. Plant architecture has an immense bearing on the resource utilization efficiency and, hence, the agronomic performance of the plant. The advent of GE has made it possible for the first time in the history of plant breeding to close in on the concept of ideotype (Huang et al. 2021). For example, *DEP1* (*Dense and erect Panicle 1*) is an important gene that plays a triple role by increasing meristematic activity and nitrogen-use efficiency and improving population canopy structure through erect panicles (Xu et al. 2016). An elite allele of this gene, *dep1*, has already been exploited during the course of *Japonica* rice domestication in China. GE can be employed to replicate the same in other rice backgrounds. However, the gene has a complex structure with five different domains, each of which causes a particular phenotypic effect. In an elegant experiment, each of the domains was individually knocked out using GE to study their individual and combined effects (Li et al. 2019). For example, by targeting the specific domain (GGL domain and all C-terminal VWFC domains) of the *DEP1* gene, grain length was reduced by 4.97%. (Li et al. 2019). Such mutations can be induced to produce short-grained rice varieties that can be released in areas where short grain is desired. Since plant architecture is under hormonal control, genes involved in hormone signaling pathways can be also targeted by gene editing using CRISPR-Cas9. For example, in the case of rice, the *CKX2* or *Gn1a* gene, which encodes the cytokinin oxidase/dehydrogenase enzyme, is a good target. The reduced expression of *CKX2* leads to cytokinin accumulation in the reproductive tissues, causing an increase in grain number (Li et al. 2016b). Editing its promoter sequence or transcriptional factor that binds to the promoter might result in the lower expression of the gene. For example, *DST* is a transcription factor, and it binds to the promoter sequence of the *CKX2* gene. *DST* gene has been edited to enhance tolerance to salinity and drought in rice (Kumar et al. 2020). Similarly, the editing of the *CKX2* (*Gn1a*) homeologs in wheat has shown increased grain number (Zhang et al. 2019b). Also in wheat, the knockout of the *TaCKX-D1* gene significantly increased the grain number per spikelet (Zhang et al. 2019b). However, altering the same gene in barley did not achieve the desired consequences, indicating that these gene effects are background-specific (Gasparis et al. 2019).

Plant yield is a highly complicated trait, and understanding the complex process behind plant yield is critical to enhancing plant production. GE has been widely used to solve the complex genetic architecture of plant yield. Different genes that have been edited in various crops to improve plant architecture are shown in Table 2. In rice, knockout studies have demonstrated the antagonistic effects of the *PRR1* (*Pseudo-response regulator 1*) and *CCA 1* (*Circadian clock associated 1*) genes on the tiller numbers. Overexpressing *CCA1* and mutating *PRR1* reduces the number of tillers, whereas the reverse combination, i.e., overexpressing *PRR1* and mutating *CCA1*, increases the number of tillers. The *CCA1* gene is upstream of other important plant architecture genes, such as *IPA1* (*Ideal plant architecture 1*), *D14* (*Dwarf 14*), and *SCM3* (*Strong culm 3*) (Wang et al. 2020). Interestingly, the *IPA1* gene is considered the new green revolution gene. Targeting the *OsmiR156* target site in the *IPA1* gene can result in larger panicles with more grain weight, resulting in better plant production. By targeting grain-filling genes, several studies have been conducted to unleash the underlying mechanism of yield. Genes such as *SWEET11*, *GFR1* (*Grain-filling rate1*), and *PK3* (*Pyruvate kinase 3*) have been studied using CRISPR/Cas9. In each of these cases, grain filling was severely impaired. For example, *SWEET11* encodes a sucrose transporter; mutation in this gene reduces the sucrose content in the developing embryo (Ma et al. 2017), whereas knocking out the *PK3* gene reduces pyruvate kinase activity, causing defects in the source-to-sink

Table 2 Genes that have been edited to improve plant architecture and yield in different crops (modified from Huang et al. 2021)

Crop	Gene targeted	Trait affected	Reference
Rice	*GW5*	Grain width and weight	Liu et al. (2017b)
	CCD7	Plant height and tillering	Butt et al. (2018)
	AAP3	Tiller number, biomass, grain yield	Lu et al. (2018)
	DEP1, *GGC2*, *GS3*, *RGB1*, *RGA1*	All the genes code for subunits of the heterotrimeric G protein that regulates grain size in rice	Sun et al. (2018)
	LGG	Grain length	Chiou et al. (2019)
	PIL15	Grain size and weight	Ji et al. (2019)
	SNB	Grain length, width, weight	Ma et al. (2019)
	AAP5	Tiller number, grain yield	Wang et al. (2019c)
	ER1	Spikelets per panicle	Guo et al. (2020)
	HOX1, *HOX28*	Tiller angle	Hu et al. (2020c)
	Cyt P450 family	Grain size	Usman et al. (2020)
	AFG1	Grain size	Yu et al. (2020)
	GLW2	Grain length and grain width	Li et al. (2016c)
	CYP734A4	Grain number	Qian et al. (2017)
	SP3	Plant height, no. of branches and spikelets per panicle	Huang et al. (2019)
	NF-YC10	Grain morphometrics	Jia et al. (2019)
	VIN2	Seed size, grain weight	Lee et al. (2019a)
	INV3, *INV2*	Grain size	Deng et al. (2020)
	AAP1	Tiller number, grain weight	Ji et al. (2020)
	DLT10	Tiller number	Wen et al. (2020)
	BG3	Grain length	Yin et al. (2020)
	DF1	No. of florets	Ren et al. (2018)
	FON4	Floret characteristics	Ren et al. (2019)
	MOF1	No. of florets	Ren et al. (2020)
	ACS1 and *ACS2*	Root length	Lee et al. (2019b)
	LPR5	Primary root length	Ai et al. (2020)
	NAC2	Primary root length, no. of crown roots	Mao et al. (2020)
Wheat	*GASR7*, *DEP1*	GASR7—Grain weight DEP1—Inflorescence architecture, panicle growth, grain yield	Zhang et al. (2016)
	GW2	Thousand-grain weight, other grain morphometrics	Wang et al. (2018a, b) and Zhang et al. (2018)
Barley	*D-hordein*	Grain size	Yang et al. (2020)
Maize	*ZmCLE7*	Affect meristem size to produce enlarged ears	Liu et al. (2021b)
Wild tomato	*WUS*, *CLV3*	Fruit size	Rodriguez-Leal et al. (2017)

translocation of sucrose (Hu et al. 2020b). Moreover, a mutation in *GFR1* has been shown to reduce the rate of grain filling during the early grain-filling stage (Liu et al. 2019a). In addition to this, legumes have been also examined to better understand the complicated regulatory network governing their productivity. In *Medicago truncatula*, CRISPR/Cas9 GE has been used to study the role of the nodule-specific PLAT domain (*NPD1–5*) gene (Trujillo et al. 2019; Sun et al. 2018). GE can also be expanded to horticultural crops to enhance yield. Relatively fewer studies have been done on horticultural crops due to their less efficient transformation protocols. Nevertheless, some model horticultural crops have been explored with the aim of enhancing their yield. For example, in *S. pimpinellifolium* (*wild tomato*), targeting the *WUS* and *CLV3* genes resulted in increased fruit size, as discussed previously (Rodriguez-Leal et al. 2017). Although it is a complex trait, several efforts have been made to understand the genetic basis of yield. GE can be used in a very effective way to unfurl the complex gene interactions that govern the yield. Supplemented with other biotechnological tools, GE can be employed to study diverse crop families, irrespective of their ploidy and growth habits.

4.1 Developing Stable Male Sterile Lines to Facilitate Hybrid Breeding

While numerous strategies are being explored to increase global crop yields over the current levels, the most remunerative approach could be the development of hybrid varieties exploiting heterosis. GE is a powerful addition to the already available toolkit of hybrid production by creating more stable sources of male sterility and self-incompatibility. Many genes that have been edited to induce male sterility are shown in Table 3. Genes involved in the pathways for production and degradation of sporopollenin, exine, and tapetum can be targeted to confer male sterility. Being crucial to proper pollen function, the impairment of the exine development pathway may lead to a male sterile phenotype. For instance, *Ms1-B*, which encodes a lipid-transfer protein that functions in pollen coat formation, became an attractive target for generating male sterile mutants using genome editing (Okada et al. 2019). Similarly, the timing of tapetum ablation should be perfectly in sync with pollen development. Hastening or delaying tapetum ablation or preventing anther ablation completely will render the male sterile phenotype. It is well known that cytoplasmic male sterility (CMS) is conditioned by mitochondrial genes and has been used at the field level for the generation of hybrids in a number of crops. However, it is plagued by several issues, such as the requirement of a separate maintainer line (in the three-line system) and restorer line and the instability of CMS phenotypes (Weider et al. 2009). As an alternative, the genic male sterility (GMS) method has been adopted, as demonstrated by the hybrid seed production technology (SPT) recently introduced by DuPont Pioneer researchers (Chen et al. 2021). The SPT system is composed of the male-sterile (MS) female line and the male-fertile maintainer line with

Table 3 Genes edited to induce male sterility in different crops (modified from Chen et al. (2021))

Crop	Gene/target	Phenotype obtained	Reference
Rice	*OsACOS12* (acyl-CoA synthetase 5, involved in sporopollenin biosynthesis)	Male sterility	Zou et al. (2017)
Rice, maize	*OsTMS5*, *ZmTMS5* (Rnase ZS1, processes the mRNAs of three ubiquitin fusion ribosomal protein L40, involved in pollen production)	Thermo-sensitive male-sterile mutants	Zhou et al. (2016), Barman et al. (2019), Chen et al. (2020a, b) and Li et al. (2017)
Rice	*OsNP1* (glucose-methanol-choline oxidoreductase, involved in tapetum degeneration and pollen exine formation)	Male sterility	Chang et al. (2016)
Wheat	*TaMs1-B* (glycosylphosphatidylinositol-anchored lipid transfer protein, involved in exine and pollen development)	Complete male sterility in homozygous T1 mutants	Okada et al. (2019)
Wheat	*TaMs45* (encodes strictosidine synthase enzyme, involved in exine and pollen development)	Male sterility in triple homozygous mutants	Singh et al. (2018)
Maize	*ZmAPV1* (cytochrome P450 monooxygenase/hydroxylates C12 fatty acids and responsible for the development of tapetum)	Male sterility	Somaratne et al. (2017)
Maize	*ZmTGA9*, *ZmDFR*, and *ZmACOS5* genes	Male sterility	Liu et al. (2022)
Maize	*ZmMS33* (Sn-2 glycerol-3-phosphate acyltransferase, involved in the biosynthesis of extracellular lipid polyesters, pollen, and anther development)	Male sterility in homozygous mutants	Xie et al. (2018)
Tomato	*ms10³⁵* (*male sterile 1035*)	Male sterility	Liu et al. (2021a)
Tomato	*SlAP3* (*APETALA3*/class B floral organ identity *MADS-box* gene for petals and stamens development)	Homogeneous mutants have male-sterile phenotypes and reduced organ numbers in petals and stamens	Liu et al. (2019b)
Brassica napus	*BnaRFL11*	Male sterility, other phenotypic abnormalities	Farooq et al. (2022)
Brassica napus	*BnCYP450*	Genetic male sterility	Wang et al. (2023)

a transgenic male-sterility restorer gene linked to a marker gene for selection based on seed color (Wu et al. 2016). The EGMS (environmentally sensitive genic male sterility) system alleviates the need for a separate maintainer (B) line, thus converting the three-line system into a two-line system by using conditional MS mutants. Several TGMS (*TMS1–TMS10*) and PGMS (*PMS1–PMS3*) genes have been identified in rice (Huang et al. 2014; Fan and Zhang 2018). The rice gene *TMS5* encodes RNase ZS1, which specifically degrades the ubiquitin fusion ribosomal protein l40 messenger RNA (mRNA) in a temperature-sensitive manner, conferring the conditional MS phenotype (Zhou et al. 2014). The gene was knocked out to produce a commercially ready, transgene-free thermo-sensitive genic male sterile (TGMS) line (Zhou et al. 2016). Floral organ identity genes have been also manipulated in fruit crops such as tomato and strawberry to cause abnormal anther development, causing a male-sterile genotype. CRISPR/Cas9 has been used in tomatoes by inducing a mutation in the *SlAP3* gene, a homolog of the *Arabidopsis APETALA3*, which led to a male-sterile phenotype (Liu et al. 2019b). Conclusively, GE can be employed to induce MS in a diverse array of crops to facilitate hybrid seed production. Generating different MS alleles that are stable in different backgrounds can be very useful in heterosis breeding programs.

4.2 Fixation of Heterosis and Developing Apomictic Hybrids

The fixation of heterosis is a long-sought-after goal. Theoretically, there are many ways to fix heterosis, but its practical manifestation is very limited. One way to fix heterosis is by inducing artificial apomixis. It has been suggested that the incorporation of apomixis can fix heterosis, which was hitherto confined only to F_1. While much of the molecular mechanism underlying apomixis is still shrouded in mystery, research has already progressed in this field. The simultaneous modulation of three genes, viz., *REC8*, *PAIR1*, and *OSD1*, essentially converts meiosis into mitosis in gametic cells; this system has therefore been termed MiMe (mitosis instead of meiosis) (d'Erfurth et al. 2009). CRISPR/Cas-based editing of the *OsSPO11–1*, *OsREC8*, *OsOSD1*, and *OsMATL* genes has been shown to produce synthetic apomictic rice (Xie et al. 2019). The quadruple mutants could overcome the first checkpoint for successful synthetic apomixis, i.e., bypassing meiosis. The gametes produced by these mutants were clonal and unreduced. However, they failed at the second checkpoint—a bypass of fertilization. As a result, no apomictic offspring could be obtained. In another study, meiosis was successfully modulated by the multiplex editing of the *REC8*, *PAIR1*, *OSD1*, and *MTL* genes to produce diploid apomictic embryos (Wang et al. 2019b). Besides *MATL*, parthenogenesis-related genes have been also used to successfully bypass sexual reproduction (Khanday et al. 2019). The ectopic expression of the *BABY BOOM* (*BBM1*) gene in the rice egg cell was used to induce parthenogenesis. Later, combining the egg cell expression of *BBM1* with CRISPR/Cas9-engineered MiMe resulted in the production of up to 29% clonal diploid offspring (Khanday et al. 2019). Furthermore, the efficiency of this *MiMe + BBM1* system of synthetic apomixis in rice has been increased to >95% by using a single gene-editing cassette to target all four genes simultaneously and using a rice native promoter for the *BBM1* gene, instead of an *Arabidopsis* promoter (Vernet et al. 2022). While much advancement has already been achieved, understanding the missing links in the molecular pathways of male sterility, haploid induction, clonal propagation, and apomixis phenomena can enable its fine-tuning via gene-editing experiments. Supplemented with an efficient plant transformation and regeneration protocol, along with a conducive regulatory framework, can enable the agriculture community to reap the benefits of these technologies.

4.3 Generating Novel Cis-Alleles in the Promoter to Generate a Gradient of Phenotypes

In addition to genic sequences, cis elements also play a crucial role in determining plant phenotype. The ease of GE has favored the study of yet unexplored cis elements. The manipulation of cis elements can help in fine-tuning several traits by producing a phenotypic gradient of the respective trait. Rodriguez-Leal et al. (2017) first developed a promoter-editing strategy to generate a range of cis-allelic variations in tomatoes. The authors developed promoter variants for genes that regulate fruit size, inflorescence branching, and plant architecture and generated tomato lines with improved yield.

Apart from this, targeting cis elements has also been known to effectively overcome the pleiotropic effects of genes. For example, *IPA1* (*IDEAL PLANT ARCHITECTURE 1*) is one of the widely studied genes in rice and wheat (Li et al. 2016b; He et al. 2018; Qin et al. 2019). However, due to its pleiotropic nature, an increase in grains per panicle comes at the cost of reduced tillers. In the search for a genic sequence that could strike the perfect trade-off between these two contrasting attributes, Song et al. (2022) identified a 54-base pair cis-regulatory region in *IPA1*. It has been found that deleting this region resolves the trade-off between grains per panicle and tiller number, leading to a substantial increase in yield. Besides this, the editing of cis elements can also be employed to generate variability for quality attributes. Grain amylose content has a significant effect on both the eating and cooking quality of rice. It is known to be governed by the *WAXY* (*Wx*) gene, which encodes granule-bound starch synthase 1 (*GBSS1*). The editing of the region near the TATA box of the *Wxb* promoter has resulted in the creation of six new *Wx* alleles. These new alleles produced a gradient of amylose content ranging from 10.66% to 16.06% (Huang et al. 2020). It has generated new variability for the amylose content, which can be incorporated in different breeding lines according to environmental conditions and consumer preference. The CLAVATA (*CLV*)-WUSCHEL (*WUL*) signaling pathway, which is known to control meristem activity, has been studied in both maize and tomato to enhance yield. In maize, the *CLE7* and *FCP1* genes are known to increase meristem size by affecting the CLAVATA (*CLV*)-WUSCHEL (*WUL*) signaling pathway. It has been observed that weak promoter alleles of the *CLE7* and *FCP1* genes have enhanced yield by producing enlarged ears (Liu et al. 2021b).

Many economically important traits in crop plants are quantitatively controlled. Therefore, generating new quantitative trait variation (QTV) in plants using genome-editing tools holds great promise for revolutionizing and fast-tracking crop breeding. While QTV has been generated even by editing the coding sequence of mostly yield-related genes, the phenotypic outcomes of promoter editing have lower predictability. This unpredictability can be attributed to the large size of regulatory elements upstream of the transcription start site. In most cases, it harbors both cis-regulatory sequences and functionally neutral sequences. Along with this, they are also affected by epigenetic factors such as chromatin accessibility, posttranscriptional modifications of histones, etc. Though difficult, it is possible to predict the relative importance of sequence stretches within the regulatory sequences using bio-computational models. Zhou et al. (2023) designed a tool called CAPE (CRISPR-Cas12a promoter editing) to greatly ease the process of guide RNA designing to target cis-regulatory elements. This model relies on five variables, viz., sequence conservation, variation effects, chromatin accessibility, transcription factor binding motifs, and histone modification. The validity of the design tool was judged by using a previously published report on tomatoes (Rodriguez-Leal et al. 2017). It was also used to induce QTV in rice for amylose content (*GRANULE-BOUND STARCH SYNTHASE 1*), rice grain size (*OsGS3*), and green revolution traits (*OsD2*, *OsD11*, *OsD10*, *OsD1*, and *OsD18*). The use of gene editing in conjunction with a biocomputational tool such as CAPE will enable plant breeders to make more informed decisions regarding the choice of plant genes and cis-regulatory elements while using gene editing as a plant breeding tool. At a more basic level, a finer dissection of cis-regulatory elements is indeed much warranted for genes of immense consequence, such as those controlling yield or the ones akin to Green Revolution genes. Saturation editing of known genes associated with agronomic traits of interest may accelerate the molecular design of desirable traits (Cui et al. 2020).

5 Expanding Breeder's Toolbox

Developing varieties or new plant types to address new emerging problems is the main goal of plant breeding. During the classical era, several methods such as mutation breeding, with physical and chemical mutagens, helped plant breeders to create genetic variations and analyze and combine them to create new plant varieties. New tools are always desirable and preferred by plant breeders, which can increase the pace and accuracy of their breeding programs. Such tools allow plant breeders to complete their programs in a short period with increased precision and uniformity. GE represents one such tool, with unprecedented power, that can help speed up breeding programs by allowing plant breeders to create and incorporate new variations rapidly in plants. We discuss below a few major ways by which GE is contributing enormously to increasing the efficiency of plant breeding (Fig. 1).

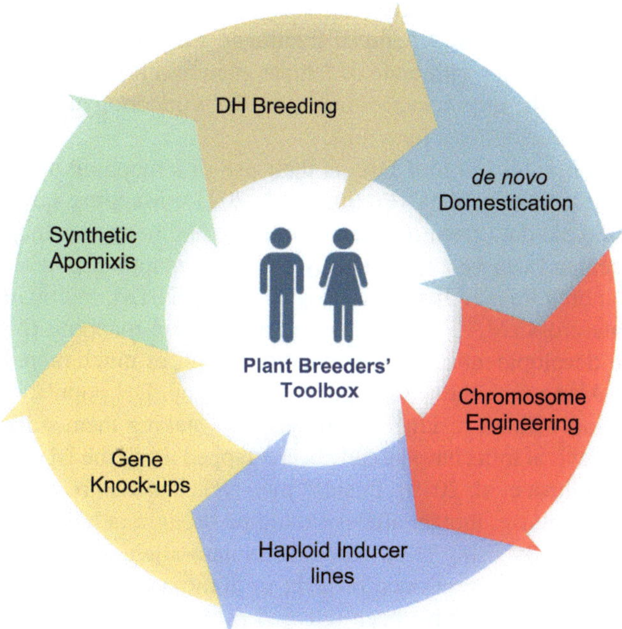

Fig. 1 Different tools to enhance the efficiency and scope of plant breeding to develop different crop varieties

5.1 Doubled Haploids (DHs) and Haploid Inducer Lines

Plants having gametic chromosome numbers in their sporophyte are called haploids. To restore their somatic genome, the chromosome of these haploids is doubled using chemical agents such as colchicine. These somatic-chromosome-restored haploids are called doubled haploids. Doubled haploids are one of the crucial and time-saving tools in a plant breeder's toolbox with multifarious benefits. It has been widely used to suffice different aspects of plant breeding. The first and foremost use of doubled haploids is in the rapid fixation of recombinants (Kyum et al. 2022). Conventional plant breeding methods broadly focused on three steps: (1) identifying and hybridizing diverse parents having desirable traits, (2) growing and screening of F_2 generation for desirable combinations of traits, and (3) advancing desired combinations by selfing to fix them through increased homozygosity. The third step is the most time-consuming, and alone it can take 6–7 years. The required homozygosity can only be achieved by five to six generations of continuous selfing. Fixing the desirable combination of genes after continued selfing is a key requirement to release any plant as a variety, whereas doubled haploid technology has the potential to achieve the same level of homozygosity in a single generation, which is otherwise achievable in five to six generations. Once hybridization is done, the gametes of F_1 plants can be captured to give rise to haploid plants. The haploidy of these plants can be doubled by using chemical agents such as colchicine. Spontaneous doubling can also happen without any chemical treatment. Capturing gametes here means generating haploid plants with male or female gametes either by in vitro or in vivo methods. Owing to the importance of haploids, the production of haploid plants has been extensively studied for a long time.

Another major application of DHs is their use as a mapping population during the study of quantitative traits (Pretini et al. 2021). Mapping quantitative traits requires phenotypic data from replicated studies of homozygous mapping populations. Such demand can be completed by perpetual mapping populations such as recombinant inbred lines (RILs), near-isogenic lines (NILs), multiparent advanced generation intercross (MAGIC), and nested association mapping (NAM) populations. But the development of these populations requires much more time as compared to DH. Along with the advantage of saving time, DH populations are highly homozygous and without residual heterozygosity, making them suitable for replicated studies. Several traits have already been mapped using the DH population (Shi et al. 2017; Kunihisa et al. 2019). Besides this, DHs can also be used for mutation studies. Being recessive, mutant alleles should be brought in homozygous form for their further study. By using DHs, recessive mutant alleles can be uncovered and can be subjected to selection once brought in homozygous form. Owing to their benefits, doubled haploids have been released as a variety in some crops, such as rice, wheat, barley, potato, tomato, *Brassica* spp., etc. (Cook 1936; Thompson 1972; Ho and Jones 1980).

Haploid inducer stocks (HISs) have been widely used in maize at the commercial level. The advantage of HIS lies in its easy *in planta* approach to producing haploids. Once developed, haploid inducer lines (HILs) can be directly used to produce haploids, omitting tissue culture and wide crossing. Before the advent of GE, this technology was mainly used in maize and was less explored in other crops, such as rice, wheat, oilseeds, and vegetables. But with the help of GE, it has been extended to other monocots and dicots as well. In the recent past, several genes have been identified underlying the phenomenon of haploid induction. It was seen that the loss of function mutation in *MTL/ZmPLA1/NLD* (Liu et al. 2017a; Kelliher et al. 2017; Gilles et al. 2017), *ZmDMP* (Zhong et al. 2019), *ZmPLD3* (Li et al. 2021b), and *ZmPOD65* (Jiang et al. 2022) genes leads to haploid induction in maize. Interestingly, most of these genes belong to different families and possess different mechanisms of inducing haploids. Unlike maize, HIS is not naturally available in other species, limiting its use in other crops. Using the GE approach, several scientists have successfully induced haploids across monocots (Yao et al. 2018; Lv et al. 2020; Wang and Ouyang 2023) and dicots (Wang et al. 2022b; Zhong et al. 2022). In most of the cases, it was found that the loss of function mutation in the above genes has led to haploid induction. A comprehensive list of potential genes for the generation of HIS in rice and wheat can be found in a recent article (Goyal et al. 2024). Moreover, GE has also helped in rectifying the other limitations associated with already available HIS in maize. The main limitation associated with HIS technology is its low haploid induction rate (HIR). A HIS with a high HIR is always desirable and required. GE has also been used to increase the efficiency of HIS by improving

HIR. For example, to increase HIR, the abovementioned genes have been knocked out in different combinations, and these genes act in a synergistic way rather than in a redundant manner. It was reported that the simultaneous knockout of three (*zmmtl* + *zmdmp* + *zmpld3*) genes led to an increase in the HIR of maize as compared to single mutants (Li et al. 2021b).

Extending its benefits to other crops, HIS has been also developed in rice (Yao et al. 2018) and wheat (Liu et al. 2020) by knocking out orthologues of the matrilineal gene (*OsMatl and TaMatl*), a phospholipase A family gene. Although with low HIR, *Matl* gene editing has enabled the development of HIS in rice and wheat, which was previously unavailable. The domain membrane protein is another important gene responsible for haploid induction. GE-mediated haploid induction has been widely explored in many dicots, such as tomato (Zhong et al. 2022), lucerne (Wang et al. 2022b), *Arabidopsis* (Zhong et al. 2020), brassica, and tobacco (Li et al. 2022). Provided a suitable tissue culture protocol is available, GE can be used to improve any crop.

Another important player in haploid induction is the *CENH3* gene (Ravi and Chan 2010). This gene differs from the previous one in that it is actively involved in the positioning of the centromere and is essential for chromosome segregation. The manipulation of this gene and its proteins has been shown to induce haploids in several crops. In recent years, it has been found that GE can be employed for haploid induction via the manipulation of *CENH3* in wheat (Lv et al. 2020), maize (Wang et al. 2021b), and *Arabidopsis* (Kuppu et al. 2020). Moreover, base editing technology, which is more targeted and precise (Molla and Yang 2019), has also been used in *Arabidopsis* for the induction of haploids (Wang and Ouyang 2023).

5.2 Haploid Inducer-Mediated Genome Editing

Recalcitrance to tissue culture and regulatory concerns for GE crops limit the use of GE up to its full potential. Different strategies have been used for conducting efficient GE in recalcitrant plant species and crop varieties. For example, morphogenic genes such as *BABYBOOM* (Boutilier et al. 2002), *WUSCHEL* (Zuo et al. 2002), and *GRF-GIF* (Debernardi et al. 2020) have been reported to overcome the problem of recalcitrance for conducting efficient GE. Interestingly, recently discussed haploid inducer stocks (HIS) can also facilitate GE in recalcitrant crops. This technique not only aided GE in recalcitrant species but also paved the way for transgene-free GE to skip regulatory issues. Briefly, this technology is based on the principle of uniparental genome elimination from the zygote derived from a cross between HIS and a host variety or species. The CRISPR/Cas9 cassette targeting the desired gene in the host variety is combined with the HIS of a respective crop. Later, haploids derived from crossing the CRISPR/Cas9 + HIS with the host variety are screened for putative edits. The artificial or spontaneous chromosome doubling of these edited haploids can convert them into edited transgene-free doubled haploids. As discussed earlier, HIS is naturally available in maize. With the help of GE, HIS has also become available in *Arabidopsis*, rice, and wheat.

This approach has been extensively explored in maize. By using HI-Edit, *ZmMatl*-based HIS has been used to carry out transgene-free GE in diverse maize backgrounds for *VLHP* and *GW2* genes. The editing frequency of haploids varied from 4.7% to 8.8% for both genes (Kelliher et al. 2019). Using a similar approach named haploid inducer-mediated genome editing (HI-IMGE), transgene-free GE was done for leaf angle (*ZmLG1*) and tassel number (*UB2*) by using CAU5 as a haploid inducer line. Interestingly, among 245 haploids, 4.1% haploids were edited for *the ZmLG1* locus, whereas for *UB2*, only one haploid plant was found to be edited (Wang et al. 2019a). Moreover, it took only two generations to obtain a homozygous DH line with intended trait improvements. The complex nature of polyploid crops has slowed down the pace of their improvement with GE. Interestingly, this technique has also been used to conduct transgene-free gene editing in complex polyploid crops such as wheat. It is well known that an intergenic cross between wheat and maize is known to induce wheat haploids. This system has been used to edit the *GRASSY TILLER1* (*TaGT1*) orthologs in wheat. The cassette targeting the *GRASSY TILLER1* gene was first transformed in maize, and the pollen of these transformed maize plants was then used to pollinate wheat. This gave rise to two edited haploids among 292 haploids for *TaGT1-4A* and *TaGT1-4B* genes (Kelliher et al. 2019). Using a similar approach, six different backgrounds of durum and bread wheat have been targeted for *BRASSINOSTEROID-INSENSITIVE 1* (*BRI1*) and *SEMI-DWARF 1* (*SD1*) genes. Homozygous mutations were observed for both genes in bread wheat, but for durum wheat, chimeric mutations were observed. The mutation frequency ranged from 3.6% to 50% across the different genotypes in both types of wheat (Budhagatapalli et al. 2020).

Furthermore, extending this approach to dicots, *GLABROUS1* (*GL1*) was targeted in *Arabidopsis*. Using *CENH3*-based paternal haploid inducer stock, 16.9% of edited haploids were generated. Moreover, the spontaneous chromosome doubling of these plants provided 100% transgene-free diploid progenies with homozygous mutations (Kelliher et al. 2019), showing that *CENH3*-based HI can also similarly work for dicots. Similarly, *Brassica* spp. has also been studied to induce transgene-free GE via a HI-mediated pathway (Li et al. 2021a). A previously reported haploid-inducing allooctaploid brassica (AAAACCCC, $2n = 8X = 76$) (Fu et al. 2018) was used for editing the *FAD2* gene in *B. napus* (*BnFAD2*) and *B. oleracea* (*BoFAD2*). CRISPR/Cas9-engineered allooctaploid brassica was used to target three homoeologues in *B. napus* and two homoeologues in *B. oleracea*. Interestingly, the editing frequency of doubled haploid offspring of *B. oleracea* and *B. napus* ranged from 1.89% to 12.97%. Conclusively, with a few modifications, different types of HIS can be used to perform transgene-free GE in both monocots and dicots irrespective of their ploidy level.

5.3 *De Novo Domestication*

The role of plant breeders is not only limited to developing new plant varieties. It also includes the development of new plant types or crops to increase diversity in the food platter of humans to maintain a well-balanced diet. Our present-day food crops are the result of the process our ancestors started many years ago. Dealing with wild species to meet their food demand, our hunter-gatherer ancestors domesticated many crops. But unfortunately, in the past, the monoculture of crops has severely impacted the environment and agricultural diversity. It is estimated that humans cultivate only 0.5% of total edible species of plants (Pattnaik et al. 2023). This dependency on a small proportion of edible species affects nutritional and food security. Furthermore, this problem is supplemented by the narrower genetic diversity of these food crops. The emerging risk of climate change and severe biotic and abiotic stresses can lead to sudden and heavy crop loss, impacting global food security. To combat these problems, it is required to explore the unexplored wild relatives of present-day crops. Moreover, it would also be desirable to focus on orphan crops such as millets and beans, which are still very less cultivated but are otherwise nutritious and resistant to biotic and abiotic stresses. Making these crops suitable to eat and developing them to maximize the economic gains from these crops are highly desirable. Wild crops possess resistant genes for biotic and abiotic stresses; domesticating such crops to make them yield high, agronomically acceptable, and nutritionally rich for human consumption is needed for the present-day scenario.

Domestication is an extremely slow process and a complex process that takes years over years. The time required to domesticate new crop species can be bypassed using different biotechnological approaches available to us. GE can serve as a key tool in the domestication of crops. Once the underlying genetic modification for domestication is known in a crop, it can be easily mimicked in its wild relative using GE. For example, in the case of rice, the loss of function mutations played a vital role in its domestication process (Gasparini et al. 2021); it has been assumed that domestication generally results from the loss of function mutations and not from the gain of function mutations (Luo et al. 2022). Imitating such mutations in wild species will be easy by using basic CRISPR/Cas9-based knockout experiments (Pattnaik et al. 2023). Moreover, in the case of mutations like single-nucleotide polymorphisms (SNPs), such domestication traits can be mimicked by using a base editing experiment, which is capable of inducing precise base substitutions, whereas mutations that include insertions or deletions can be mimicked by prime editing—a more advanced version of GE.

As reviewed previously, several attempts have been made to domesticate wild relatives of several cultivated crops, proving the practical application of GE for the rapid domestication of crops (Pattnaik et al. 2023). For example, efforts have been made to domesticate crops belonging to grass families, such as *Oryza alta* (Yu et al. 2021), *Hordeum marianum* (Kuang et al. 2022), *Oryza glaberrima* (Lacchini et al. 2020), and *Setaria virdis* (Mamidi et al. 2020), whereas a number of dicots, including *Solanum pimpinellifolium* (Li et al. 2018b), *Physalis pruinosa* (Lemmon et al.

2018), and *Solanum peruvianum* (Lin et al. 2022), have been also attempted for domestication.

Oryza alta is an allotetraploid rice variety with a CCDD genome. Polyploid crops come with their advantages, such as higher biomass, enhanced genome buffering, and resistance to biotic and abiotic stresses. Instead of inducing polyploidy in a domesticated crop, domesticating a polyploid crop can serve many advantages. Besides optimizing a tissue culture protocol and generating a high-quality genome sequence, multiplexed GE and base editing were used to domesticate *Oryza alta*. Several traits, such as seed shattering (*qSH1* homologs), awn length (*An-1*), plant height (*sd1*), and grain length (*GS3*) were edited to improve the plant type of *O. alta* (Yu et al. 2021). Moreover, base editing was also used to edit the miRNA target sites in the homoeologues of the *IPA1* gene (improved plant architecture). Later, showing the potential of multiplexed GE for domestication, *Ghd7* and *DTH7* were simultaneously edited to expand their cultivation to other regions. Different versions of GE can be employed to speed up the process of domestication.

In two independent studies, several genes have been edited to improve the plant type of *S. pimpinellifolium* to mold them according to human needs. In one of the studies, tweaking six important genes led to a multifold increase in fruit number, size, and lycopene content (Zsögön et al. 2018). This was done by targeting the genes for plant architecture (*SELF PRUNING*, *SP*), fruit shape (*OVATE*), fruit size (*FASCIATED*, *FAS* and *FRUIT WEIGHT 2.2*, *FW2.2*), fruit number (*MULTIFLORA*, *MULT*), and nutritional quality (*LYCOPENE BETA CYCLASE CycB*). Moreover, in other studies, efforts have been directed toward improved plant architecture (*Self-pruning*, *SP*), day neutrality (*Self-pruning 5G*, *SP5G*), increased flower and fruit number (*Clavata 3*, *SlCLV3* and *Wuschel*, *SlWUS*), and nutritional value (*GDP-L-galactose phosphorylase*, *SlGGP1*) (Li et al. 2018b). Interestingly, this study shows that GE can also be employed for editing different regulatory elements to improve domestication-related traits. Besides, in *SP* and *SP5G*, where exon targeting was used, promoter and transcription repressor elements were edited for *SlCLV3* and *SlWUS* genes. Furthermore, in the case of *SlGGP1*, the targeting of an upstream open reading frame (uORF) was done to enhance ascorbic acid content (Li et al. 2018b). Sometimes, a complete knockout of genes may not be desirable, and a weakening of gene expression is required; in such cases, the editing of regulatory elements can be rewarding. Similarly, *SlCLV3* and *SP5G* in *Physalis pruinosa* (Lemmon et al. 2018); *SvLes1* (seed shattering) in *Setaria virdis* (Mamidi et al. 2020); *HTD1*, *GS3*, *GW2*, and *GN1A* (plant height, grain size, grain weight, grain number) in *Oryza glaberrima* (Lacchini et al. 2020); *SpRDR6*, *SpSGS3* (plant RNA silencing), *pPR-1*, *SpProSys* (pathogenesis-related genes), and *SpMlo1*(fungal resistance) genes in *Solanum peruvianum* were edited to facilitate the domestication of these wild and orphan plants.

5.4 Herbicide Tolerance

Being a long-time noxious problem in agriculture, weeds harm crop plants in multifarious ways. Due to their competitive behavior, they can reduce plant yield by supplementing other problems, such as cohosting fungal pathogens and insects. Though several approaches have been used to control weeds since the beginning of the agricultural revolution, it remains a limiting factor in increasing crop production. Besides the use of several mechanical and cultural methods to control weeds, the chemical control of weeds always remains a method of choice due to its easiness and higher efficiency. One key requirement for applying this approach is the availability of new herbicides, along with new herbicide-tolerant (HT) varieties. Over the years, plant scientists keep on developing new HT varieties across the plant species using different approaches. Mainly, it includes artificial mutagenesis (physical and chemical) and transgenics. Due to regulatory concerns in many countries, developing HT varieties using transgenic approaches is becoming less preferred, whereas artificial mutagenesis is a technique of choice due to its fewer public concerns. However, developing herbicide tolerance through mutagenesis is highly uncertain and is a matter of chance. Recently, an EMS-generated mutant in the *OsALS* gene has been transferred in basmati rice, making it amenable to direct-seeded rice (DSR) conditions (Grover et al. 2020). Ideally, a desirable HT variety should be resistant to multiple herbicides with low residuals over the seasons and years without affecting the next crops in rotations. In such conditions, employing GE for targeting multiple genes can help in developing a plant that can be tolerant to multiple herbicides.

Several genes involved in different pathways have been targeted to develop HT crops. It includes 5-enolpyruvoylshikimate-3-phosphate synthase (*EPSPS*), glutamine synthetase, acetolactate synthase (*ALS*), acetyl-CoA carboxylase (*ACCase*), hydroxyphenylpyruvate dioxygenase (*HPPD*), and tubulin genes (*TubA2*) (Molla et al. 2021). The above mentioned genes are directly and indirectly involved in different vital pathways, such as the synthesis of branched and aromatic amino acids (AAs), plant nitrogen metabolism, fatty acid synthesis, carotenoid biosynthesis, etc. Only some specific types of mutations at some specific sites can provide tolerance toward herbicides. Sometimes, natural mutations in weeds provide the ability to withstand various herbicides. Weeds showing tolerance toward such herbicides can be used to mine alleles, which can be further imitated to develop novel tolerant alleles in food crops. Breeding for HT demands specific mutations at specific sites, which makes GE the most suited among various tools.

Though less, some studies have reported the use of HDR for generating HT crops. Using this strategy, the *EPSPS* gene has been modified to induce TIPS (T102I + P106S) mutation in rice (Li et al. 2016a) and TIPA mutations (T102I/P106A) in cassava (Hummel et al. 2018), whereas the *ALS* gene has also been targeted in many crops for imparting HT. The *ALS* gene has been mutated in rice for *G628W* substitution (Wang et al. 2021a), basmati rice for W548L (Zafar et al. 2023), sugarcane for W574L and S653I (Oz et al. 2021), soybean for P178S (Li et al. 2015), and maize for P165S (Svitashev et al. 2015) substitution mutations. For example, to imitate the already known substitutions, NHEJ cannot be used since it generates random mutations. Such substitution mutations can be induced by using HDR approaches, but the lower efficiency and difficult delivery of donor fragments are the major bottlenecks in using such an approach (Shimatani et al. 2017; Molla et al. 2022).

To overcome the above mentioned limitations of NHEJ and HDR in breeding HT crops, base editing has been widely used for this purpose. The advent of base editing (BE) has revolutionized the breeding of HT crops. This section focuses on different ways to employ BE for imparting herbicide tolerance in crops. Owing to ancestral similarity between genes, a mutation in one gene can behave similarly at the corresponding site in another ancestrally related gene. Similarly, an HT mutation in a particular gene can behave in the same manner as the mutation in an ancestrally related gene in another crop. Following this principle, the *ALS* gene (P197S) was mutated to develop HT oilseed rape based on information derived from the *ALS* gene of *Arabidopsis thaliana* (Wu et al. 2020). Similarly, the *ALS* gene was also mutated to develop HT watermelon (P190S and P190L) (Tian et al. 2018) and HT *B. napus* (P197F) (Hu et al. 2020a) using BE approaches. As discussed previously, information derived from herbicide-resistant weeds can help develop HT crops. For example, *Lolium rigidum* is a weed that was found to be resistant to *ACCase-inhibiting* herbicides. The underlying resistance was attributed to C2088R mutation, which imparts HT for *ACCase-inhibiting* herbicides (Yu et al. 2007). Using this information, a similar mutation was induced in the rice *ACCase* gene. The mutation at C2186R in rice corresponding to C2088R in *Lolium rigidium* imparted HT against *ACCase* herbicides (Li et al. 2018a). This shows that information derived from weed genetics can be replicated in cultivated crops to develop HT traits. Furthermore, developing crop varieties tolerant to multiple herbicides is always desirable. It can be done by manipulating multiple genes or targeting different sites in a single gene. For example, the mutation of two sites in the rice *ALS* gene, viz., P171 and G628, confers its tolerance to sulfonylurea (SU) herbicides as well as imidazolinone (IMI) herbicides (Zhang et al. 2021), expanding its resistance to diverse herbicides. Additionally, mutating two different genes can also expand the scope of HT in crop varieties, as seen in the case of wheat. It has been observed that the coediting of both *ALS* and *ACCase* genes has been shown to impart tolerance to multiple herbicides (Zhang et al. 2019a).

Like other traits, new alleles for HT can boost breeding for HT varieties. Briefly, it can be done by generating several alleles in the target gene and subsequent phenotyping for the putative trait. To accomplish this, several approaches have been developed based on BE. Saturated targeted endogenous mutagenesis editor (STEME) is one of its kind and has been specially designed to combine the function of both adenine base editors (ABEs) and cytosine base editors (CBEs) (Li et al. 2020). STEMEs can be used to saturate the target genic region with both transition and transversion mutations. Using this approach, novel alleles (P1927F, S1866F, and A1884P) imparting HT have been generated by targeting the *ACCase* gene in rice. It was also concluded that P1927F can be employed for combating weeds in rice improvement programs. Similarly in another study, the *ALS* gene in rice was targeted using the base-editing-mediated gene evolution (BEMGE) approach (Kuang et al. 2020). New alleles were generated (P171F, P171L, and P171S) at the same position with varying levels of tolerance to bispyribac-sodium (BS) herbicide. In another approach focusing on both CBEs and ABEs, the *ACCase* gene was targeted to generate several mutations in rice (Wang et al. 2022a), showing the feasibility of the method to generate novel HT alleles. Furthermore, lowering reliance on *ALS*, *ACCase*, and *EPSPS* genes for HT, other important genes can be base edited to impart HT, such as *OsTubA2* (Met-268-Thr) for dinitroaniline herbicides in rice (Liu et al. 2021c) and *OsGS1* (S59G) in rice for glufosinate herbicides (Ren et al. 2023).

Despite its so many applications for herbicide tolerance, base editing suffers from some limitations. It has been mainly used to induce transition mutations. Although some transversion base editors have been reported recently (Molla et al. 2020; Tong et al. 2023), their use in inducing transversion mutations is still limited. To overcome these limitations, prime editing has been employed to induce herbicide tolerance. Though this technology is still in its infancy, it is proving effective to induce precise edits to impart HT. In diverse studies, different genes have been tweaked to provide HT in rice. *OsALS* (Xu et al. 2020b; Liang et al. 2023) and *OsACC1* (Xu et al. 2020a, b) genes have been edited in rice using prime editing (PE). Analogous to saturation mutagenesis through base editing, the prime-editing-library-mediated saturation mutagenesis (PLSM) approach has been developed for a similar purpose (Xu et al. 2021). Using PLSM, 16 different HT mutations were induced at six conserved sites of the *OsACC1* gene in rice, yielding a vast array of alleles in the *OsACC1* gene. Though PLSM is more comprehensive as compared to BE-mediated approaches, BE can still play an important role in generating HT alleles. Using BE, it is better to first identify the key mutable sites in genes to induce HT alleles. Once the key sites are identified, they can be saturated with different mutations using PLSM approaches, generating new HT alleles (Xu et al. 2021).

5.5 Chromosome Engineering

Chromosome rearrangements (CRs) played an important role in the history of plant genome evolution. The advent of advanced cytogenetic tools in the classical era facilitated the study of CRs and revealed their role in plant genome evolution. CRs, including inversions, translocations, deletions, and duplications, helped unprecedently in studying genes and genomes when there were no molecular markers available. Different types of translocations, inversions, and other CRs played a significant role in shaping the genomes of the crops we have today. It is well known that translocations have extensively contributed to trait diversity, speciation, and genome evolution. Moreover, inversions are known to suppress crossing over (CO) and limit recombination, resulting in establishing an unbreakable linkage between genes present on an inverted segment. This leads to inhibiting recombination, impeding the stacking of genes and QTLs through classical plant breeding approaches

(Keilwagen et al. 2019), and slowing down the process of varietal development. Inversions can be used to release yet unexplored variability by converting a dead recombination region into a recombination active region. Interestingly, GE can be employed to generate CRs in a targeted manner. In the recent past, GE has been used to mimic both inversions and translocations, though with low frequency. Mimicking CRs, such as inversions, will enable plant breeders to establish and break linkage groups at their will in a targeted way.

In one of the pioneer studies, four loci were targeted to induce inversions, viz., ADH1, TT4, ECA3, and PSDG in *Arabidopsis* (Schmidt et al. 2019). Using GE, inversions of 8 to 18 kb size could be generated with a variable frequency of 0.64%–5%. It was also observed that the frequency of inversions could be increased by using an egg-cell-specific promoter and knocking out a KU70 regulatory protein. Furthermore, to expand the scope of GE to induce big inversions, a study was focused on reinverting an hk4S inversion in *Arabidopsis*. As previously mentioned, inversions can be used to restore recombinations in an inverted region. This has been practically shown in *Arabidopsis* by Schmidt et al. (2020). In this study, the reinversion of a 1.17 Mb long inverted segment leads to converting a dead recombination region into an active recombination region by enhancing CO. This characteristic inversion is present on chromosome number 4 of the Col-0 variety (Columbian accession) and is also known as hk4S. On the crossing of reinverted Col-0 with Ler1 (Landsberg accession; a variety with no hk4S inversion), recombination was observed, though with low frequency (0.5%). Besides *Arabidopsis*, maize has also been explored for targeted CRs. Using an integrated set of advanced approaches, including pangenomics and morphogenic genes (*Baby Boom* (*BBM*) and *Wuschel2* (*Wus2*)), a big pericentric inversion of 75.5 Mb was induced (Schwartz et al. 2020). This shows that targeted CRs can also be expanded to other commercially important crops using the GE approach. These customized manipulations can help in unlocking variations that are yet unexplored due to CO inhibiting CRs such as inversions.

CO and recombination have their pros and cons according to the breeder's requirements. Segregation and variability are not always required in breeding programs. For example, in heterosis breeding, segregation and recombination are the main reasons for the breakdown of heterosis. There are many approaches to fixing heterosis, but they have not been used commercially. Hypothetically, heterosis can also be fixed by using inversions via the suppression of segregation and recombination. Recently, GE-mediated inversions have been used to reduce CO and recombination in *Arabidopsis* (Rönspies et al. 2022). By inverting a fragment of 17.1 Mb on chromosome 2 in *Arabidopsis*, recombination events were reduced by 92%. Interestingly, a nine-tenth part of chromosome 2 was inverted, which means only a meager part of the telomeric region had been excluded for inversion. Chromosome restructuring played a crucial role in the development and evolution of new crops via the disruption and creation of new linkage groups. Mimicking or reverting such restructuring can unveil the history of the origin of crop plants, further enhancing their research. Translocations can help in such restructuring of chromosomes. GE has been used to induce such heritable translocation in *Arabidopsis* (Beying et al.

2020). Two translocations were induced in *Arabidopsis*, one between chromosomes 1 and 2 and the other between chromosomes 1 and 5.

The targeted induction of CRs via GE is a very crucial tool for crop improvement programs. As most of the studies have been done in *Arabidopsis* and other model crops, its extension to other commercial crops is highly desirable. The induction of CRs with GE also suffers from various limitations, including its low frequency and demand for high throughput screening. Moreover, different modifications of GE can be used to carry out targeted CRs. Recently, prime editing has been used to induce targeted inversions in human cells using twin prime editing (twinPE) (Anzalone et al. 2022). Such tools can also be replicated in plants to conduct efficient CRs at targeted sites.

6 Conclusion

GE can greatly suffice multiple needs of plant breeding. By aiding the development of new tools, it can solve yet unsolved problems. It has helped fill the void created by the lack of natural variations for many traits. By combining different tools, a series of targeted variations can be created for a gene that can be later screened for the desired phenotypic variation. A combination of GE and DH technologies can hasten plant breeding programs in an unprecedented way. Employing techniques like HI-mediated GE will not only facilitate the GE of recalcitrant species, but it can also enhance the development of transgene-free GE varieties. To combat the undesirable effects of climate change, new crops can be developed via GE-mediated de novo domestication. But the progress of GE is limited by low efficiency in different systems and their confinement to model crops. Increasing efficiency and the extension of these systems to commercial crops are the major goals of GE-mediated crop improvement. The availability of a regeneration protocol for important germplasms will enable us to directly edit their genome to improve them for multiple traits to rapidly bring them into mainstream agriculture. In addition, a conducive regulatory environment for genome-edited crops is a must to realize the full potential of GE.

References

Acquaah G (2012) Principles of plant genetics and breeding. Wiley-Blackwell, Chichester

Ai H, Cao Y, Jain A, Wang X, Hu Z, Zhao G, Hu S, Shen X, Yan Y, Liu X, Sun Y, Lan X, Xu G, Sun S (2020) The ferroxidase *LPR5* functions in the maintenance of phosphate homeostasis and is required for normal growth and development of rice. J Exp Bot 71:4828–4842

Allard RW (1999) Principles of plant breeding. Wiley

Anzalone AV, Randolph PB, Davis JR, Sousa AA, Koblan LW, Levy JM, Chen PJ, Wilson C, Newby GA, Raguram A, Liu DR (2019) Search-and-replace genome editing without double-strand breaks or donor DNA. Nature 576(7785):149–157

Anzalone AV, Gao XD, Podracky CJ, Nelson AT, Koblan LW, Raguram A, Levy JM, Mercer JA, Liu DR (2022) Programmable deletion, replacement, integration and inversion of large DNA sequences with twin prime editing. Nat Biotechnol 40(5):731–740

Ashikari M, Sakakibara H, Lin S, Yamamoto T, Takashi T, Nishimura A, Angeles ER, Qian Q, Kitano H, Matsuoka M (2005) Cytokinin oxidase regulates rice grain production. Science 309(741):745

Aury JM, Engelen S, Istace B, Monat C, Lasserre-Zuber P, Belser C, Cruaud C, Rimbert H, Leroy P, Arribat S, Dufau I (2022) Long-read and chromosome-scale assembly of the hexaploid wheat genome achieves high resolution for research and breeding. GigaScience 11:giac034

Barman HN, Sheng Z, Fiaz S, Zhong M, Wu Y, Cai Y, Wang W, Jiao G, Tang S, Wei X, Hu P (2019) Generation of a new thermo-sensitive genic male sterile rice line by targeted mutagenesis of TMS5 gene through CRISPR/Cas9 system. BMC Plant Biol 19:109

Beying N, Schmidt C, Pacher M, Houben A, Puchta H (2020) CRISPR–Cas9-mediated induction of heritable chromosomal translocations in *Arabidopsis. Nat Plants* 6(6):638–645

Boutilier K, Offringa R, Sharma VK, Kieft H, Ouellet T, Zhang L, Hattori J, Liu CM, van Lammeren AA, Miki BL, Custers JB (2002) Ectopic expression of *BABY BOOM* triggers a conversion from vegetative to embryonic growth. Plant Cell 14(8):1737–1749

Budhagatapalli N, Halbach T, Hiekel S, Büchner H, Müller AE, Kumlehn J (2020) Site-directed mutagenesis in bread and durum wheat via pollination by cas9/guide RNA-transgenic maize used as haploidy inducer. Plant Biotechnol J 18(12):2376

Butt H, Jamil M, Wang JY, Al-Babili S, Mahfouz M (2018) Engineering plant architecture via CRISPR/Cas9-mediated alteration of strigolactone biosynthesis. BMC Plant Biol 18:174

Chang Z, Chen Z, Wang N, Xie G, Lu J, Yan W, Zhou J, Tang X, Deng XW (2016) Construction of a male sterility system for hybrid rice breeding and seed production using a nuclear male sterility gene. Proc Natl Acad Sci USA 113:14145–14150

Chen HQ, Liu HY, Wang K, Zhang SX, Ye XG (2020a) Development and innovation of haploid induction technologies in plants. Hereditas (Beijing) 42:466–482

Chen RR, Zhou YB, Wang DJ, Zhao XH, Tang XD, Xu SC, Tang QY, Fu XX, Wang K, Liu XM, Yang YZ (2020b) CRISPR/Cas9-mediated editing of the thermo-sensitive genic male-sterile gene TMS5 in rice. Acta Agron Sin 46:1157–1165

Chen G, Zhou Y, Kishchenko O, Stepanenko A, Jatayev S, Zhang D, Borisjuk N (2021) Gene editing to facilitate hybrid crop production. Biotechnol Adv 46:107676

Chiou WY, Kawamoto T, Himi E, Rikiishi K, Sugimoto M, Hayashi- Tsugane M, Tsugane K, Maekawa M (2019) *LARGE GRAIN* encodes a putative RNA-binding protein that regulates spikelet hull length in rice. Plant Cell Physiol 60:503–515

Choi K (2017) Advances towards controlling meiotic recombination for plant breeding. Mol Cells 40:814–822

Cook RR (1936) A haploid Marglobe tomato. J Hered 27:433–435

Cui Y, Hu X, Liang G, Feng A, Wang F, Ruan S, Dong G, Shen L, Zhang B, Chen D, Zhu L, Hu J, Lin Y, Guo L, Matsuoka M, Qian Q (2020) Production of novel beneficial alleles of a rice yield-related QTL by CRISPR/Cas9. Plant Biotechnol J 18:1987–1989

Debernardi JM, Tricoli DM, Ercoli MF, Hayta S, Ronald P, Palatnik JF, Dubcovsky J (2020) A GRF–GIF chimeric protein improves the regeneration efficiency of transgenic plants. Nat Biotechnol 38(11):1274–1279

Deng X, Han X, Yu S, Liu Z, Guo D, He Y, Li W, Tao Y, Sun C, Xu P, Liao Y, Chen X, Zhang H, Wu X (2020) *OsINV3* and its homolog, *OsINV2*, control grain size in rice. Int J Mol Sci 21:2199

d'Erfurth I, Jolivet S, Froger N, Catrice O, Novatchkova M, Mercier R (2009) Turning meiosis into mitosis. PLoS Biol 7(6):e1000124

Dhugga KS (2022) Gene editing to accelerate crop breeding. Front Plant Sci 13:889995

Fan Y, Zhang Q (2018) Genetic and molecular characterization of photoperiod and thermo-sensitive male sterility in rice. Plant Reprod 31:3–14

Fan C, Xing Y, Mao H, Lu T, Han B, Xu C, Li X, Zhang Q (2006) *GS3*, a major QTL for grain length and weight and minor QTL for grain width and thickness in rice, encodes a putative transmembrane protein. Theor Appl Genet 112:1164–1171

Farooq Z, Nouman Riaz M, Farooq MS, Li Y, Wang H, Ahmad M, Tu J, Ma C, Dai C, Wen J, Shen J (2022) Induction of male sterility by targeted mutation of a restorer-of-fertility gene with CRISPR/Cas9-mediated genome editing in *Brassica napus* L. Plan Theory 11(24):3501

Filler Hayut S, Melamed Bessudo C, Levy AA (2017) Targeted recombination between homologous chromosomes for precise breeding in tomato. Nat Commun 8(1):15605

Fu S, Yin L, Xu M, Li Y, Wang M, Yang J, Fu T, Wang J, Shen J, Ali A, Zou Q (2018) Maternal doubled haploid production in interploidy hybridization between *Brassica napus* and *Brassica allooctaploids*. Planta 247:113–125

Gasparini K, dos Reis Moreira J, Peres LEP, Zsögön A (2021) *De novo* domestication of wild species to create crops with increased resilience and nutritional value. Curr Opin Plant Biol 60:102006

Gasparis S, Przyborowski M, Kala M, Nadolska-Orczyk A (2019) Knockout of the *HvCKX1* or *HvCKX3* gene in barley (*Hordeum vulgare* L.) by RNA-guided Cas9 nuclease affects the regulation of cytokinin metabolism and toot morphology. Cell 8:782

Gaudelli NM, Komor AC, Rees HA, Packer MS, Badran AH, Bryson DI, Liu DR (2017) Programmable base editing of A• T to G• C in genomic DNA without DNA cleavage. Nature 551(7681):464–471

Gilles LM, Khaled A, Laffaire JB, Chaignon S, Gendrot G, Laplaige J, Bergès H, Beydon G, Bayle V, Barret P, Comadran J (2017) Loss of pollen-specific phospholipase NOT LIKE DAD triggers gynogenesis in maize. EMBO J 36(6):707–717

Goff SA, Ricke D, Lan TH, Presting G, Wang R, Dunn M, Glazebrook J, Sessions A, Oeller P, Varma H, Hadley D (2002) A draft sequence of the rice genome (*Oryza sativa* L. ssp. japonica). Science 296(5565):92–100

Goyal L, Kaur M, Mandal M et al (2024) Potential gene editing targets for developing haploid inducer stocks in rice and wheat with high haploid induction frequency. 3 Biotech 14:14. https://doi.org/10.1007/s13205-023-03857-9

Goyal L, Intwala CG, Modha KG, Acharya VR (2022) Genetic variability analysis and characterization for MYMV resistance in mungbean. Indian J Ecol 49(5):1757–1763

Grover N, Kumar A, Yadav AK, Gopala Krishnan S, Ellur RK, Bhowmick PK, Vinod KK, Bollinedi H, Nagarajan M, Viswanathan C, Sevanthi AMV (2020) Marker assisted development and characterization of herbicide tolerant near isogenic lines of a mega basmati rice variety, "Pusa basmati 1121". Rice 13(1):1–13

Guo T, Lu Z, Shan J, Ye W, Dong N, Lin H (2020) *ERECTA1* acts upstream of the OsMKKK10-*OsMKK4-OsMPK6* cascade to control spikelet number by regulating cytokinin metabolism in rice. Plant Cell 2:2763–2779

Han Y, Teng K, Nawaz G, Feng X, Usman B, Wang X, Luo L, Zhao N, Liu Y, Li R (2019) Generation of semi-dwarf rice (*Oryza sativa* L.) lines by CRISPR/Cas9-directed mutagenesis of *OsGA20ox2* and proteomic analysis of unveiled changes caused by mutations. 3 Biotech 9:387

He X, Singh PK, Schlang N, Duveiller E, Dreisigacker S, Payne T, He Z (2014) Characterization of Chinese wheat germplasm for resistance to fusarium head blight at CIMMYT, Mexico. Euphytica 195:383–395

He F, Zhang F, Sun W, Ning Y, Wang GL (2018) A versatile vector toolkit for functional analysis of rice genes. Rice 11(1):1–10

Hilton AJ, Jenkinson P, Hollins TW, Parry DW (1999) Relationship between cultivar height and severity of fusarium ear blight in wheat. Plant Pathol 48(2):202–208

Ho KM, Jones GE (1980) Mingo barley. Can J Plant Sci 60(1):279–280

Hu L, Amoo O, Liu Q, Cai S, Zhu M, Shen X, Yu K, Zhai Y, Yang Y, Xu L, Fan C (2020a) Precision genome engineering through cytidine base editing in rapeseed (*Brassica napus L*). Front Genome Ed 2:605768

Hu L, Tu B, Yang W, Yuan H, Li J, Guo L, Zheng L, Chen W, Zhu X, Wang Y, Qin P, Ma B, Li S (2020b) Mitochondria-associated pyruvate kinase complexes regulate grain filling in rice. Plant Physiol 183:1073–1087

Hu Y, Li S, Fan X, Song S, Zhou X, Weng X, Xiao J, Li X, Xiong L, You A, Xing Y (2020c) *OsHOX1* and *OsHOX28* redundantly shape rice tiller angle by reducing *HSFA2D* expression and auxin content. Plant Physiol 184:1424–1437

Huang JZ, Zhi-Guo E, Zhang HL, Shu QY (2014) Workable male sterility systems for hybrid rice: genetics, biochemistry, molecular biology, and utilization. Rice (NY) 7:13

Huang Y, Bai X, Luo M, Xing Y (2019) *Short panicle 3* controls panicle architecture by upregulating *APO2/RFL* and increasing cytokinin content in rice. J Integr Plant Biol 61:987–999

Huang L, Li Q, Zhang C, Chu R, Gu Z, Tan H, Zhao D, Fan X, Liu Q (2020) Creating novel Wx alleles with fine-tuned amylose levels and improved grain quality in rice by promoter editing using CRISPR/Cas9 system. Plant Biotechnol J 18(11):2164

Huang X, Hilscher J, Stoger E, Christou P, Zhu C (2021) Modification of cereal plant architecture by genome editing to improve yields. Plant Cell Rep 40:953–978

Hummel AW, Chauhan RD, Cermak T, Mutka AM, Vijayaraghavan A, Boyher A, Starker CG, Bart R, Voytas DF, Taylor NJ (2018) Allele exchange at the *EPSPS* locus confers glyphosate tolerance in cassava. Plant Biotechnol J 16(7):1275–1282

International Wheat Genome Sequencing Consortium (IWGSC), Appels R, Eversole K, Stein N, Feuillet C, Keller B, Rogers J, Pozniak CJ, Choulet F, Distelfeld A, Poland J (2018) Shifting the limits in wheat research and breeding using a fully annotated reference genome. Science 361(6403):eaar7191

Ji X, Du Y, Li F, Sun H, Zhang J, Li J, Peng T, Xin Z, Zhao Q (2019) The basic helix-loop-helix transcription factor, *OsPIL15*, regulates grain size via directly targeting a purine permease gene *OsPUP7* in rice. Plant Biotechnol 17:1527–1537

Ji Y, Huang W, Wu B, Fang Z, Wang X (2020) The amino acid transporter *AAP1* mediates growth and grain yield by regulating neutral amino acid uptake and reallocation in *Oryza sativa*. J Exp Bot 71:4763–4777

Jia S, Xiong Y, Xiao P, Wang X, Yao J (2019) *OsNF-YC10*, a seed preferentially expressed gene regulates grain width by affecting cell proliferation in rice. Plant Sci 280:219–227

Jiang C, Sun J, Li R, Yan S, Chen W, Guo L, Qin G, Wang P, Luo C, Huang W, Zhang Q (2022) A reactive oxygen species burst causes haploid induction in maize. Mol Plant 15(6):943–955

Karmakar S, Das P, Panda D, Xie K, Baig MJ, Molla KA (2022) A detailed landscape of CRISPR-Cas-mediated plant disease and pest management. Plant Sci 323:111376

Keilwagen J, Lehnert H, Berner T, Beier S, Scholz U, Himmelbach A, Stein N, Badaeva ED, Lang D, Kilian B, Hackauf B (2019) Detecting large chromosomal modifications using short read data from genotyping-by-sequencing. Front Plant Sci 10:1133

Kelliher T, Starr D, Richbourg L, Chintamanani S, Delzer B, Nuccio ML, Green J, Chen Z, McCuiston J, Wang W, Liebler T (2017) *MATRILINEAL*, a sperm-specific phospholipase, triggers maize haploid induction. Nature 542(7639):105–109

Kelliher T, Starr D, Su X, Tang G, Chen Z, Carter J, Wittich PE, Dong S, Green J, Burch E, McCuiston J (2019) One-step genome editing of elite crop germplasm during haploid induction. Nat Biotechnol 37(3):287–292

Khanday I, Skinner D, Yang B, Mercier R, Sundaresan V (2019) A male-expressed rice embryogenic trigger redirected for asexual propagation through seeds. Nature 565:91–95

Komor AC, Kim YB, Packer MS, Zuris JA, Liu DR (2016) Programmable editing of a target base in genomic DNA without double-stranded DNA cleavage. Nature 533(7603):420–424

Kuang Y, Li S, Ren B, Yan F, Spetz C, Li X, Zhou X, Zhou H (2020) Base-editing-mediated artificial evolution of *OsALS1* in planta to develop novel herbicide-tolerant rice germplasms. Mol Plant 13(4):565–572

Kuang L, Shen Q, Chen L, Ye L, Yan T, Chen ZH, Waugh R, Li Q, Huang L, Cai S, Fu L (2022) The genome and gene editing system of sea barleygrass provide a novel platform for cereal domestication and stress tolerance studies. Plant Commun 3(5):100333

Kumar SVV, Verma RK, Yadav SK, Yadav P, Watts A, Rao MV, Chinnusamy V (2020) CRISPR-Cas9 mediated genome editing of *drought and salt tolerance* (*OsDST*) gene in *indica* mega rice cultivar MTU1010. Physiol Mol Biol Plants 26:1099–1110

Kunihisa M, Takita Y, Yamaguchi N, Okada H, Sato M, Komori S, Nishitani C, Terakami S, Yamamoto T (2019) The use of a fertile doubled haploid apple line for QTL analysis of fruit traits. Breed Sci 69(3):410–419

Kuppu S, Ron M, Marimuthu MP, Li G, Huddleson A, Siddeek MH, Terry J, Buchner R, Shabek N, Comai L, Britt AB (2020) A variety of changes, including CRISPR/Cas9-mediated deletions, in *CENH3* lead to haploid induction on outcrossing. Plant Biotechnol J 18(10):2068–2080

Kyum M, Kaur H, Kamboj A, Goyal L, Bhatia D (2022) Strategies and prospects of haploid induction in rice (*Oryza sativa*). Plant Breed 141(1):1–11

Lacchini E, Kiegle E, Castellani M, Adam H, Jouannic S, Gregis V, Kater MM (2020) CRISPR-mediated accelerated domestication of African rice landraces. PLoS One 15(3):e0229782

Lee DW, Lee SK, Rahman MM, Kim YJ, Zhang D, Jeon JS (2019a) The role of rice *vacuolar invertase2* in seed size control. Mol Cells 42:711–720

Lee HY, Chen Z, Zhang C, Yoon GM (2019b) Editing of the *OsACS* locus alters phosphate deficiency-induced adaptive responses in rice seedlings. J Exp Bot 70:1927–1940

Lemmon ZH, Reem NT, Dalrymple J, Soyk S, Swartwood KE, Rodriguez-Leal D, Van Eck J, Lippman ZB (2018) Rapid improvement of domestication traits in an orphan crop by genome editing. Nat Plants 4(10):766–770

Li Z, Liu ZB, Xing A, Moon BP, Koellhoffer JP, Huang L, Ward RT, Clifton E, Falco SC, Cigan AM (2015) Cas9-guide RNA directed genome editing in soybean. Plant Physiol 169(2):960–970

Li J, Meng X, Zong Y, Chen K, Zhang H, Liu J, Li J, Gao C (2016a) Gene replacements and insertions in rice by intron targeting using CRISPR–Cas9. Nat Plants 2(10):1–6

Li M, Li X, Zhou Z, Wu P, Fang M, Pan X, Lin Q, Luo W, Wu G, Li H (2016b) Reassessment of the four yield-related genes *Gn1a*, *DEP1*, *GS3*, and *IPA1* in rice using a CRISPR/Cas9 system. Front Plant Sci 7:377

Li S, Gao F, Xie K, Zeng X, Cao Y, Zeng J, He Z, Ren Y, Li W, Deng Q, Wang S, Zheng A, Zhu J, Liu H, Wang L, Li P (2016c) The *OsmiR396c-OsGRF4-OsGIF1* regulatory module determines grain size and yield in rice. Plant Biotechnol J 14:2134–2146

Li J, Zhang H, Si X, Tian Y, Chen K, Liu J, Chen H, Gao C (2017) Generation of thermosensitive male-sterile maize by targeted knockout of the ZmTMS5 gene. J Genet Genomics 44:465–468

Li C, Zong Y, Wang Y, Jin S, Zhang D, Song Q, Zhang R, Gao C (2018a) Expanded base editing in rice and wheat using a Cas9-adenosine deaminase fusion. Genome Biol 19:1–9

Li T, Yang X, Yu Y, Si X, Zhai X, Zhang H, Dong W, Gao C, Xu C (2018b) Domestication of wild tomato is accelerated by genome editing. Nat Biotechnol 36(12):1160–1163

Li X, Tao Q, Miao J, Yang Z, Gu M, Liang G, Zhou Y (2019) Evaluation of differential *qPE9-1/DEP1* protein domains in rice grain length and weight variation. Rice (N Y) 12:5

Li C, Zhang R, Meng X, Chen S, Zong Y, Lu C, Qiu JL, Chen YH, Li J, Gao C (2020) Targeted, random mutagenesis of plant genes with dual cytosine and adenine base editors. Nat Biotechnol 38(7):875–882

Li C, Sang S, Sun M, Yang J, Shi Y, Hu X, Li Y, Hao M, Chu W, Zhang H, Wang H (2021a) Direct modification of multiple gene homoeologs in *Brassica oleracea* and *Brassica napus* using doubled haploid inducer-mediated genome-editing system. Plant Biotechnol J 19(10):1889

Li Y, Lin Z, Yue Y, Zhao H, Fei X, Lizhu E, Liu C, Chen S, Lai J, Song W (2021b) Loss-of-function alleles of *ZmPLD3* cause haploid induction in maize. Nat Plants 7(12):1579–1588

Li Y, Li D, Xiao Q, Wang H, Wen J, Tu J, Shen J, Fu T, Yi B (2022) An in planta haploid induction system in *Brassica napus*. J Integr Plant Biol 64(6):1140–1144

Liang Z, Wu Y, Guo Y, Wei S (2023) Addition of the T5 exonuclease increases the prime editing efficiency in plants. J Genet Genomics 50:582

Lin CS, Hsu CT, Yuan YH, Zheng PX, Wu FH, Cheng QW, Wu YL, Wu TL, Lin S, Yue JJ, Cheng YH (2022) DNA-free CRISPR-Cas9 gene editing of wild tetraploid tomato *Solanum peruvianum* using protoplast regeneration. Plant Physiol 188(4):1917–1930

Liu C, Li X, Meng D, Zhong Y, Chen C, Dong X, Xu X, Chen B, Li W, Li L, Tian X (2017a) A 4-bp insertion at *ZmPLA1* encoding a putative phospholipase A generates haploid induction in maize. Mol Plant 10(3):520–522

Liu J, Chen J, Zheng X, Wu F, Lin Q, Heng Y, Tian P, Cheng Z, Yu X, Zhou K, Zhang X, Guo X, Wang J, Wang H, Wan J (2017b) *GW5* acts in the brassinosteroid signalling pathway to regulate grain width and weight in rice. Nat Plants 3:17043

Liu E, Zeng S, Zhu S, Liu Y, Wu G, Zhao K, Liu X, Liu Q, Dong Z, Dang X, Xie H, Li D, Hu X, Hong D (2019a) Favourable alleles of *GRAIN-FILLING RATE1* increase the grain-filling RATE and yield of rice. Plant Physiol 181:1207–1222

Liu Y, Qiu S, Jin M, Deng H, Yin M, Chen Z, Zhou X, Tang X (2019b) Study on the application of CRISPR/Cas9 technology in development of tomato (*Solanum lycopersicum*) male sterile line. J Agric Biotechnol 27:951–960

Liu C, Zhong Y, Qi X, Chen M, Liu Z, Chen C, Tian X, Li J, Jiao Y, Wang D, Wang Y (2020) Extension of the in vivo haploid induction system from diploid maize to hexaploid wheat. Plant Biotechnol J 18(2):316

Liu J, Wang S, Wang H, Luo B, Cai Y, Li X, Zhang Y, Wang X (2021a) Rapid generation of tomato male-sterile lines with a marker use for hybrid seed production by CRISPR/Cas9 system. Mol Breed 41:1–12

Liu L, Gallagher J, Arevalo ED, Chen R, Skopelitis T, Wu Q, Bartlett M, Jackson D (2021b) Enhancing grain-yield-related traits by CRISPR–Cas9 promoter editing of maize *CLE* genes. Nat Plants 7(3):287–294

Liu L, Kuang Y, Yan F, Li S, Ren B, Gosavi G, Spetz C, Li X, Wang X, Zhou X, Zhou H (2021c) Developing a novel artificial rice germplasm for dinitroaniline herbicide resistance by base editing of *OsTubA2*. Plant Biotechnol J 19(1):5

Liu X, Zhang S, Jiang Y, Yan T, Fang C, Hou Q, Wu S, Xie K, An X, Wan X (2022) Use of CRISPR/Cas9-based gene editing to simultaneously mutate multiple homologous genes required for pollen development and male fertility in maize. Cell 11(3):439

Lu K, Wu B, Wang J, Zhu W, Nie H, Qian J, Huang W, Fang Z (2018) Blocking amino acid transporter *OsAAP3* improves grain yield by promoting outgrowth buds and increasing tiller number in rice. Plant Biotechnol J 16:1710–1722

Luo G, Najafi J, Correia PM, Trinh MDL, Chapman EA, Østerberg JT, Thomsen HC, Pedas PR, Larson S, Gao C, Poland J (2022) Accelerated domestication of new crops: yield is key. Plant Cell Physiol 63(11):1624–1640

Lv J, Yu K, Wei J, Gui H, Liu C, Liang D, Wang Y, Zhou H, Carlin R, Rich R, Lu T (2020) Generation of paternal haploids in wheat by genome editing of the centromeric histone *CENH3*. Nat Biotechnol 38(12):1397–1401

Lyzenga WJ, Pozniak CJ, Kagale S (2021) Advanced domestication: harnessing the precision of gene editing in crop breeding. Plant Biotechnol J 19(4):660–670

Ma L, Zhang D, Miao Q, Yang J, Xuan Y, Hu Y (2017) Essential role of sugar transporter *OsSWEET11* during the early stage of rice grain filling. Plant Cell Physiol 58:863–873

Ma X, Feng F, Zhang Y, Elesawi IE, Xu K, Li T, Mei H, Liu H, Gao N, Chen C, Luo L, Yu S (2019) A novel rice grain size gene *OsSNB* was identified by genome-wide association study in natural population. PLoS Genet 15:e1008191

Mamidi S, Healey A, Huang P, Grimwood J, Jenkins J, Barry K, Sreedasyam A, Shu S, Lovell JT, Feldman M, Wu J (2020) A genome resource for green millet *Setaria viridis* enables discovery of agronomically valuable loci. Nat Biotechnol 38(10):1203–1210

Mao Y, Botella JR, Liu Y, Zhu JK (2019) Gene editing in plants: progress and challenges. Natl Sci Rev 6(3):421–437

Mao C, He J, Liu L, Deng Q, Yao X, Liu C, Qiao Y, Li P, Ming F (2020) *OsNAC2* integrates auxin and cytokinin pathways to modulate rice root development. Plant Biotechnol J 18:429–442

Molla KA, Yang Y (2019) CRISPR/Cas-mediated base editing: technical considerations and practical applications. Trends Biotechnol 37(10):1121–1142

Molla KA, Qi Y, Karmakar S, Baig MJ (2020) Base editing landscape extends to perform transversion mutation. Trends Genet 36(12):899–901
Molla KA, Sretenovic S, Bansal KC, Qi Y (2021) Precise plant genome editing using base editors and prime editors. Nat Plants 7(9):1166–1187
Molla KA, Shih J, Wheatley MS, Yang Y (2022) Predictable NHEJ insertion and assessment of HDR editing strategies in plants. Front Genome Ed 4:825236
Neeraja CN, Maghirang-Rodriguez R, Pamplona A, Heuer S, Collard BC, Septiningsih EM, Vergara G, Sanchez D, Xu K, Ismail AM, Mackill DJ (2007) A marker-assisted backcross approach for developing submergence-tolerant rice cultivars. Theor Appl Genet 115:767–776
Okada A, Arndell T, Borisjuk N, Sharma N, Watson-Haigh NS, Tucker EJ, Baumann U, Langridge P, Whitford R (2019) CRISPR/Cas9-mediated knockout of *Ms1* enables the rapid generation of male-sterile hexaploid wheat lines for use in hybrid seed production. Plant Biotechnol J 17:1905–1913
Oz MT, Altpeter A, Karan R, Merotto A, Altpeter F (2021) CRISPR/Cas9-mediated multi-allelic gene targeting in sugarcane confers herbicide tolerance. Front Genome Ed 3:673566
Pattnaik D, Avinash SP, Panda S, Bansal KC, Chakraborti M, Kar MK, Baig MJ, Molla KA (2023) Accelerating crop domestication through genome editing for sustainable agriculture. J Plant Biochem Biotechnol:1–17
Pretini N, Vanzetti LS, Terrile II, Donaire G, González FG (2021) Mapping QTL for spike fertility and related traits in two doubled haploid wheat (*Triticum aestivum* L.) populations. BMC Plant Biol 21:1–18
Qian W, Wu C, Fu Y, Hu G, He Z, Liu W (2017) Novel rice mutants overexpressing the brassinosteroid catabolic gene *CYP734A4*. Plant Mol Biol 93:197–208
Qin R, Li J, Li H, Zhang Y, Liu X, Miao Y, Zhang X, Wei P (2019) Developing a highly efficient and wildly adaptive CRISPR-SaCas9 toolset for plant genome editing. Plant Biotechnol J 17(4):706
Ravi M, Chan SW (2010) Haploid plants produced by centromere-mediated genome elimination. Nature 464(7288):615–618
Ren D, Yu H, Rao Y, Xu Q, Zhou T, Hu J, Zhang Y, Zhang G, Zhu L, Gao Z, Chen G, Guo L, Zeng D, Qian Q (2018) 'Two-floret spikelet' as a novel resource has the potential to increase rice yield. Plant Biotechnol J 16:351–353
Ren D, Xu QK, Qiu ZN, Cui YJ, Zhou TT, Zeng DL, Guo LB, Qian Q (2019) *FON4* prevents the multi-floret spikelet in rice. Plant Biotechnol J 17:1007–1009
Ren D, Rao Y, Yu H, Xu Q, Cui Y, Xia S, Yu X, Liu H, Hu H, Xue D, Zeng D, Hu J, Zhang G, Gao Z, Zhu L, Zhang Q, Shen L, Guo L, Qian Q (2020) *MORE FLORET1* encodes a MYB transcription factor that regulates spikelet development in rice. Plant Physiol 184:251–265
Ren B, Kuang Y, Xu Z, Wu X, Zhang D, Yan F, Li X, Zhou X, Wang G, Zhou H (2023) Three novel alleles of *OsGS1* developed by base-editing-mediated artificial evolution confer glufosinate tolerance in rice. Crop J 11(2):661–665
Rodriguez-Leal D, Lemmon ZH, Man J, Bartlett ME, Lippman ZB (2017) Engineering quantitative trait variation for crop improvement by genome editing. Cell 171:470–480
Rönspies M, Schmidt C, Schindele P, Lieberman-Lazarovich M, Houben A, Puchta H (2022) Massive crossover suppression by CRISPR–Cas-mediated plant chromosome engineering. Nat Plants 8(10):1153–1159
Sadhu MJ, Bloom JS, Day L, Kruglyak L (2016) CRISPR-directed mitotic recombination enables genetic mapping without crosses. Science 352(6289):1113–1116
Sarno R, Vicq Y, Uematsu N, Luka M, Lapierre C, Carroll D, Bastianelli G, Serero A, Nicolas A (2017) Programming sites of meiotic crossovers using Spo11 fusion proteins. Nucleic Acids Res 45:e164
Schmidt C, Pacher M, Puchta H (2019) Efficient induction of heritable inversions in plant genomes using the CRISPR/Cas system. Plant J 98(4):577–589

Schmidt C, Fransz P, Rönspies M, Dreissig S, Fuchs J, Heckmann S, Houben A, Puchta H (2020) Changing local recombination patterns in *Arabidopsis* by CRISPR/Cas mediated chromosome engineering. Nat Commun 11(1):4418

Schwartz C, Lenderts B, Feigenbutz L, Barone P, Llaca V, Fengler K, Svitashev S (2020) CRISPR–Cas9-mediated 75.5-Mb inversion in maize. Nat Plants 6(12):1427–1431

Shen L, Wang C, Fu Y, Wang J, Liu Q, Zhang X, Yan C, Qian Q, Wang K (2018) QTL editing confers opposing yield performance in different rice varieties. J Integr Plant Biol 60:89–93

Shi Z, Song W, Xing J, Duan M, Wang F, Tian H, Xu L, Wang S, Su A, Li C, Zhang R (2017) Molecular mapping of quantitative trait loci for three kernel-related traits in maize using a double haploid population. Mol Breed 37:1–10

Shimatani Z, Kashojiya S, Takayama M, Terada R, Arazoe T, Ishii H, Teramura H, Yamamoto T, Komatsu H, Miura K, Ezura H (2017) Targeted base editing in rice and tomato using a CRISPR-Cas9 cytidine deaminase fusion. Nat Biotechnol 35(5):441–443

Singh M, Kumar M, Albertsen MC, Young JK, Cigan AM (2018) Concurrent modifications in the three homeologs of *Ms45* gene with CRISPR-Cas9 lead to rapid generation of male sterile bread wheat (*Triticum aestivum* L.). Plant Mol Biol 97:371–383

Somaratne Y, Tian Y, Zhang H, Wang M, Huo Y, Cao F, Zhao L, Chen H (2017) Abnormal pollen Vacuolation1 (APV1) is required for male fertility by contributing to anther cuticle and pollen exine formation in maize. Plant J 90:96–110

Song X, Meng X, Guo H, Cheng Q, Jing Y, Chen M, Liu G, Wang B, Wang Y, Li J, Yu H (2022) Targeting a gene regulatory element enhances rice grain yield by decoupling panicle number and size. Nat Biotechnol 40(9):1403–1411

Srinivasachary, Gosman N, Steed A, Simmonds J, Leverington-Waite M, Wang Y, Snape J, Nicholson P (2008) Susceptibility to fusarium head blight is associated with the Rht-D1b semi-dwarfing allele in wheat. Theor Appl Genet 116:1145–1153

Sun S, Wang L, Mao H, Shao L, Li X, Xiao J, Ouyang Y, Zhang Q (2018) A G-protein pathway determines grain size in rice. Nat Commun 9:851

Svitashev S, Young JK, Schwartz C, Gao H, Falco SC, Cigan AM (2015) Targeted mutagenesis, precise gene editing, and site-specific gene insertion in maize using Cas9 and guide RNA. Plant Physiol 169(2):931–945

Thompson KF (1972) Oil-seed rape. In: Reports of the plant breeding institute. Cambridge University Press, Cambridge, pp 94–96

Tian S, Jiang L, Cui X, Zhang J, Guo S, Li M, Zhang H, Ren Y, Gong G, Zong M, Liu F (2018) Engineering herbicide-resistant watermelon variety through CRISPR/Cas9-mediated base-editing. Plant Cell Rep 37(9):1353–1356

Tong H, Wang X, Liu Y, Liu N, Li Y, Luo J, Ma Q, Wu D, Li J, Xu C, Yang H (2023) Programmable A-to-Y base editing by fusing an adenine base editor with an N-methylpurine DNA glycosylase. Nat Biotechnol 41:1–5

Trujillo DI, Silverstein KAT, Young ND (2019) Nodule-specific PLAT domain proteins are expanded in the Medicago lineage and required for nodulation. New Phytol 222:1538–1550

Usman B, Nawaz G, Zhao N, Liu Y, Li R (2020) Generation of high yielding and fragrant rice (*Oryza sativa* L.) lines by CRISPR/ Cas9 targeted mutagenesis of three homoeologs of cytochrome P450 gene family and *OsBADH2* and transcriptome and proteome profiling of revealed changes triggered by mutations. Plants (Basel) 9:788

Vernet A, Meynard D, Lian Q, Mieulet D, Gibert O, Bissah M, Rivallan R, Autran D, Leblanc O, Meunier AC, Frouin J (2022) High-frequency synthetic apomixis in hybrid rice. Nat Commun 13(1):7963

Wang S, Ouyang K (2023) Rapid creation of *CENH3*-mediated haploid induction lines using a cytosine base editor (CBE). Plant Biol 25(1):226–230

Wang W, Pan Q, He F, Akhunova A, Chao S, Trick H, Akhunov E (2018a) Transgenerational CRISPR-Cas9 activity facilitates multiplex gene editing in allopolyploid wheat. CRISPR J 1:65–74

Wang W, Simmonds J, Pan Q, Davidson D, He F, Battal A, Akhunova A, Trick HN, Uauy C, Akhunov E (2018b) Gene editing and mutagenesis reveal inter-cultivar differences and additivity in the contribution of *TaGW2* homoeologues to grain size and weight in wheat. Theor Appl Genet 131:2463–2475

Wang B, Zhu L, Zhao B, Zhao Y, Xie Y, Zheng Z, Li Y, Sun J, Wang H (2019a) Development of a haploid-inducer mediated genome editing system for accelerating maize breeding. Mol Plant 12(4):597–602

Wang C, Liu Q, Shen Y, Hua Y, Wang J, Lin J, Wu M, Sun T, Cheng Z, Mercier R, Wang K (2019b) Clonal seeds from hybrid rice by simultaneous genome engineering of meiosis and fertilization genes. Nat Biotechnol 37:283–286

Wang J, Wu B, Lu K, Wei Q, Qian J, Chen Y, Fang Z (2019c) The amino acid permease 5 (*OsAAP5*) regulates tiller number and grain yield in rice. Plant Physiol 180:1031–1045

Wang F, Han T, Song Q, Ye W, Song X, Chu J, Li J, Chen ZJ (2020) Rice circadian clock regulates tiller growth and panicle development through strigolactone signaling and sugar sensing. Plant Cell 32:3124

Wang F, Xu Y, Li W, Chen Z, Wang J, Fan F, Tao Y, Jiang Y, Zhu QH, Yang J (2021a) Creating a novel herbicide-tolerance *OsALS* allele using CRISPR/Cas9-mediated gene editing. Crop J 9(2):305–312

Wang N, Gent JI, Dawe RK (2021b) Haploid induction by a maize *cenh3* null mutant. Sci Adv 7(4):eabe2299

Wang H, He Y, Wang Y, Li Z, Hao J, Song Y, Wang M, Zhu JK (2022a) Base editing-mediated targeted evolution of ACCase for herbicide-resistant rice mutants. J Integr Plant Biol 64(11):2029–2032

Wang N, Xia X, Jiang T, Li L, Zhang P, Niu L, Cheng H, Wang K, Lin H (2022b) In planta haploid induction by genome editing of *DMP* in the model legume *Medicago truncatula*. Plant Biotechnol J 20(1):22

Wang Z, Zhang Y, Song M, Tang X, Huang S, Linhu B, Jin P, Guo W, Li F, Xing L, An R (2023) Genome-wide identification of the cytochrome P450 superfamily genes and targeted editing of *BnCYP704B1* confers male sterility in rapeseed. Plan Theory 12(2):365

Weider C, Stamp P, Christov N, Hüsken A, Foueillassar X, Camp K-H, Munsch M (2009) Stability of cytoplasmic male sterility in maize under different environmental conditions. Crop Sci 49:77–84

Wen X, Sun L, Chen Y, Xue P, Yang Q, Wang B, Yu N, Cao Y, Zhang Y, Gong K, Wu W, Chen D, Cao L, Cheng S, Zhang Y, Zhan X (2020) Rice *dwarf and low tillering 10* (*OsDLT10*) regulates tiller number by monitoring auxin homeostasis. Plant Sci 297:110502

Wu Y, Fox TW, Trimnell MR, Wang L, Xu R, Cigan AM, Huffman GA, Garnaat CW, Hershey H, Albertsen MC (2016) Development of a novel recessive genetic male sterility system for hybrid seed production in maize and other cross-pollinating crops. Plant Biotechnol J 14:1046–1054

Wu J, Chen C, Xian G, Liu D, Lin L, Yin S, Sun Q, Fang Y, Zhang H, Wang Y (2020) Engineering herbicide-resistant oilseed rape by CRISPR/Cas9-mediated cytosine base-editing. Plant Biotechnol J 18(9):1857

Xie K, Wu S, Li Z, Zhou Y, Zhang D, Dong Z, An X, Zhu T, Zhang S, Liu S, Li J, Wan X (2018) Map-based cloning and characterization of *Zea mays male sterility33* (*ZmMs33*) gene, encoding a glycerol-3-phosphate acyltransferase. Theor Appl Genet 131:1363–1378

Xie E, Li Y, Tang D, Lv Y, Shen Y, Cheng Z (2019) A strategy for generating rice apomixis by gene editing. J Integr Plant Biol 61:911–916

Xu R, Yang Y, Qin R, Li H, Qiu C, Li L, Wei P, Yang J (2016) Rapid improvement of grain weight via highly efficient CRISPR/Cas9- mediated multiplex genome editing in rice. J Genet Genomics 43:529–532

Xu R, Li J, Liu X, Shan T, Qin R, Wei P (2020a) Development of plant prime-editing systems for precise genome editing. Plant Commun 1(3):100043

Xu W, Zhang C, Yang Y, Zhao S, Kang G, He X, Song J, Yang J (2020b) Versatile nucleotides substitution in plant using an improved prime editing system. Mol Plant 13(5):675–678

Xu R, Liu X, Li J, Qin R, Wei P (2021) Identification of herbicide resistance *OsACC1* mutations via in planta prime-editing-library screening in rice. Nat Plants 7(7):888–892

Yang Q, Zhong X, Li Q, Lan J, Tang H, Qi P, Ma J, Wang J, Chen G, Pu Z, Li W, Lan X, Deng M, Harwood W, Li Z, Wei Y, Zheng Y, Jiang Q (2020) Mutation of the D-hordein gene by RNA-guided Cas9 targeted editing reducing the grain size and changing grain compositions in barley. Food Chem 311:125892

Yao L, Zhang Y, Liu C, Liu Y, Wang Y, Liang D, Liu J, Sahoo G, Kelliher T (2018) *OsMATL* mutation induces haploid seed formation in indica rice. Nat Plants 4(8):530–533

Yin W, Xiao Y, Niu M, Meng W, Li L, Zhang X, Liu D, Zhang G, Qian Y, Sun Z, Huang R, Wang S, Liu CM, Chu C, Tong H (2020) *ARGONAUTE2* enhances grain length and salt tolerance by activating *BIG GRAIN3* to modulate cytokinin distribution in rice. Plant Cell 32:2292–2306

Yu Q, Collavo A, Zheng MQ, Owen M, Sattin M, Powles SB (2007) Diversity of acetyl-coenzyme A carboxylase mutations in resistant *Lolium* populations: evaluation using clethodim. Plant Physiol 145(2):547–558

Yu X, Xia S, Xu Q, Cui Y, Gong M, Zeng D, Zhang Q, Shen L, Jiao G, Gao Z, Hu J, Zhang G, Zhu L, Guo L, Ren D, Qian Q (2020) *Abnormal flower and grain 1* encodes *OsMADS6* and determines Palea identity and affects rice grain yield and quality. Sci China Life Sci 63:228–238

Yu H, Lin T, Meng X, Du H, Zhang J, Liu G, Chen M, Jing Y, Kou L, Li X, Gao Q (2021) A route to de novo domestication of wild allotetraploid rice. Cell 184(5):1156–1170

Zafar K, Khan MZ, Amin I, Mukhtar Z, Zafar M, Mansoor S (2023) Employing template directed CRISPR-based editing of the *OsALS* gene to create herbicide tolerance in basmati rice. AoB Plants 15:plac059

Zhang Y, Liang Z, Zong Y, Wang Y, Liu J, Chen K, Qiu JL, Gao C (2016) Efficient and transgene-free genome editing in wheat through transient expression of CRISPR/Cas9 DNA or RNA. Nat Commun 7:12617

Zhang Y, Li D, Zhang D, Zhao X, Cao X, Dong L, Liu J, Chen K, Zhang H, Gao C, Wang D (2018) Analysis of the functions of *TaGW2* homoeologs in wheat grain weight and protein content traits. Plant J 94:857–866

Zhang R, Liu J, Chai Z, Chen S, Bai Y, Zong Y, Chen K, Li J, Jiang L, Gao C (2019a) Generation of herbicide tolerance traits and a new selectable marker in wheat using base editing. Nat Plants 5(5):480–485

Zhang Z, Hua L, Gupta A, Tricoli D, Edwards KJ, Yang B, Li W (2019b) Development of an agrobacterium-delivered CRISPR/Cas9 system for wheat genome editing. Plant Biotechnol J 17:1623–1635

Zhang J, Zhang X, Chen R, Yang L, Fan K, Liu Y, Wang G, Ren Z, Liu Y (2020) Generation of transgene-free semidwarf maize plants by gene editing of gibberellin-oxidase20-3 using CRISPR/Cas9. Front Plant Sci 11:1048

Zhang R, Chen S, Meng X, Chai Z, Wang D, Yuan Y, Chen K, Jiang L, Li J, Gao C (2021) Generating broad-spectrum tolerance to *ALS*-inhibiting herbicides in rice by base editing. Sci China Life Sci 64:1624–1633

Zhong Y, Liu C, Qi X, Jiao Y, Wang D, Wang Y, Liu Z, Chen C, Chen B, Tian X, Li J (2019) Mutation of *ZmDMP* enhances haploid induction in maize. Nat Plants 5(6):575–580

Zhong Y, Chen B, Li M, Wang D, Jiao Y, Qi X, Wang M, Liu Z, Chen C, Wang Y, Chen M (2020) A *DMP*-triggered in vivo maternal haploid induction system in the dicotyledonous *Arabidopsis*. Nat Plants 6:466–472

Zhong Y, Chen B, Wang D, Zhu X, Li M, Zhang J, Chen M, Wang M, Riksen T, Liu J, Qi X (2022) In vivo maternal haploid induction in tomato. Plant Biotechnol J 20(2):250–252

Zhou H, Zhou M, Yang Y, Li J, Zhu L, Jiang D, Dong J, Liu Q, Gu L, Zhou L, Feng M, Qin P, Hu X, Song C, Shi J, Song X, Ni E, Wu X, Deng Q, Liu Z, Chen M, Liu Y-G, Cao X, Zhuang C (2014) RNase ZS1 processes UbL40 mRNAs and controls thermosensitive genic male sterility in rice. Nat Commun 5:4884

Zhou H, He M, Li J, Chen L, Huang Z, Zheng S, Zhu L, Ni E, Jiang D, Zhao B, Zhuang C (2016) Development of commercial thermo-sensitive genic male sterile rice accelerates hybrid rice breeding using the CRISPR/Cas9-mediated TMS5 editing system. Sci Rep 6:37395

Zhou J, Liu G, Zhao Y, Zhang R, Tang X, Li L, Jia X, Guo Y, Wu Y, Han Y, Bao Y (2023) An efficient CRISPR–Cas12a promoter editing system for crop improvement. Nat Plants 9(4):588–604

Zou T, He Z, Qu L, Liu M, Zeng J, Liang Y, Wang T, Chen D, Xiao Q, Zhu J, Liang Y, Deng Q, Wang S, Zheng A, Wang L, Li P, Li S (2017) Knockout of *OsACOS12* caused male sterility in rice. Mol Breed 37:1–12

Zsögön A, Čermák T, Naves ER, Notini MM, Edel KH, Weinl S, Freschi L, Voytas DF, Kudla J, Peres LEP (2018) *De novo* domestication of wild tomato using genome editing. Nat Biotechnol 36(12):1211–1216

Zuo J, Niu QW, Frugis G, Chua NH (2002) The *WUSCHEL* gene promotes vegetative-to-embryonic transition in *Arabidopsis*. Plant J 30(3):349–359

Plant Breeding Using the CRISPR-Cas9 System for Food Security and Facing Climate Change

Ambika, Sharmista Bhati, and Rajendra Kumar

Abstract With the rapid expansion of the global population and the influence of climate change on agriculture, crops with higher yields and greater resistance to abiotic stress are in demand. However, traditional crop development is a time-consuming procedure that cannot keep up with the demand for new crop varieties. The most popular of the new gene editing technologies, CRISPR/Cas9, may make it possible to quickly improve crops. The CRISPR/Cas9 genome editing technique specifically uses a guide RNA and a Cas9 protein that can cut the genome at particular loci. The CRISPR/Cas9 system has quickly emerged as the most popular method for editing plant genomes due to its effectiveness and simplicity. It is optimal for developing novel germplasm resources and for altering the traits of various plants, especially food crops. This chapter focuses on the application of CRISPR/Cas systems to knockout genes in crops to improve attributes including yield, quality, stress tolerance, and disease resistance. De novo domestication and hybrid breeding using the CRISPR/Cas9 system are also discussed. Additionally, the system's drawbacks are described for CRISPR/Cas9. This chapter will give researchers crucial knowledge on the use of CRISPR/Cas9 gene editing technology for crop improvement and plant breeding.

Keywords CRISPR/Cas9 · Climate change · Food security · Plant breeding · Crop improvement

Ambika
Department of Genetics and Plant Breeding, UAS, GKVK, Bangalore, Karnataka, India

S. Bhati
School of Biotechnology, Gautam Buddha University,
Greater Noida, Gautam Budh Nagar, Uttar Pradesh, India

R. Kumar (✉)
Division of Genetics, ICAR-Indian Agricultural Research Institute, New Delhi, India

J.-T. Chen, S. Ahmar (eds.), *Plant Genome Editing Technologies*, Interdisciplinary Biotechnological Advances,
https://doi.org/10.1007/978-981-99-9338-3_6

1 Introduction

By 2050, there will be 9.7 billion people on Earth, a staggering rise from the current rate of growth (FAOSTAT 2023). Additionally, altered weather patterns brought on by climate change affect agricultural output and also increase environmental pressures like drought, soil salinity, and the emergence of new diseases and insect pests. During the last two decades (1990–2013), the amount of salinity in irrigated farmland has increased by 37% (Ghassemi et al. 1995; Qadir et al. 2014). Drought stress has become more common and severe because of changes in precipitation patterns and a rise in evapotranspiration caused by global warming (Dai 2011). According to a new meta-analysis study, the global average temperature will rise by 2.0–4.9 °C by 2100 (Raftery et al. 2017). Increased heavy metal poisoning of agricultural areas is restricting food output while also posing major health dangers to humans (Rehman et al. 2018). Further, food availability and food quality access are also impacted by climate change. Food security challenges may eventually arise as a result of the growing world population and the detrimental effects of the climate. According to a recent estimate, food production will need to be expanded by 70–100% by 2050 to keep up with the rate of population growth and properly feed the global population (Jones et al. 2014). Variable crops with higher yields biofortified with protein (Kumar et al. 2021; Misra et al. 2016; Rashmi et al. 2012; Ramani et al. 2021), macro- and micronutrients (Mittal et al. 2023), better climatic adaptation, and more tolerance to biotic and abiotic (Singh et al. 2022) stresses utilizing various naturally available germplasm will need to be created urgently over the next 10 years to fulfill this anticipated demand.

Although it is difficult to accurately measure the impacts of abiotic stress on crop output, it is clear that abiotic stress continues to have a significant impact on plants, as evidenced by the proportion of land area affected (Cramer et al. 2011). According to an FAO report published in 2007 (http://www.fao.org/docrep/010/a1075e/a1075e00.htm), abiotic stress affects roughly 96.5% of worldwide rural land area, and over 35,000 articles published during 2001–2011 dealt directly with abiotic stress. Crop yields in lower latitude regions are currently declining, whereas yields in higher latitude regions are increasing (Iizumi et al. 2018). However, as a direct effect of climate change, worldwide yield and crop adaptability are expected to fall over the century. Extreme weather occurrences will disrupt and reduce global food supply, resulting in higher food costs. Climate change and desertification, in particular, are predicted to diminish agricultural production in arid areas of the world and effects have already begun to manifest and will surely intensify.

Traditional breeding strategies for crop improvement heavily rely on genetic variation that arises either naturally (Kumar and Singh 1986; Kumar et al. 1999, 2000, 2002; Singh et al. 2001a, b, 2002; Tiwari et al. 2001; Yadav et al. 2002, 2003a, b, c) or through hybridization, domestication, diversification (Ambika and Kumar 2023; Ambika et al. 2022; Singh et al. 2022) or induced artificially through chemical and physical mutagenesis or cutting-edge techniques like insertional mutagenesis by T-DNA insertions or transposon tagging (Songstad et al. 2017). Natural genetic variation may be constrained, and artificial mutagenesis techniques have clear drawbacks, including the randomness of the produced mutations, their low effectiveness, as well as the fact that they are time-consuming, expensive, and labor-intensive.

2 Challenges in Plant Breeding for Crop Improvement

Since their inception, land plants have existed in an essentially hostile environment. Low or high temperature, deficient or excessive water, high salinity, heavy metals, and ultraviolet (UV) radiation are only a few of the physical or chemical elements that are harmful to them. Abiotic stressors, as a group, are posing a danger to agriculture and the ecosystem, accounting for significant crop output losses (Wang et al. 2003). The greatest persistent stress is salt stress, which is exacerbated by the global salinization of arable land (Munns and Tester 2008). Because of ionic toxicity, osmotic pressure, oxidative damage, and nutritional shortage, most plants cannot survive when NaCl concentrations are above 200 mM (Flowers and Colmer 2008). More significantly, it is linked to drought, a global problem that can be exacerbated by excessive temperatures (Ashraf and Foolad 2007; Slama et al. 2015).

Plants must meet challenges and develop effective adaptive methods to avoid or tolerate their negative effects to survive and grow due to their sessile nature. There are numerous cellular, physiological, and morphological protections in place. The cuticle, a ubiquitous outermost shield, is the most visible (Shepherd and Wynne Griffiths 2006; Yeats and Rose 2013; Fich et al. 2016). It is also noteworthy that halophytes have evolved a specific organ to excrete salt, as seen in *Limonium bicolor*'s epidermal salt gland. Due to forward and reverse genetic approaches as well as genome-wide analyses conducted on various model species like the classical model Arabidopsis thaliana and its extremophile relative thellungiella salsuginea, which has exceptional multi-stress resistance, tremendous progress has been made toward understanding the biochemical and molecular mechanisms underpinning the defenses.

Desaturation of membrane lipids, activation of reactive species (RS) scavengers, induction of molecular chaperones, and accumulation of suitable solutes are therefore emerging as more universal and conserved cellular defense mechanisms. This is consistent with the notion that a variety of abiotic stimuli can cause membrane injury, RS damage, protein denaturation, and osmotic stress (mainly dehydration). Upstream signaling molecules such as stress hormones [e.g., abscisic acid (ABA)], reactive oxygen species (ROS), hydrogen sulfide (H2S), nitric oxide (NO), polyamines (PAs), phytochromes, and calcium (Ca2+), as well as downstream gene regulation factors, particularly transcription factors, orchestrate these defenses in the stress response (TFs).

Understanding the changes in cellular, metabolic, and molecular machinery that occur in response to stress has been critical to progress in breeding superior crops under stress. The use of modern molecular methods entails the discovery and application of molecular markers that can help improve breeding programs. Introgression of genomic sections (QTLs) implicated in stress tolerance, however, frequently brings with it unfavorable agronomic traits from the donor parents. This is due to a lack of detailed knowledge of the main genes that underpin QTLs. As a result, the introduction and/or overexpression of selected genes into genetically altered plants appears to be a promising alternative for hastening the breeding of "better" plants. Genetic engineering, on the surface, appears to be a speedier approach to insert beneficial genes than conventional or molecular breeding. When genes of interest come from cross-barrier species, distant cousins, or non-plant sources, it would be the sole alternative. Several attributes have been studied in transgenic plants to see if they correlate with resistance.

The transgenic technique is a strong way of adding a wide range of genes with the ability to up- or downregulate certain metabolic pathways linked to various abiotic stress responses. Although the genetic transformation of a single gene that encodes either biochemical pathways or the endpoint of signaling pathways is considered difficult because abiotic stress tolerance is a complex polygenic trait, the results obtained from the transfer of a single gene that encodes either biochemical pathways or the endpoint of signaling pathways have been encouraging. Transgenics that produce and store osmoprotectants and molecular chaperons, reactive oxygen scavenging/detoxification enzymes, late-embryo genesis abundant protein genes, ion/proton transporters, and water channels/aquaporins have been found in crop plants.

The first phase in a plant's reaction to abiotic stress is signal perception, in which a sensor detects external stimuli and transmits signals to cellular destinations. In a plant, the two conserved proteins, response regulator (RR) protein and histidine protein kinase are critical for receiving and transducing intracellular and extracellular signals. Stress signaling in plants has been linked to the CBL-CIPK, CDPK, and MAPK pathways. Furthermore, ABA signaling and transduction stimulate SnRK2s, which then phosphorylate and induce the expression of *AREB/ABF/ABI5* target genes (Chae et al. 2007). The entire molecular mechanism of signal transmission from stress perception to gene expression involves two types of proteins: functional proteins that protect cells from dehydration (functional proteins) and protein factors that regulate gene expression and stress tolerance (functional proteins) (Shinozaki et al. 2003). Functional proteins include transporters, chaperones, ROS (reactive oxygen species) scavengers, osmolytes including fructan, trehalose, proline, glycine betaine, mannitol, and polyamines, as well as protective proteins like LEA and heat-shock proteins. Biotechnological applications in three broad fields are used in molecular breeding strategies for enhancing agricultural yields against abiotic stresses: (1) DNA marker technology for improving breeding efficiency and precision; (2) genetic engineering (GE) for easily transferring agronomically valuable features across species; and (3) genomic techniques for discovering new and useful genes/alleles. Genetic engineering, often known as genetically modified crop technology, allows scientists to transfer valuable genes from a completely separate gene pool into crop plants with the least amount of disturbance to the plant genome. Gene modification (GM) or gene technology is frequently advocated as a solution for raising crop yields around the world, particularly in less-developed areas where food insecurity and low crop productivity are a concern (Nelson et al. 2007). Transgenic techniques for abiotic stress tolerance in diverse crops are widely employed around the world (Ashraf 2010). Transgenic plants with diverse genes encoding transcription factors (TFs), heat-shock proteins (Hsp), late embryogenesis abundant proteins (LEA), and suitable organic osmolytes have been developed in recent investigations.

In order to get over these constraints, using epigenomic (Chandana et al. 2022) and biotechnological technologies (Aggarwal et al. 2003; Bhardwaj et al. 2014; Harshavardhana et al. 2019; Kumar et al. 2022a; Prabha et al. 2017; Sharma et al. 2002; Singh et al. 2012, 2022; Soi et al. 2014; Srivastava et al. 2016) including CRISPR/Cas9 is very crucial for crop development. With the use of sequence-specific nucleases, genome editing is the most effective method for changing the structure of the plant genome. Genome editing for crop enhancement has the extraordinary potential to reduce global food insecurity and create a system of agriculture that is climate-smart (Mahto et al. 2022; Singh et al. 2023; Gupta et al. 2023; Yadav et al. 2023a, b). Genome editing technologies allow for exact modifications in an organism's DNA by introducing targeted mutations, insertion/deletion (indel), and particular sequence alteration through the recruitment of certain nucleases. Mega nucleases (Puchta et al. 1993), transcription activator-like nucleases (TALENs) (Zhang et al. 2013a), zinc-finger nucleases (ZFNs) (Zhang et al. 2010), and more recently CRISPR-Cas9 (Jiang et al. 2013) have all been developed and are currently being employed for genome editing. Given its capacity to induce desired mutations in an effective, affordable, quick, and accurate manner, the clustered regularly interspaced short palindromic repeats (CRISPR)/CRISPR-associated protein 9 (CRISPR/Cas9) system have shown to be the most successful genome editing system across a wide range of organisms, including plants (Chen et al. 2019).

3 CRISPR/Cas9 System as a Facilitating Tool for Plant Breeding

CRISPR/Cas9 system has been applied in several plant species such as *Nicotiana benthamiana* (Li et al. 2013), *Nicotiana tabacum* (Kumar et al. 2022b), Arabidopsis (Li et al. 2013), wheat (Shan et al. 2013), maize (Liang et al. 2014), rice (Shan et al. 2013), liverwort (Sugano et al. 2014), tomato (Brooks et al. 2014), potato (Wang et al. 2015), soybean (Jacobs et al. 2015), sweet orange (Jia and Nian 2014), banana (Tripathi et al. 2019), pepper (Park et al. 2021), barley (Garcia-Gimenez and Jobling 2021), peanut (Wei et al. 2021), foxtail millet (Lin et al. 2018), and sugarcane (Oz et al. 2021). Additionally, CRISPR/Cas9-based multiplexing by targeting multiple genes in a single organism has also been carried out successfully across a range of crop species such as wheat (Wang et al. 2018a), rice (Miao et al. 2013), cotton (Gao et al. 2017), and maize (Char et al. 2017). Therefore, by simultaneously targeting a large number of stress-sensitive genes in an elite, high-yielding, but sensitive cultivar, this technology has enormous potential to produce genome-edited crop plants tolerant to multiple stresses. Tolerance genes can also be over-expressed using CRISPR-mediated gene activation (Zafar et al. 2020). To create crop plants that are tolerant to changing climates, CRISPR/Cas9 technologies may be used to target the genes involved in signal transduction, metabolite production, and gene regulatory networks linked with stress (Jain 2015).

As of right now, CRISPR/Cas-based genome engineering has been effectively applied to study the genetic mechanism underpinning tolerance against a variety of abiotic stresses, including drought, salt, heat, and nutritional values in diverse crop plants (Li et al. 2019). In this study, we summarize the majority of possible uses of the CRISPR/Cas9-mediated genome editing strategy in crop plants for mitigating climate change effects. We also talk about potential future uses of CRISPR/Cas-based systems to create crop types that can withstand stress.

4 History of the CRISPR-Cas System

The discovery of CRISPR cluster repeats occurred in the DNA sequences of E. coli while sequencing the *iap* gene encoding alkaline phosphatase isozyme conversion enzyme (Ishino et al. 1987). In 1993, CRISPR clustered repeats were found in Mycobacterium tuberculosis (Mojica et al. 1993). Later, it was found that CRISPR families are widespread in prokaryotes (Mojica et al. 2000). In 2002, Cas genes were identified and described and coined the term "CRISPR," an acronym for short DNA repeats (Jansen et al. 2002). These non-repetitive sequences were found to be homologous with foreign DNA sequences derived from plasmids and phages (Mojica et al. 2005). In 2007 this CRISPR system was established as an adaptive immunity of bacteria from bacteriophage (Barrangou et al. 2007). In 2008, CRISPR system type 3-A targeting DNA was discovered (Brouns et al. 2008). In 2009, a Cmr complex that cleaves ssRNA was discovered (Moscou and Bogdanove 2009). In 2010, the target region of Cas9 protein for the cleavage of DNA sequence that resides within a protospacer (Garneau et al. 2010). The discovery of the CRISPR/Cas9 gene editing system has revolutionized research in animal and plant biology with its utility in genome editing being first demonstrated in 2012 in mammalian cells (Jinek et al. 2012; Ann Ran et al. 2013). In 2014, the crystal structure of Cas9 protein was discovered (Nishimasu et al. 2014). The first CRISPR germline editing in implanted human embryos and the first CRISPR clinical trial for treatment against HIV-1 was conducted (Ormond et al. 2017). In 2019, the first in vivo CRISPR clinical trial for treatment against blindness was done. In 2020, for the first time in history, a Nobel prize was awarded to two women, Emmanuelle Charpentier and Jennifer Doudna, who made key discoveries in the field of DNA manipulation with the CRISPR/Cas9 system. The summary of CRISPR history is presented in Fig. 1.

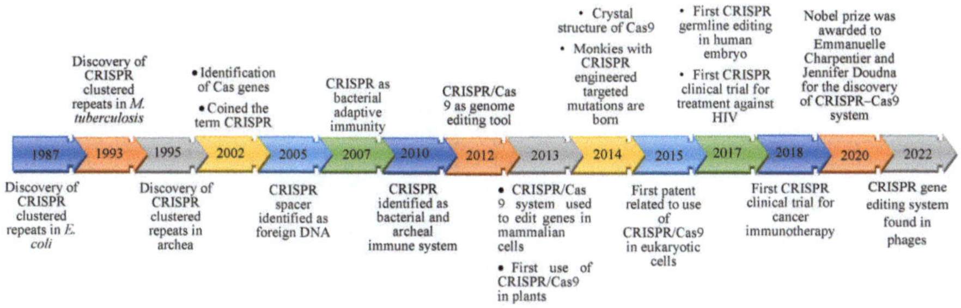

Fig. 1 The timeline of the CRISPR/Cas system indicates critical milestones

5 Phases of the CRISPR-Cas System

The CRISPR system's key functional phases are adaptation, crRNA processing, and interference.

5.1 Adaptation

The process of incorporating foreign DNA sequences into CRISPR arrays is known as adaptation, sometimes known as insertion or acquisition. Cas1 and Cas2 are needed to capture and catalytically insert the spacer in known adaptation pathways. Spacers can also be produced from RNA by type III systems that contain reverse transcriptase-Cas1 fusion proteins (Silas et al. 2016).

5.2 crRNA Processing

CRISPR RNA (crRNA) biogenesis, also known as crRNA processing, is the transcription of a CRISPR array into a lengthy precursor CRISPR RNA (pre-crRNA) which associates with Cas proteins for further processing into mature crRNAs. Cas6 is required in class 1 systems in order to convert pre-crRNA into distinct crRNA molecules (Taylor et al. 2019). While type II systems rely on RNase III (Charpentier et al. 2015), type V-A and type VI signature nucleases have a unique active site for processing crRNA, and some type V subtypes depend on host nucleases (Liu et al. 2017). Using individual promoters from the CRISPR array, certain type II systems directly generate mature crRNAs (Zhang et al. 2013b).

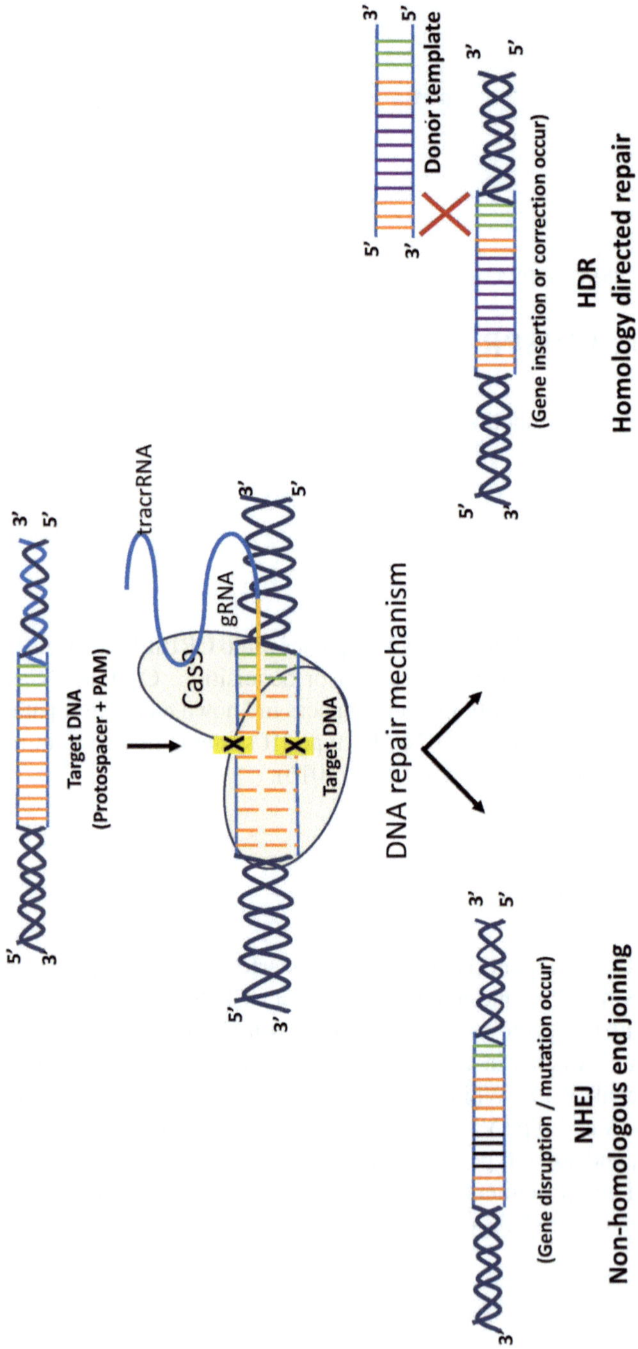

Fig. 2 Molecular mechanisms of the CRISPR/Cas system

Table 1 Summary of CRISPR/Cas systems applicable in plants

Class	Type	Subtype	Effector	Target	Nuclease domains	TracrRNA requirement	PAM/PFS
1 Multi-Cas proteins	I	A, B, C, D, E, F, U	Cascade	dsDNA	HD fused to Cas3	No	–
1 Multi-Cas proteins	III	A, B, C, D	Cascade	ssRNA	HD fused to Cas10	No	–
1 Multi-Cas proteins	III	A, B, C, D	Cascade	ssRNA	HD fused to Cas10	No	–
1 Multi-Cas proteins	IV	A, B	Cascade	dsDNA	Unknown	No	–
2 Single-Cas protein	II	A	SpCas9	dsDNA	RuvC, HNH	Yes	NGG
2 Single-Cas protein	II	A	SaCas9	dsDNA	RuvC, HNH	Yes	NNGRRT
2 Single-Cas protein	II	B	FnCas9	dsDNA/ ssRNA	RuvC, HNH	Yes	NGG
2 Single-Cas protein	II	C	NmCas9	dsDNA	RuvC, HNH	Yes	NNNNGATT
2 Single-Cas protcin	V	A	Cas12a (Cpf1)	dsDNA	RuvC, Nuc	No	5^0 AT-rich PAM
2 Single-Cas protein	V	B	Cas12b (C2c1)	dsDNA	RuvC	Yes	5^0 AT-rich PAM
2		V	C	Cas12c (C2c3)	dsDNA	RuvC	Yes
2		VI	A	Cas13a (C2c2)	ssRNA	2xHEPN	No
2		VI	B	Cas13b (C2c4)	ssRNA	2xHEPN	No
2		VI	C	Cas13c (C2c7)	ssRNA	2xHEPN	No
2		VI	D	Cas13d	ssRNA	2xHEPN	No

5.3 *Interference*

This stage involves the sequence-specific targeting and cleavage of foreign DNA and/or RNA. The development of a multiprotein effector complex or a single effector protein is required for interference. Interference involves R-loop formation as the crRNA guide region hybridizes to target DNA or base-pairing between the crRNA guide region and target RNA. This is followed by cleavage/degradation of the target.

6 Structure of CRISPR Locus and Classification of Cas Proteins

The chromosomes of prokaryotic species generally contain one CRISPR locus (Jansen et al. 2002). The CRISPR locus consists of three major parts: The CRISPR array consists of short, direct repeats bordered with spacers. The direct repeats are nucleotide sequences in the genome with identical sequence and length. The average size of the repeats is 32 bp; however, the size may vary from 21 to 47 bp. The spacers are nucleotide sequences with a fixed length, but they are highly variable in sequences. The average size of spacers is from 20 to 72 bp (Karginov and Hannon 2010). CRISPR-associated (Cas) genes represent a cluster of genes in varying orientations and order that code corresponding Cas proteins. In summary, 93 different cas genes have been identified until now. These genes were classified into 35 families based on the sequence similarities (Makarova et al. 2015). Cas proteins play a major role in the acquisition and destruction of foreign sequences as presented in Fig. 2.

7 Classification of the CRISPR-Cas System

CRISPR-Cas systems are divided into two categories: class I and class II. Each of the two classes is divided into three subtypes, with type I, III, and IV in class I and type II, V, and VI in class II based on differences in Cas protein composition and sequence divergence among effector complexes as presented in Table 1 (Makarova et al. 2015). The type I system is divided into seven subtypes (I-A, I-B, I-C, I-D, I-E, I-F, and I-G), each with its own set of variants. Cas3, a gene that generates a protein with both helicase and nuclease domains, is the only one of type I (Brouns et al. 2008). In the interference stage, this class uses an effector complex called cascade made up of several Cas proteins and crRNA. This recruits a *trans*-activating nuclease-helicase (Cas3) for unwinding and DNA cleavage. The type III system is divided into six subtypes (III-A, III-B, III-C, III-D, III-E, and III-F), with reverse transcriptase included in some of them. For type III, the signature gene is cas10, encoding a multidomain protein with nuclease activity called cmr complex involved in the interference step (Ramia et al. 2014). A major functional difference between the Cascade and Cmr complexes is that the cascade complex targets DNA (Brouns et al. 2008), while the Cmr complex targets both RNA (Hale et al. 2009) and DNA (Samai et al. 2015). There are three subtypes in a type IV system (IV-A, IV-B, and IV-C). The proposed signature gene for type IV is csf1 (Makarova et al. 2018). Type IV is different from the two other types in that it does not contain cas1 and cas2 genes. Moreover, its Cas genes are usually not found close to a CRISPR locus, but the associated proteins are nevertheless predicted to form an effector complex (Makarova et al. 2018). In the adaptation process and target cleavage of foreign genomes, these systems frequently lack Cas proteins.

Unlike class 1, class 2 systems have a single, large, multidomain Cas protein coupled to crRNA. For DNA targeting and degradation, type II systems use the multifunctional protein Cas9 under the control of a naturally occurring dual-RNA heteroduplex made up of a crRNA and a tracrRNA (Jinek et al. 2012). By joining the 3′ end of crRNA to the 5′ end of tracrRNA with a linker sequence, the dual crRNA:tracrRNA can be created as a chimeric single-guide RNA that effectively instructs Cas9 to cleave target DNA sequences that match the RNA's 20 nucleotide (nt) guide sequence (Jinek et al. 2012). In the genomes of eukaryotic cells, site-specific double-stranded (ds) DNA breaks have been created using Cas9 and single-guide RNAs. These breaks can then be repaired using non-homologous end joining (NHEJ) or homology-directed repair (HDR), leading to site-specific and long-lasting genome modifications (Cong et al. 2013). The Cas12 protein is a single effector complex in a type V system, which has ten subtypes (V-A, V-B, V-C, V-D, V-E, V-F, V-G, V-H, V-I, and V-K). There are four subtypes in a type VI system (VI-A, VI-B, VI-C, and VI-D). Cas13 proteins distinguish single effector complexes in type VI systems from other effector complexes in class 2.

8 Application of CRISPR/Cas9 System in Plant Breeding for Crop Improvement

CRISPR-Cas system has been used to edit variety of crops, including rice, maize, wheat, soybean, barley, sorghum, potato, tomato, flax, rapeseed, camelina, cotton, cucumber, lettuce, grapes, grapefruit, apple, oranges, and watermelon for various traits including yield improvement, biotic and abiotic stress management (Ricroch et al. 2017; Zhang et al. 2016a). Here, we propose an extensive study of gene editing-based possibilities in response to climate change-related restrictions on agricultural productivity. The proof of concept is demonstrated by these instances.

8.1 Improvement in Yield and Yield-Related Traits

Increasing yield is a critical challenge in crop breeding. Due to the quantitative and multigenic nature of crop output, traditional breeding to increase yield is laborious and time-consuming. Only removing one factor may not raise yield in the field significantly because the majority of yield-associated variables are quantitative and controlled by numerous loci. Three genes (*OsGW2, OsGW5*, and *OsTGW6*) involved in grain weight were knocked out at the same time in rice to induce trait pyramiding, which considerably enhanced grain weight by up to 30% (Viana et al. 2019). In rice, the *LAZY1* gene was knocked out using CRISPR/Cas9, resulting in a tiller-spreading phenotype that, under certain conditions, could increase crop output (Miao et al. 2013).

The rice cultivar Zhonghua11's four yield-related genes, *Gn1a, DEP1, IPA1*, and *GS3*, were altered using the CRISPR/Cas9 system, and the mutants had more grains, densely erect panicles, and larger grains, all of which led to enhanced yield (Li et al. 2016). Wheat grain length and width are increased, which enhances grain yield when the *TaGW2* gene—which encodes a RING E3 ligase—is deleted (Wang et al. 2018b; Zhang et al. 2018a, b). Modulating cytokinin homeostasis may be a successful tactic for increasing grain yield among other parameters. The grain yield is specifically increased in wheat by knocking down *TaCKX2-D1*, which is a cytokinin oxidase/dehydrogenase (Dahan-Meir et al. 2018). Tomato plants with early flowering, determinate flowering, and early yields were produced by CRISPR/Cas9-mutations in the flowering repressor gene *SP5G* (Soyk et al. 2017).

The gene *CAROTENOID CLEAVAGE DIOXYGENASE 7* (*CCD7*), which governs a crucial step in strigolactone production, was recently silenced in rice using the CRISPR/Cas9 technology. The rice plant phenotypes of the *ccd7* mutants were perfect, with high tillering and a dwarf habit (Butt et al. 2018). *OsGS3* and *OsGL3.1* were modified using CRISPR technology, which resulted in larger grains, larger grains overall, and a higher yield per plant for rice (Yuyu et al. 2020). In addition, grain production was increased in altered plants relative to wild type by multiplex gene editing of three genes (*GS3, GW2*, and *Gn1a*) (Lacchini et al. 2020). In a different study, *OsPIN5b* (which regulates panicle length), *GS3*, and *OsMYB30* (which affects drought tolerance in rice) were simultaneously knocked out using CRISPR-Cas9, resulting in high-yielding lines of rice that are resistant to drought (Zeng et al. 2020b).

By raising the quantity of mevalonic acid, a precursor to the plant hormone gibberellin, a rice double mutant was created by multiplex gene editing of the microRNA genes *MIR396e* and *MIR396f*. This mutant displayed enhanced grain yield and flag leaf area (Miao et al. 2020). Furthermore, *OsLOGL5*, a gene that encodes a cytokinin-activation enzyme, is negatively regulated by CRISPR-based editing, which results in increased root growth, tiller numbers, and rice production (Wang 2020). The *teosinte branched 1* (*tb1*) gene was knocked out using CRISPR/Cas9 in the switchgrass biofuel crop, increasing the number of plant tillers and fresh biomass (Liu and Zhang 2020). Simultaneous knockout of all four *BnaMAX1* alleles in rapeseed led to semi-dwarf and more silique-rich branching phenotypes. These traits contributed to increased yield compared with their wild-type controls (Zheng et al. 2020).

8.2 CRISPR/Cas9 in Plant Hybrid Breeding

Hybrid breeding is a quick, simple, and effective method for increasing agricultural yields. A maternal line that is male-sterile is a need for creating high-quality hybrid varieties. CRISPR/Cas9 has achieved significant success in establishing male-sterile lines by gene deletion. By deleting the *TMS5* gene in rice, thermogenic male sterility was recently created (Barman et al. 2019). Similarly, the CRISPR-Cas system was used to eliminate a gene, *TaNP1*, which affects the development of the tapetum and the production of pollen exine in wheat and encodes a putative glucose-methanol-choline oxidoreductase, at three homeologous loci, resulting in male-sterile wheat plants (Li et al. 2020b). The *ZmMTL* (*ZmPLA1*) gene was modified using the CRISPR/Cas9 system to produce maternal haploid inducers with robust haploid identification markers that can be exploited for the breeding of doubled-haploid cereals, including maize (Dong et al. 2018).

8.3 CRISPR/Cas9 in Apomictic Breeding

Apomixis is a practical method for fixing hybrid vigor and preserving hybrid lineages. By simultaneously editing four genes, including *OsPAIR1* (which prevents meiotic recombination), *OsREC8* (facilitates the separation of sister chromatids in the first meiotic division), *OsOSD1* (skipping of the second meiotic division), and *OsMTL* (affects fertilization), heterosis was fixed in rice using the mitosis instead of meiosis (MiMe) system (Wang 2020). The ectopic production of *Baby Boom1* (*BBM*) gene by CRISPR-Cas to accelerate embryogenesis in MiMe rice egg cells is another instance in rice-produced offspring that were similar to the female parent (Khanday et al. 2019). In another investigation, rice *OsMTL* and *OsMiMe* genes were knocked out to create clonal diploid embryos. This process results in paternal genome deletion after fertilization and fixing of hybrid vigor (Wang et al. 2019). Because of the reduced germination rates of the haploid seeds and problems with pollen viability, the hybrid seeds created using these technologies have not yet been commercialized. These findings show that crop plants can be successfully modified using CRISPR/Cas9 to increase crop yields.

9 Development of Climate-Resilient Crops

Climate change has increased drought, warmth, and soil salinization, resulting in lower crop yields. Cas9 has been used to combat several abiotic stresses in important crops like rice, wheat, maize, cotton, and potatoes as given below.

9.1 Drought Stress Tolerance/Resistance

The maize gene *ARGOS8* was subjected to CRISPR/Cas mutation, which increased hybrid yield under stress (Shi et al. 2017). Recently, CRISPR/Cas9 editing in chickpea protoplast targeted two candidate genes *4-coumarate ligase* (*4CL*) and *Reveille 7* (*RVE7*) genes, both of which are connected with drought tolerance. Future kinds of chickpea that can withstand drought were created by the new method of knocking these genes out using Cas9 (Badhan et al. 2021). Tomatoes with *slmapk3* protein gene mutations caused by CRISPR/Cas9 have better defense responses to drought stress (Wang et al. 2017).

By altering the protein expression pattern and scavenging ROS in rice, the Cas9-induced mutation of *Leaf1,2* conferred drought resistance (Liao et al. 2019). In comparison to plants of the wild type, *AREB-1*-activated Arabidopsis via CRISPR recently demonstrated greater drought tolerance (de Melo et al. 2020). Similar to this, the Cas-based mutation of the *Drb2a* and *Drb2b* genes, which controlled drought and salt tolerance in soybeans, was investigated (Curtin et al. 2018). To improve fruit sets under heat stress conditions, CRISPR/Cas9-based genome editing of the heat-sensitive gene *SlAGAMOUS-LIKE 6* (*SlAGL6*) in tomatoes was developed (Klap et al. 2017). In lettuce (*Lactuca sativa*), employing CRISPR to disrupt the *NCED4* gene enhances seed germination at high temperatures, with >70% germination efficiency at 37 °C (Bertier et al. 2018).

9.2 Salt Stress Tolerance/Resistance

Salt stress in plants results in a variety of physiological and morphological alterations, including necrosis, the premature death of old leaves, and the abrupt stoppage of ions in cells (Julkowska and Testerink 2015). To increase plant salt tolerance, several genes have been discovered and characterized using CRISPR/Cas-based genome editing. In soybean (*Glycine max*) plants, salt stress tolerance was imparted using CRISPR/Cas9-based knockout mutants of the ABA-induced transcription repressors (AITRs) genes (Wang et al. 2021). Rice that had SnRK2 and the ABA-activated protein kinases *SAPK-1* and *SAPK-2* genes deleted by CRISPR showed tolerance to salt stress (Lau and Suh 2017). Rice's ability to tolerate salt was improved by the targeted mutation of the *OsRR22* gene (Zhang et al. 2019).

9.3 Cold Stress Tolerance/Resistance

According to Hannah et al. (2006), cold stress tolerance in plants is a very complex feature involving a variety of different cell compartments and metabolic processes. The tomato plant is guarded against chilling/cold damage and has less electrolyte leakage, as demonstrated by CRISPR/Cas9-based cbf1 mutants (Li et al. 2018a). Overexpression of the *CsWRKY6* gene in cucumbers increased cold stress tolerance while decreasing proline buildup and ABA sensitivity (Zhang et al. 2016b). Three genes, *OsPIN5b, GS3*, and *OsMYB30*, that were subjected to CRISPR/Cas9-based genome editing changed simultaneously and displayed improved yield and resistance to cold stress (Zeng et al. 2020b).

The japonica rice *annexin* gene (*Osann3*), a Ca^{2+}–dependent phospholipid binding protein, was knocked down using CRISPR/Cas9, and the resulting mutants had much higher relative electrical conductivities. *OsANN3* could be a possible target for engineering rice with improved cold resistance as a result of this result, which highlights the crucial function this gene plays in rice cold resistance (Shen et al. 2017).

10 Heavy Metals Tolerance/Resistance

One of the main issues that has a negative impact on crop productivity in agriculture is heavy metal stress (Jha and Bohra 2016; Yadav et al. 2006, 2008, 2019). The evolution of the oxp1/CRISPR mutant in Arabidopsis recently demonstrated resistance to cadmium, indicating better heavy metal detoxification in mutants (Baeg et al. 2021). The introduction of point mutations into the rice *OsALS* gene using CRISPR/Cas9-mediated homologous recombination conferred herbicide tolerance on rice plants (Sun et al. 2017). *OsNramp5* and *OsLCT1* gene mutants created using CRISPR/Cas9 allowed rice to have a permissible level of cadmium (Lu et al. 2017).

10.1 Herbicide Tolerance

Weeds are a global agricultural problem that endangers crop output by fiercely competing with the primary crop for resources like nutrients, soil moisture, light, space, and CO_2. The best method for controlling weeds right now is CRISPR/Cas9-based genome editing to create herbicide-resistant crop plants (Toda and Okamoto 2020). Three genes, *ALS (acetolactate synthase), EPSPS (5-enolpyruvylshikimate-3-phosphate synthase*), and *pds (phytoene desaturase)*, were specifically modified using CRISPR/Cas9 to give herbicide resistance in *Solanum lycopersicum cv. Micro-Tom* (Yang et al. 2022). These results demonstrated that Cas9 might modify genes to promote resistance to numerous abiotic stressors, including salt, low temperatures, and drought.

Recently, *acetolactate synthase 1* (*ZmALS1*) and *ZmALS2* were modified in maize to increase herbicide tolerance using CRISPR-Cas9 base editing and prime editing systems (Nuccio et al. 2021). Herbicide resistance in rice has also risen as a result of precise nicking of the *OsALS* gene and HDR-mediated repair employing prime editing technology (Ali et al. 2020; Butt et al. 2020). In order to create point mutations in rice plants, the nuclease-deficient Cas9 (dCas9) fused to Petromyzon marinus cytidine deaminase (*PmCDA1*), which led to a herbicide resistance phenotype in the mutant lines (Shimatani et al. 2017). Base editing was used to modify the *OsTubA2* sequence in order to create novel artificial rice germplasm that is resistant to dinitroaniline herbicides (Liu et al. 2021).

10.2 Development of Disease-Resistant Crops

Crop production is significantly hampered by biological stressors such as bacterial, viral, fungal, and insect pests, and the continual emergence of novel virulent pathogens makes the battle against pathogens increasingly challenging. The ability of plants to withstand various biotic stresses is being severely affected by the changing climate. Increased resistance to illness and insects can be achieved by employing CRISPR/Cas technology to knockout susceptibility genes. Rice plants were found to be resistant to the bacterial blight caused by *Xanthomonas oryzae pv oryzae* following CRISPR-Cas9 targeted mutation of the TALE binding site in the promoter of *OsSWEET11* (*Xa13*) (Li et al. 2020a), *OsSWEET13* (Zhou et al. 2015), and *OsSWEET14* (Zeng et al. 2020a). Specific rice genes, namely *OsERF92231* and *Pi21*, have been knocked out to create rice lines resistant to rice blast disease (Nawaz et al. 2020). Similar to this, targeting the *Mildew Locus O* (*MLO*) gene by CRISPR-Cas increased wheat's resistance to powdery mildew, a disease brought on by the fungus, *Podosphaera xanthii* (Wang et al. 2014).

By simultaneously targeting all three homologs of *MLO* (Wang et al. 2014) and *enhanced disease resistance 1* (Zhang et al. 2017) in bread wheat using CRISPR/Cas9, wheat lines resistant to powdery mildew were created. It is interesting to note that the disease-resistant plants had increased yield-related features, indicating a potential chance to increase output in order to fulfill the aim of ending world hunger by creating disease-resistant plants using CRISPR technology. To provide resistance against leaf curl virus and powdery mildew, the targeted mutation of *SlPeLo* and *SlMlO1* was carried out for trait introgression of tomato utilizing Cas9 (Pramanik et al. 2021). The citrus canker-causing gene *CsLOBI*'s promoter region was altered using Cas9, and the resulting mutations might increase resistance to citrus canker (Jia et al. 2017). Thus, Cas9 is a powerful tool to enhance the genetic makeup of crops to increase their resistance against viruses and other casual agents.

In order to develop virus resistance against cucumber vein yellowing virus (CVYV), zucchini yellow mosaic virus (ZYMV), and papaya ringspot mosaic virus type-W (PRSV-W), the CRISPR/Cas9 genome editing system was used to alter the recessive gene *eukaryotic translation initiation factor 4E* (*eIF4E*) in cucumber (*Cucumis sativus*) (Chandrasekaran et al. 2016). Editing the cocoa non-expressor of pathogenesis-related *PR3* gene (*TcNPR3*) in leaves increases resistance to infection with the cacao pathogen *Phytophthora tropicalis* and upstream defense gene expression (Fister et al. 2018).

Table 2 Application of CRISPR-based genome editing approaches in plants for various traits

Crop	Method	Target gene	Stress/trait	Reference
A. thaliana	NHEJ	dsDNA of virus (A7, B7, and C3	Beet severe curly top virus resistance	Ji et al. (2015)
Rice	NHEJ	*OsERF922* (ethylene responsive)	Blast resistance	Wang et al. (2016)
Bread wheat	NHEJ	*TaMLO-A1*, *TaMLO-B1*	Powdery mildew resistance	Wang et al. (2014)
Cucumber	NHEJ	*eIF4E*		Chandrasekaran et al. (2016)
Abiotic stress				
Maize	HDR	*ARGOS8*	Increased grain yield under drought stress	Shi et al. (2017)
Tomato	NHEJ	*SlMAPK3*	Drought tolerance	Wang et al. (2017)
A. thaliana	NHEJ	*UGT79B2*, *UGT79B3*	Susceptibility to cold, salt, and drought stresses	Li et al. (2017)
Rice	HDR, NHEJ	*OsPDS, OsMPK2, OsBADH2*	Involved in various abiotic stress tolerance	Xie and Yang (2013)
Rice	NHEJ, HDR	*OsMPK2, OsDEP1*	Yield under stress	Shan et al. (2014)
Nutritional traits				
Rice	NHEJ	25,604 gRNA for 12,802 genes	Creating genome-wide mutant library	Meng et al. (2017)
Maize	NHEJ	*ZmIPK1A, ZmIPK* and *ZmMRP4*	Phytic acid synthesis	Liang et al. (2014)
Wheat	HDR	*TaVIT2*	Fe content	Connorton et al. (2017)
Soybean	NHEJ	*GmPDS11 and GmPDS18*	Carotenoid biosynthesis	Du et al. (2016)
Tomato	NHEJ	*Rin*	Fruit ripening	Ito et al. (2015)
Potato	HDR	*ALS1*	Herbicide resistance	Butler et al. (2016)
Cassava	NHEJ	*MePDS*	Carotenoid biosynthesis	Odipio et al. (2017)

Progress in insect pest resistance has lagged behind the widespread use of gene editing technologies for disease resistance, possibly due to the former's more intricate resistance mechanism. Recently, cytochrome *P450* gene *CYP71A1*, encoding tryptamine 5-hydroxylase, which catalyzes the conversion of tryptamine to serotonin, was knocked out using CRISPR/Cas9 technology (Lu et al. 2018). This resulted in rice that was resistant to brown planthopper (BPH) via inhibition of serotonin production. This study proposes a way for using gene editing technology to create insect-resistant cultivars in other crops and shows promise for developing insect-resistant rice as presented in Table 2.

10.3 De Novo Domestication of Crop Wild Relatives and Orphan Crops

De novo domestication aids in reaping the rewards of wild relatives and orphan crops, which are powerful sources of biotic and abiotic stress resistance and have high nutritional value. The pace of de novo domestication has been further accelerated using multiplex genome editing (MGE) technologies. Only a few successful cases of using wild relatives for de novo domestication through genome editing of the distinctive loci responsible for some agronomic features exist as of yet, serving as proof of concept. Using the CRISPR-Cas9 genome editing technique, valuable features from wild lines were combined with agronomically favorable traits. De novo domestication of the wild *Solanum pimpinellifolium* was made possible by editing six loci that are critical for yield and productivity in today's tomato crop lines: general plant growth habit (*SELF-PRUNING*), fruit shape and size (*OVATE and FASCIATED and FRUIT WEIGHT 2.2*), fruit number (*MULTIFLORA*), and nutritional quality (LYCOPENE BETA-CYCLASE). The size, number, and nutritional content of the fruits were also modified in engineered *S. pimpinellifolium* morphology. Fruit size and fruit number increased threefold and tenfold, respectively, in engineered lines. A notable improvement in fruit lycopene accumulation is a 500% increase over the commonly grown *S. lycopersicum* (Li et al. 2018b; Zsögön et al. 2018).

Oryza alta was domesticated into cultivated rice by using a CRISPR-Cas mediated MGE system to target *A1An-1, An-2/LAB, GAD1/RAE2, qSH1, Sh4, OsLG1, Rc, Bh4, GW5/GSE5/qSW, OsSD1, OsGS3,* and *PROG1*. Awn length, shattering resistance, panicle form, pericarp color, hull color, grain width, plant height, grain size, and tiller angle were among the interesting attributes mutant plants displayed (Yu et al. 2021). The findings demonstrated the de novo domestication of novel allotetraploid rice, which can improve food security. De novo domestication was also used to increase adaptability, productivity, and consumer acceptability in the orphan crop ground cherry (*Physalis pruinosa*) by targeting multiple genes, including *Ppr-SP5G, Ppr-CLV3*, and *Ppr-SP*, which control flower production, fruit size, and plant architecture, respectively (Lemmon et al. 2018). These researches emphasize the importance of CRISPR-assisted de novo domestication of wild relatives and orphan crops for achieving global food security.

11 Limitations of CRISPR-Cas

The major limitations of CRISPR-Cas are presented in Fig. 3 and discussed as follows:

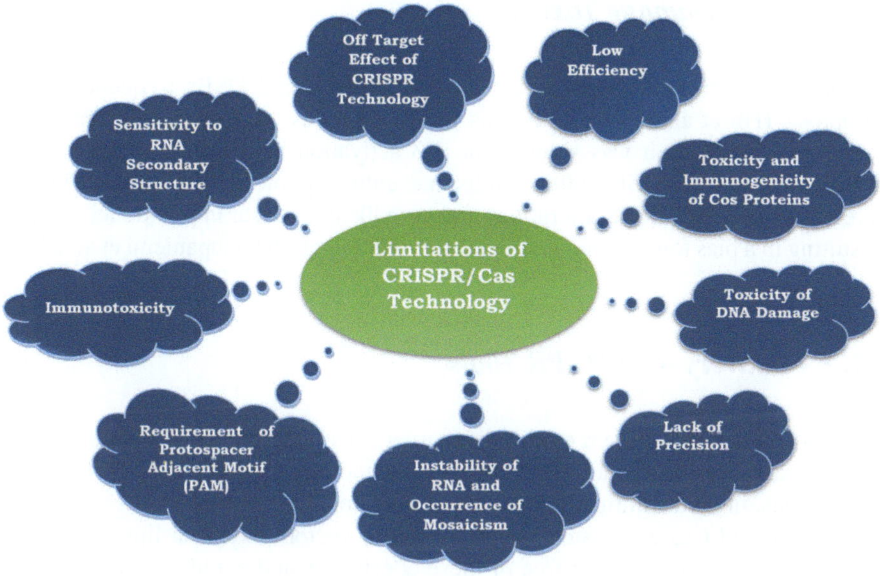

Fig. 3 Limitations of the CRISPR/Cas system

11.1 Off-Target Effects of CRISPR Technology

A key problem for using CRISPR/Cas9 for gene therapy is the relatively high frequency of off-target effects (OTEs), which have been reported at a rate of up to 50% (Zhang et al. 2015). The delivery of the CRISPR system to the target cells has been the most difficult issue thus far. The CRISPR system's non-specific targeting of nucleic acids (with as many as 3–5 mismatches) has been a source of worry, especially when used for therapeutic and diagnostic purposes (Jin et al. 2019; Zuo et al. 2019).

11.2 Protospacer Adjacent Motif Requirement

The requirement for a PAM near the target site is another drawback of the method. Cas9 from *Streptococcus pyogenes* (SpCas9) is one of the most widely utilized Cas9s, with a short canonical PAM recognition site of 5′NGG3′, where N can be any nucleotide. SpCas9, however, is quite big and difficult to bundle into AAV vectors (Lino et al. 2018; Lou et al. 2017).

11.3 DNA-Damage Toxicity

Rather than the intended gene change, CRISPR-induced DSBs frequently cause apoptosis (Hu et al. 2014). When this technology was used in human pluripotent stem cells (hPSCs), it was shown that p53 activation in response to the harmful DSBs induced by CRISPR often results in eventual apoptosis (Ihry et al. 2018). As a result, effective CRISPR alterations are more likely to occur in p53-deficient cells, resulting in a bias toward oncogenic cell survival selection (Haapaniemi et al. 2018).

11.4 Delivery of CRISPR Tools

The distribution of different tools inside target cells under in vivo conditions is one of the therapeutic applications of CRISPR technology. Though viral vectors have advantages, in vivo administration with viral vectors has drawbacks such as immunogenicity and Cas expression length. Nonviral vectors based on lipids and inorganic particles, however, are free of these constraints and could be examined as a therapeutic delivery option for CRISPR-cas systems (Wilbie et al. 2019).

11.5 Toxicity and Immunogenicity of Cas Proteins

Because Cas proteins are derived from prokaryotic sources, in vivo administration of these proteins may cause toxicity in the human cells that contain them, as well as immunological activation and the creation of Cas protein-specific antibodies (Charlesworth et al. 2019).

11.6 Target Site Restriction

PAM sequence is required for targeting a specific area in the target genome by several Cas proteins, including Cas9 and Cas12a. The PAM sequences are highly particular for the type of CRISPR-Cas system, and the lack of a specific PAM sequence near the target region severely limits its application for therapeutic purposes (Moon et al. 2019). To improve their efficiency, these CRISPR technologies must be able to choose targets with more flexibility.

11.7 Sensitivity to RNA Secondary Structure

In most cells, RNA gains secondary structure, which can reduce the efficacy of RNA-specific Cas proteins like Cas13. The discovery of the processes underlying Cas13 proteins' sensitivity could aid in the development of Cas13 proteins that are less sensitive to RNA secondary structure (Wolter and Puchta 2018).

11.8 RNA Instability and Occurrence of Mosaicism

The fragile nature of RNA due to the widespread presence of RNase may have a substantial impact on the diagnostic system based on CRISPR. Other drawbacks to using the CRISPR system for therapeutic purposes include mosaicism, which occurs when transduced cells divide before editing or cleaving target nucleic acids (Yen et al. 2014).

11.9 Immunotoxicity

CRISPR/Cas9, like classical gene therapy, has technical limitations and raises concerns about immunogenic toxicity. More than half of the human participants in investigations had anti-Cas9 antibodies against the two most often employed bacterial orthologs, SaCas9 and SpCas9 (Charlesworth et al. 2019).

11.10 Precision Gene Editing with CRISPR

For CRISPR gene therapy to work, precise genome editing is required. Although HDR pathways can assist a desired edit, their low efficiency limits their utility for precise gene editing in clinical intervention, with NHEJ being the default repair process for human cells. The regulation of the NHEJ pathway through pharmacological inhibition of key NHEJ regulating enzymes like Ku has improved HDR efficiency (Li et al. 2018a).

11.11 Delivery of CRISPR Gene Therapy

The way CRISPR tools are delivered has a big impact on their safety and therapeutic efficacy. While classical gene therapy based on viruses has been scrutinized due to the potential of immunotoxicity and insertional oncogenesis, AAV vectors have remained a crucial delivery vehicle for CRISPR gene therapy and are still widely employed due to their high delivery efficiency (Xu et al. 2019).

12 Future Perspectives

The CRISPR-Cas system is a next-generation technology that aids in the development of novel highly specific antiviral treatments and molecular diagnostics that may be used quickly, accurately, and inexpensively at the point of care. Biofuels, new materials, and other industries will be impacted by the CRISPR revolution. CRISPR technology can bring back extinct species and possibly generate entirely new ones in the future. CRISPR-Cas, a sequence-specific nuclease capable of editing the exact gene sequence, has revolutionized biology and created a new era in genetic engineering and site-specific editing of nucleotide(s) within a faulty gene. CRISPR/Cas9 offers a wide range of programmable gene editing options and has the potential to be a valuable tool in modern medicine. The method may be used to change bacterial, fungal, and yeast strains to qualitatively and quantitatively manipulate secondary metabolite pathways. Finally, the CRISPR-Cas system is a one-of-a-kind gene editing technique. We can improve our quality of life by using CRISPR-Cas, a very precise genome editing technique. Without the presence of poisons or pathogens, our food will become more nutrient-dense. Several crops have already shown CRISPR-mediated improvements in quality and quantity, as well as tolerance to viruses, herbicides, drought, salt, and cold. However, the technique will usher in a whole new crop generation, complete with fresh types. Misuse of CRISPR-Cas for gene editing, however, could pose a risk and danger; thus, ethical discussions concerning CRISPR in the scientific community are critical. Despite the concerns, we believe that the use of CRISPR will be a huge benefit to humanity.

New breeding techniques allow scientists to implant desired features more precisely and quickly than traditional breeding. CRISPR/Cas9-based genome editing is a game-changing technology. In the future, crop development using genome editing methods to boost yield, nutritional value, disease resistance, and other qualities will be a major focus. Future clinical developments will require a review of potential dangers connected with CRISPR use.

13 Conclusion

Although genome engineering for crop enhancement is still in its early stages, it has been used to generate some crop species. The advent and widespread use of gene editing tools in crops and livestock has permitted a recent surge in gene editing for climate change. Despite its success in the lab, gene editing for climate change has yet to have a significant impact in the real world. Regulations, societal hurdles, and prohibitive policies, among other externalities outside the technical limits stated, have hampered the adoption of these technological advancements. The majority of developments in gene editing applications for agriculture have occurred recently, which helps to explain why agricultural output has a low throughput. Current technical advances are rapidly expanding, thanks to the continued efforts of both public and commercial organizations. Even having many limitations in gene editing tools, all the gene editing tools help in the development of a better life in agriculture.

References

Aggarwal RAK, Sharma R, Kumar R, Mohapatra T, Sharma RP (2003) Molecular mapping of loci for contents of three major fatty acids in Indian mustard (*Brassica juncea* L.). J Plant Biochem Biotechnol 12(2):131–137

Ali Z, Shami A, Sedeek K, Kamel R, Alhabsi A, Tehseen M, Hassan N, Butt H, Kababji A, Hamdan SM, Mahfouz MM (2020) Fusion of the Cas9 endonuclease and the VirD2 relaxase facilitates homology-directed repair for precise genome engineering in rice. Commun Biol 3:44

Ambika, Kumar R (2023) Genetic analysis of yield traits and identification of markers for stem growth habit in chickpea. The registered company Eliva Press Global Ltd. part of Eliva Press S.R.L. Publishing Group, Chisinau, Moldova, Europe, pp 1–82

Ambika, Aski MS, Gayacharan, Hamwieh A, Talukdar A, Kumar Gupta S, Sharma BB, Joshi R, Upadhyaya HD, Singh K, Kumar R (2022) Unraveling origin, history, genetics, and strategies for accelerated domestication and diversification of food legumes. Front Genet 13:932430. https://doi.org/10.3389/fgene.2022.932430

Ann Ran F, Hsu PD, Wright J, Agarwala V, Scott DA, Zhang F (2013) Genome engineering using the CRISPR/Cas9 system. Nat Protoc 8(11):2281–2308. https://doi.org/10.1038/nprot.2013.143

Ashraf M (2010) Inducing drought tolerance in plants: recent advances. Biotechnol Adv 28(1):169–183

Ashraf MFMR, Foolad MR (2007) Roles of glycine betaine and proline in improving plant abiotic stress resistance. Environ Exp Bot 59(2):206–216

Badhan S, Ball AS, Mantri N (2021) First report of CRISPR/Cas9 mediated DNA-free editing of *4CL* and *RVE7* genes in chickpea protoplasts. Int J Mol Sci 22:1. https://doi.org/10.3390/ijms22010396

Baeg GJ, Kim SH, Choi DM, Tripathi S, Han YJ, Kim J (2021) CRISPR/Cas9-mediated mutation of 5-oxoprolinase gene confers resistance to sulfonamide compounds in Arabidopsis. Plant Biotechnol Rep 15:753–764. https://doi.org/10.1007/s11816-021-00718-w

Barman HN, Sheng Z, Fiaz S, Zhong M, Wu Y, Cai Y, Wang W, Jiao G, Tang S, Wei X, Hu P (2019) Generation of a new thermo-sensitive genic male sterile rice line by targeted mutagenesis of *TMS5* gene through CRISPR/Cas9 system. BMC Plant Biol 19:109

Barrangou R, Fremaux C, Deveau H, Richards M, Boyaval P, Moineau S, Romero DA, Horvath P (2007) CRISPR provides acquired resistance against viruses in prokaryotes. Science 315:1709–1712

Bertier LD, Ron M, Huo H, Bradford KJ, Britt AB, Michelmore RW (2018) High-resolution analysis of the efficiency, heritability, and editing outcomes of CRISPR/Cas9 -induced modifications of *NCED4* in lettuce (*Lactuca sativa*). G3 8:1513–1521. https://doi.org/10.1534/g3.117.300396

Bhardwaj J, Kumari N, Ford R, Yadav R, Choi I, Kumar R (2014) In silico development and validation of EST derived new SSR markers for drought tolerance in *Cicer arietinum* L. Ind J Genet Plant Breed 74(2):254–256. https://doi.org/10.5958/0975-6906.2014.00164.3

Brooks C, Nekrasov V, Lipppman ZB, van Eck J (2014) Efficient gene editing in tomato in the first generation using the clustered regularly interspaced short palindromic repeats/CRISPR-associated cas9 system. Plant Physiol 166:1292–1297. https://doi.org/10.1104/pp.114.247577

Brouns SJ, Jore MM, Lundgren M, Westra ER, Slijkhuis RJ, Snijders AP, Dickman MJ, Makarova KS, Koonin EV, Van Der Oost J (2008) Small CRISPR RNAs guide antiviral defense in prokaryotes. Science 321(5891):960–964

Butler NM, Baltes NJ, Voytas DF, Douches DS (2016) Geminivirus-mediated genome editing in potato (*Solanum tuberosum* L.) using sequence-specific nucleases. Front Plant Sci 7:1045. https://doi.org/10.3389/fpls.2016.01045

Butt H, Jamil M, Wang JY, Al-Babili S, Mahfouz M (2018) Engineering plant architecture via CRISPR/Cas9-mediated alteration of strigolactone biosynthesis. BMC Plant Biol 18(1):1–9

Butt H, Rao GS, Sedeek K, Aman R, Kamel R, Mahfouz M (2020) Engineering herbicide resistance via prime editing in rice. Plant Biotechnol J 18(12):2370–2372

Chae MJ, Lee JS, Nam MH, Cho K, Hong JY, Yi SA, Suh SC, Yoon IS (2007) A rice dehydration-inducible SNF1-related protein kinase 2 phosphorylates an abscisic acid responsive element-binding factor and associates with ABA signaling. Plant Mol Biol 63:151–169

Chandana BS, Mahto RK, Singh RK, Ford R, Vaghefi N, Gupta SK, Yadav HK, Manohar M, Kumar R (2022) Epigenomics as potential tools for enhancing magnitude of breeding approaches for developing climate resilient chickpea. Front Genet 13:900253. https://doi.org/10.3389/fgene.2022.900253

Chandrasekaran J, Brumin M, Wolf D, Leibman D, Klap C, Pearlsman M, Sherman A, Arazi T, Gal-On A (2016) Development of broad virus resistance in non-transgenic cucumber using CRISPR/Cas9 technology. Mol Plant Pathol 17(7):1140–1153

Char SN, Neelakandan AK, Nahampun H, Frame B, Main M, Spalding MH, Becraft PW, Meyers BC, Walbot V, Wang K, Yang B (2017) An agrobacterium-delivered CRISPR/Cas9 system for high-frequency targeted mutagenesis in maize. Plant Biotechnol J 15(2):257–268

Charlesworth CT, Deshpande PS, Dever DP, Camarena J, Lemgart VT, Cromer MK, Vakulskas CA, Collingwood MA, Zhang L, Bode NM, Behlke MA (2019) Identification of preexisting adaptive immunity to Cas9 proteins in humans. Nat Med 25(2):249–254

Charpentier E, Richter H, van der Oost J, White MF (2015) Biogenesis pathways of RNA guides in archaeal and bacterial CRISPR/Cas adaptive immunity. FEMS Microbiol Rev 39:428–441. https://doi.org/10.1093/femsre/fuv023

Chen K, Wang Y, Zhang R, Zhang H, Gao C (2019) CRISPR/Cas genome editing and precision plant breeding in agriculture. Annu Rev Plant Biol 70:667–697. https://doi.org/10.1146/annurev-arplant-050718-100049

Cong L, Ran FA, Cox D, Lin S, Barretto R, Habib N, Hsu PD, Wu X, Jiang W, Marraffini LA, Zhang F (2013) Multiplex genome engineering using CRISPR/Cas systems. Science 339:819–823

Connorton JM, Jones ER, Rodriguez-Ramiro I, Fairweather-Tait S, Uauy C, Balk J (2017) Wheat vacuolar iron transporter *TaVIT2* transports Fe and Mn and is effective for biofortification. Plant Physiol 174:2434–2444. https://doi.org/10.1104/pp.17.00672

Cramer GR, Urano K, Delrot S, Pezzotti M, Shinozaki K (2011) Effects of abiotic stress on plants: a systems biology perspective. BMC Plant Biol 11(1):1–14

Curtin SJ, Xiong Y, Michno JM, Campbell BW, Stec AO, Cermák T, Starker C, Voytas DF, Eamens AL, Stupar RM (2018) Crispr/cas9 and TALENS generate heritable mutations for genes involved in small RNA processing of *Glycine max* and *Medicago truncatula*. Plant Biotechnol J 16:1125–1137

Dahan-Meir T, Filler-Hayut S, Melamed-Bessudo C, Bocobza S, Czosnek H, Aharoni A, Levy AA (2018) Efficient in planta gene targeting in tomato using geminiviral replicons and the CRISPR/Cas9 system. Plant J 95:5–16

Dai A (2011) Drought under global warming: a review. Wiley Interdiscip Rev Clim Chang 2(1):45–65

de Melo BP, Lourenço-Tessutti IT, Paixão JFR, Noriega DD, Silva MCM, de Almeida-Engler J, Fontes EPB, Grossi-deSa MF (2020) Transcriptional modulation of *AREB-1* by CRISPRa improves plant physiological performance under severe water deficit. Sci Rep 10:16231

Dong L, Xie C, Li L, Liu C, Liu C, Geng S, Li X, Huang C, Mao L, Chen S (2018) Genome editing and double-fluorescence proteins enable robust maternal haploid induction and identification in maize. Mol Plant 11:1214–1217

Du H, Zeng X, Zhao M, Cui X, Wang Q, Yang H, Cheng H, Yu D (2016) Efficient targeted mutagenesis in soybean by TALENs and CRISPR/Cas9. J Biotechnol 217:90–97

FAO (2007). http://www.fao.org/docrep/010/a1075e/a1075e00.htm

FAOSTAT (2023). https://www.fao.org/faostat/en/#home. Accessed 11 Jul 2023

Fich EA, Segerson NA, Rose JK (2016) The plant polyester cutin: biosynthesis, structure, and biological roles. Annu Rev Plant Biol 67:207–233

Fister AS, Landherr L, Maximova SN, Guiltinan MJ (2018) Transient expression of CRISPR/cas9 machinery targeting *TCNPR3* enhances defense response in *Theobroma cacao*. Front Plant Sci 9:268. https://doi.org/10.3389/fpls.2018.00268

Flowers TJ, Colmer TD (2008) Salinity tolerance in halophytes. New Phytol 179:945–963

Gao W, Long L, Tian X, Xu F, Liu J, Singh PK, Botella JR, Song C (2017) Genome editing in cotton with the CRISPR/Cas9 system. Front Plant Sci 8:1364

Garcia-Gimenez G, Jobling SA (2021) Gene editing for barley grain quality improvement. J Cereal Sci:103394. https://doi.org/10.1016/j.jcs.2021.103394

Garneau JE, Dupuis MÈ, Villion M, Romero DA, Barrangou R, Boyaval P, Fremaux C, Horvath P, Magadán AH, Moineau S (2010) The CRISPR/Cas bacterial immune system cleaves bacteriophage and plasmid DNA. Nature 468:67. https://doi.org/10.1038/nature09523

Ghassemi F, Jakeman AJ, Nix HA (1995) Salinisation of land and water resources: human causes, extent, management and case studies. CAB International, Wallingford

Gupta A, Sharma T, Singh SP, Bhardwaj A, Srivastava D, Kumar R (2023) Prospects of microgreens as budding living functional food: breeding and biofortification through OMICS and other approaches for nutritional security. Front Genet 14:1053810

Haapaniemi E, Botla S, Persson J, Schmierer B, Taipale J (2018) CRISPR/Cas9 genome editing induces a p53-mediated DNA damage response. Nat Med 24:927–930. https://doi.org/10.1038/s41591-018-0049-z

Hale CR, Zhao P, Olson S, Duff MO, Graveley BR, Wells L et al (2009) RNA-guided RNA cleavage by a CRISPR RNA-Cas protein complex. Cell 139:945–956. https://doi.org/10.1016/j.cell.2009.07.040

Hannah MA, Wiese D, Freund S, Fiehn O, Heyer AG, Hincha DK (2006) Natural genetic variation of freezing tolerance in Arabidopsis. Plant Physiol 142:98–112. https://doi.org/10.1104/pp.106.081141

Harshavardhana Y, Hegde V, Tripathi S, Raje R, Jain P, Gaikwad K, Bharadwaj C, Kumar R, Singh R, Sharma M, Chauhan S (2019) Genetics of semi-determinacy and identification of molecular marker linked to *Dt1* locus in chickpea (*Cicer arietinum* L.). Ind J Genet Plant Breed 79(01S):270–275. https://doi.org/10.31742/ijgpb.79s.1.18

Hu Z, Yu L, Zhu D, Ding W, Wang X, Zhang C, Wang L, Jiang X, Shen H, He D, Li K (2014) Disruption of *HPV16-E7* by CRISPR/Cas system induces apoptosis and growth inhibition in HPV16 positive human cervical cancer cells. Biomed Res Int 2014:612823

Ihry RJ, Worringer KA, Salick MR, Frias E, Ho D, Theriault K, Kommineni S, Chen J, Sondey M, Ye C, Randhawa R (2018) p53 inhibits CRISPR–Cas9 engineering in human pluripotent stem cells. Nat Med 24(7):939–946

Iizumi T, Shin Y, Kim W, Kim M, Choi J (2018) Global crop yield forecasting using seasonal climate information from a multi-model ensemble. Clim Serv 11:13–23

Ishino Y, Shinagawa H, Makino K, Amemura M, Nakata A (1987) Nucleotide sequence of the *Iap* gene, responsible for alkaline phosphatase isozyme conversion in *Escherichia coli*, and identification of the gene product. J Bacteriol 169:5429–5433

Ito Y, Nishizawa-Yokoi A, Endo M, Mikami M, Toki S (2015) CRISPR/Cas9-mediated mutagenesis of the *RIN* locus that regulates tomato fruit ripening. Biochem Biophys Res Commun 467:76–82. https://doi.org/10.1016/j.bbrc.2015.09.117

Jacobs TB, LaFayette PR, Schmitz RJ, Parrott WA (2015) Targeted genome modifications in soybean with CRISPR/Cas9. BMC Biotechnol 15:1–10. https://doi.org/10.1186/s12896-015-0131-2
Jain M (2015) Function genomics of abiotic stress tolerance in plants: A CRISPR approach. Front Plant Sci 6:e93806. https://doi.org/10.3389/fpls.2015.00375
Jansen R, van Embden JDA, Gaastra W, Schouls LM (2002) Identification of genes that are associated with DNA repeats in prokaryotes. Mol Microbiol 43:1565–1575
Jha UC, Bohra A (2016) Genomics enabled breeding approaches for improving cadmium stress tolerance in plants. Euphytica 208:1–31. https://doi.org/10.1007/s10681-015-1580-3
Ji X, Zhang H, Zhang Y, Wang Y, Gao C (2015) Establishing a CRISPR/Cas-like immune system conferring DNA virus resistance in plants. Nat Plants 1:15144. https://doi.org/10.1038/nplants.2015.144
Jia H, Nian W (2014) Targeted genome editing of sweet orange using Cas9/sgRNA. PLoS One 9:e93806. https://doi.org/10.1371/journal.pone.0093806
Jia H, Zhang Y, Orbovi'c V, Xu J, White FF, Jones JB, Wang N (2017) Genome editing of the disease susceptibility gene *CsLOB1* in citrus confers resistance to citrus canker. Plant Biotechnol J 15:817–823
Jiang W, Zhou H, Bi H, Fromm M, Yang B, Weeks DP (2013) Demonstration of CRISPR/Cas9/sgRNA-mediated targeted gene modification in Arabidopsis, tobacco, sorghum and rice. Nucleic Acids Res 41:e188–e188. https://doi.org/10.1093/nar/gkt780
Jin S, Zong Y, Gao Q, Zhu Z, Wang Y, Qin P, Liang C, Wang D, Qiu JL, Zhang F, Gao C (2019) Cytosine, but not adenine, base editors induce genome-wide off-target mutations in rice. Science 364(6437):292–295
Jinek M, Chylinski K, Fonfara I, Hauer M, Doudna JA, Charpentier E (2012) A programmable dual-RNA-guided DNA endonuclease in adaptive bacterial immunity. Science 337:816–821. https://doi.org/10.1126/science.1225829
Jones JD, Witek K, Verweij W, Jupe F, Cooke D, Dorling S, Tomlinson L, Smoker M, Perkins S, Foster S (2014) Elevating crop disease resistance with cloned genes. Philos T R Soc B 369(1639):20130087
Julkowska MM, Testerink C (2015) Tuning plant signaling and growth to survive salt. Trends Plant Sci 20:586–594. https://doi.org/10.1016/j.tplants.2015.06.008
Karginov FV, Hannon GJ (2010) The CRISPR system: small RNA-guided defense in bacteria and archaea. Mol Cell 37:7
Khanday I, Skinner D, Yang B, Mercier R, Sundaresan V (2019) A male-expressed rice embryogenic trigger redirected for asexual propagation through seeds. Nature 565(7737):91–95
Klap C, Yeshayahou E, Bolger AM, Arazi T, Gupta SK, Shabtai S, Usadel B, Salts Y, Barg R (2017) Tomato facultative parthenocarpy results from *SlAGAMOUS-LIKE 6* loss of function. Plant Biotechnol J 15:634–647
Kumar R, Singh KP (1986) Genetic variability, heritability and genetic advance in grain sorghum. Farm Sci J 1:94
Kumar R, Rathi AS, Singh A (1999) Association analysis in *Vigna mungo* (L.). Ann Agric Res 20(4):468–471
Kumar R, Singh A, Rathi AS (2000) Estimating genetic parameters in Urdbean. Ann Agric Res 21(3):335–337
Kumar R, Rathi AS, Singh A, Krishna R (2002) Genetic divergence in relation to breeding for grain yield in urdbean (*Vigna Mungo* L.). Legume Res 25(2):109–112
Kumar R, Singh RK, Misra J, Yadav A, Kumar A, Yadav R, Hegde V, Kumar S, Yadav N (2021) Dissecting proteomic estimates for enhanced bioavailable nutrition during varied stages of germination and identification of potential genotypes in chickpea. Legume Res 45(9):1082–1087. https://doi.org/10.18805/LR-4531
Kumar A, Yadav A, Yadav R, Misra JP, Yadav RS, Upadhyaya HD, Kumar R (2022a) Identification of highly polymorphic molecular markers and potential genotypes for harnessing chickpea breeding strategies. Legume Res 45(7):804–814. https://doi.org/10.18805/LR-4379

Kumar M, Ayzenshtat D, Marko A, Bocobza S (2022b) Optimization of TDNA configuration with *UBIQUITIN10* promoters and tRNA–sgRNA complexes promotes highly efficient genome editing in allotetraploid tobacco. Plant Cell Rep 41:1–20. https://doi.org/10.1007/s00299-021-02796-0

Lacchini E, Kiegle E, Castellani M, Adam H, Jouannic S, Gregis V, Kater MM (2020) CRISPR-mediated accelerated domestication of African rice landraces. PLoS One 15(3):e0229782

Lau CH, Suh Y (2017) In vivo genome editing in animals using AAV-CRISPR system: applications to translational research of human disease. F1000Res 6:2153. https://doi.org/10.12688/f1000research.11243.1

Lemmon ZH, Reem NT, Dalrymple J, Soyk S, Swartwood KE, Rodriguez-Leal D, Van Eck J, Lippman ZB (2018) Rapid improvement of domestication traits in an orphan crop by genome editing. Nat Plants 4(10):766–770

Li JF, Norville JE, Aach J, McCormack M, Zhang D, Bush J, Church GM, Sheen J (2013) Multiplex and homologous recombination–mediated genome editing in Arabidopsis and *Nicotiana benthamiana* using guide RNA and Cas9. Nat Biotechnol 31(8):688–691

Li M, Li X, Zhou Z, Wu P, Fang M, Pan X, Lin Q, Luo W, Wu G, Li H (2016) Reassessment of the four yield-related genes *Gn1a, DEP1, GS3*, and *IPA1* in rice using a CRISPR/Cas9 system. Front Plant Sci 7:377. https://doi.org/10.3389/fpls.2016.00377

Li P, Li YJ, Zhang FJ, Zhang GZ, Jiang XY, Yu HM, Hou BK (2017) The Arabidopsis UDP-glycosyltransferases UGT79B2 and UGT79B3, contribute to cold, salt and drought stress tolerance via modulating anthocyanin accumulation. Plant J 89(1):85–103

Li G, Liu D, Zhang X, Quan R, Zhong C, Mo J, Huang Y, Wang H, Ruan X, Xu Z, Zheng E (2018a) Suppressing *Ku70/Ku80* expression elevates homology-directed repair efficiency in primary fibroblasts. Int J Biochem Cell Biol 99:154–160

Li T, Yang X, Yu Y, Si X, Zhai X, Zhang H, Dong W, Gao C, Xu C (2018b) Domestication of wild tomato is accelerated by genome editing. Nat Biotechnol 36:1160–1163

Li R, Liu C, Zhao R, Wang L, Chen L, Yu W, Zhang S, Sheng J, Shen L (2019) CRISPR/Cas9-mediated *SlNPR1* mutagenesis reduces tomato plant drought tolerance. BMC Plant Biol 19(1):1–13

Li C, Li W, Zhou Z, Chen H, Xie C, Lin Y (2020a) A new rice breeding method: CRISPR/Cas9 system editing of the *Xa13* promoter to cultivate transgene-free bacterial blight-resistant rice. Plant Biotechnol J 18(2):313

Li J, Wang Z, He G, Ma L, Deng XW (2020b) CRISPR/Cas9- mediated disruption of *TaNP1* genes results in complete male sterility in bread wheat. J Genet Genomics 47(5):263–272

Liang Z, Zhang K, Chen K, Gao C (2014) Targeted mutagenesis in *Zea mays* using TALENs and the CRISPR/Cas system. J Genet Genomics 41:63–68. https://doi.org/10.1016/j.jgg.2013.12.001

Liao S, Qin X, Luo L, Han Y, Wang X, Usman B, Nawaz G, Zhao N, Liu Y, Li R (2019) CRISPR/Cas9-induced mutagenesis of semi-rolled *leaf1, 2* confers curled leaf phenotype and drought tolerance by influencing protein expression patterns and ros scavenging in rice (*Oryza sativa* L.). Agronomy 9:728

Lin CS, Hsu CT, Yang LH, Lee LY, Fu JY, Cheng QW, Wu FH, Hsiao HCW, Zhang Y, Zhang R, Chang WJ (2018) Application of protoplast technology to CRISPR/Cas9 mutagenesis: from single-cell mutation detection to mutant plant regeneration. Plant Biotechnol J 16(7):1295–1310

Lino CA, Harper JC, Carney JP, Timlin JA (2018) Delivering CRISPR: a review of the challenges and approaches. Drug Deliv 25:1234–1257. https://doi.org/10.1080/10717544.2018.1474964

Liu HW, Zhang BH (2020) Virus-based CRISPR/Cas9 genome editing in plants. Trends Genet 36(11):810–813. https://doi.org/10.1016/j.tig.2020.08.002

Liu L, Li X, Wang J, Wang M, Chen P, Yin M, Li J, Sheng G, Wang Y (2017) Two distant catalytic sites are responsible for *C2c2* RNase activities. Cell 168(1):121–134

Liu L, Kuang Y, Yan F, Li S, Ren B, Gosavi G, Spetz C, Li X, Wang X, Zhou X, Zhou H (2021) Developing a novel artificial rice germplasm for dinitroaniline herbicide resistance by base editing of *OsTubA2*. Plant Biotechnol J 19(1):5

Lou D, Wang H, Liang G, Yu D (2017) *OsSAPK2* confers abscisic acid sensitivity and tolerance to drought stress in rice. Front Plant Sci 8:1371. https://doi.org/10.3389/fpls.2017.00993

Lu HP, Liu SM, Xu SL, Chen WY, Zhou X, Tan YY, Huang JZ, Shu QY (2017) CRISPR-S: an active interference element for a rapid and inexpensive selection of genome-edited, transgene-free rice plants. Plant Biotechnol J 15(11):1371

Lu HP, Luo T, Fu HW, Wang L, Tan YY, Huang JZ, Wang Q, Ye GY, Gatehouse AM, Lou YG, Shu QY (2018) Resistance of rice to insect pests mediated by suppression of serotonin biosynthesis. Nat Plants 4(6):338–344

Mahto RK, Ambika, Singh C, Chandana BS, Singh RK, Verma S, Gahlaut V, Manohar M, Yadav N, Kumar R (2022) Chickpea biofortification for Cytokinin dehydrogenase via genome editing to enhance abiotic-biotic stress tolerance and food security. Front Genet 13:900324. https://doi.org/10.3389/fgene.2022.900324

Makarova KS, Wolf YI, Alkhnbashi OS, Costa F, Shah SA, Saunders SJ, Barrangou R, Brouns SJ, Charpentier E, Haft DH, Horvath P (2015) An updated evolutionary classification of CRISPR–Cas systems. Nat Rev Microbiol 13(11):722–736

Makarova KS, Wolf YI, Koonin EV (2018) Classification and nomenclature of CRISPR/Cas systems: where from here? CRISPR J 1(5):325–336. https://doi.org/10.1089/crispr.2018.0033

Meng Y, Hou Y, Wang H, Ji R, Liu B, Wen J, Niu L, Lin H (2017) Targeted mutagenesis by CRISPR/Cas9 system in the model legume *Medicago truncatula*. Plant Cell Rep 36:371–374

Miao J, Guo D, Zhang J, Huang Q, Qin G, Zhang X, Wan J, Gu H, Qu LJ (2013) Targeted mutagenesis in rice using CRISPR/Cas system. Cell Res 23(10):1233–1236

Miao C, Wang D, He R, Liu S, Zhu JK (2020) Mutations in *MIR 396e* and *MIR 396f* increase grain size and modulate shoot architecture in rice. Plant Biotechnol J 18(2):491–501

Misra JP, Yadav A, Kumar A, Yadav R, Vaishali, Kumar R (2016) Bio-chemical characterization of chickpea genotypes with special reference to protein. Res J Chem Environ 20(8):38–43

Mittal N, Bhardwaj J, Verma S, Singh RK, Yadav R, Kaur D, Talukdar A, Yadav N, Kumar R (2023) Disentangling potential polymorphic markers and genotypes for macro-micro nutrients in chickpea. Sci Rep 13:10731. https://doi.org/10.1038/s41598-023-37602-2

Mojica FJ, Juez G, Rodríguez-Valera F (1993) Transcription at different salinities of *Haloferax mediterranei* sequences adjacent to partially modified PstI sites. Mol Microbiol 9:613–621

Mojica FJ, Díez-Villaseñor C, Soria E, Juez G (2000) Biological significance of a family of regularly spaced repeats in the genomes of archaea, bacteria and mitochondria. Mol Microbiol 36:244–246

Mojica FJM, Díez-Villaseñor C, García-Martínez J, Soria E (2005) Intervening sequences of regularly spaced prokaryotic repeats derive from foreign genetic elements. J Mol Evol 60:174–182

Moon SB, Kim DY, Ko J, Kim YS (2019) Recent advances in the CRISPR genome editing tool set. Exp Mol Med 51:1–11. https://doi.org/10.1038/s12276-019-0339-7

Moscou MJ, Bogdanove AJ (2009) A simple cipher governs DNA recognition by TAL effectors. Science 326:150. https://doi.org/10.1126/science.1178817

Munns R, Tester M (2008) Mechanisms of salinity tolerance. Annu Rev Plant Biol 59:651–681

Nawaz G, Usman B, Peng H, Zhao N, Yuan R, Liu Y, Li R (2020) Knockout of *Pi21* by CRISPR/Cas9 and iTRAQ-based proteomic analysis of mutants revealed new insights into *M. oryzae* resistance in Elite Rice Line. Gene 11(7):735

Nelson DR, Adger WN, Brown K (2007) Adaptation to environmental change: contributions of a resilience framework. Annu Rev Env Resour 32:395–419

Nishimasu H, Ann Ran F, Hsu PD, Konermann S, Shehata SI, Dohmae N, Ishitani R, Zhang F, Nureki O (2014) Crystal structure of Cas9 in complex with guide RNA and target DNA. Cell 156(5):935–949. https://doi.org/10.1016/j.cell.2014.02.001

Nuccio ML, Claeys H, Heyndrickx KS (2021) CRISPR/Cas technology in corn: a new key to unlock genetic knowledge and create novel products. Mol Breed 41:11

Odipio J, Alicai T, Ingelbrecht I, Nusinow DA, Bart R, Taylor NJ (2017) Efficient CRISPR/Cas9 genome editing of phytoene desaturase in cassava. Front Plant Sci 8:1780. https://doi.org/10.3389/fpls.2017.01780

Ormond KE, Mortlock DP, Scholes DT, Bombard Y, Brody LC, Faucett WA, Nanibaa AG, Hercher L, Isasi R, Middleton A, Musunuru K (2017) Human germline genome editing. Am J Hum Genet 101(2):167–176

Oz MT, Altpeter A, Karan R, Merotto A, Altpeter F (2021) CRISPR/ Cas9-mediated multi-allelic gene targeting in sugarcane confers herbicide tolerance. Front Genome 3:673566. https://doi.org/10.3389/fgeed.2021.673566

Park S II, Kim HB, Jeon HJ, Kim H (2021) Agrobacterium-mediated *Capsicum annuum* gene editing in two cultivars, hot pepper *CM334* and bell pepper dempsey. Int J Mol Sci 22:3921. https://doi.org/10.3390/ijms22083921

Prabha S, Yadav A, Yadav HK, Kumar S, Kumar R (2017) Importance of molecular marker in linseed (*Linum usitatissimum*) genome analysis–a review. Crop Res 52(1, 2 & 3):61–66

Pramanik D, Shelake RM, Park J, Kim MJ, Hwang I, Park Y, Kim JY (2021) CRISPR/Cas9-mediated generation of pathogen resistant tomato against tomato yellow leaf curl virus and powdery mildew. Int J Mol Sci 22:1878

Puchta H, Dujon B, Hohn B (1993) Homologous recombination in plant cells is enhanced by in vivo induction of double strand breaks into DNA by a site specific endonuclease. Nucleic Acids Res 21:5034–5040. https://doi.org/10.1093/nar/21.22.5034

Qadir M, Quillérou E, Nangia V, Murtaza G, Singh M, Thomas RJ, Drechsel P, Noble AD (2014) Economics of salt-induced land degradation and restoration. Nat Resour Forum 8(4):282–295

Raftery AE, Zimmer A, Frierson DM, Startz R, Liu P (2017) Less than 2 C warming by 2100 unlikely. Nat Clim Change 7(9):637–641

Ramani A, Kushwaha R, Malaviya R, Kumar R, Yadav N (2021) Molecular, functional and nutritional properties of chickpea (*Cicer arietinum* L.) protein isolates prepared by modified Solubilization methods. J Food Meas Charact 15(3):2352–2368. https://doi.org/10.1007/s11694-020-00778-6

Ramia NF, Tang L, Cocozaki AI, Li H (2014) *Staphylococcus epidermidis* Csm1 is a 3=−5= exonuclease. Nucleic Acids Res 42(2):1129–1138. https://doi.org/10.1093/nar/gkt914

Rashmi V, Kumar R, Yadav N (2012) Protein quantification of chickpea genotypes at different stress conditions. Prog Agric Int J 12(1):169–174

Rehman K, Fatima F, Waheed I, Akash MSH (2018) Prevalence of exposure of heavy metals and their impact on health consequences. J Cell Biochem 119(1):157–184

Ricroch A, Clairand P, Harwood W (2017) Use of CRISPR systems in plant genome editing: toward new opportunities in agriculture. Emerg Top Life Sci 1:169. https://doi.org/10.1042/ETLS20170085

Samai P, Pyenson N, Jiang W, Goldberg GW, Hatoum-Aslan A, Marraffini LA (2015) Co-transcriptional DNA and RNA cleavage during type III CRISPR/Cas immunity. Cell 161(5):1164–1174. PMID:25959775. https://doi.org/10.1016/j.cell.2015.04.027

Shan Q, Wang Y, Li J, Zhang Y, Chen K, Liang Z, Zhang K, Liu J, Xi JJ, Qiu JL, Gao C (2013) Targeted genome modification of crop plants using a CRISPR/Cas system. Nat Biotechnol 31(8):686–688

Shan Q, Wang Y, Li J, Gao C (2014) Genome editing in rice and wheat using the CRISPR/Cas system. Nat Protoc 9(10):2395–2410

Sharma R, Agarwal RAK, Kumar R, Mohapatra T, Sharma RP (2002) Construction of RAPD linkage map and localization of QTLs for oleic acid level using recombinant inbreds in mustard. Genome 45(3):467–472

Shen C, Que Z, Xia Y, Tang N, Li D, He R, Cao M (2017) Knock out of the annexin gene *OsAnn3* via CRISPR/Cas9-mediated genome editing decreased cold tolerance in rice. J Plant Biol 60:539–547

Shepherd T, Wynne Griffiths D (2006) The effects of stress on plant cuticular waxes. New Phytol 171(3):469–499

Shi J, Gao H, Wang H, Lafitte HR, Archibald RL, Yang M, Hakimi SM, Mo H, Habben JE (2017) *ARGOS8* variants generated by CRISPR/Cas9 improve maize grain yield under field drought stress conditions. Plant Biotechnol J 15(2):207. https://doi.org/10.1111/pbi.12603

Shimatani Z, Kashojiya S, Takayama M, Terada R, Arazoe T, Ishii H, Teramura H, Yamamoto T, Komatsu H, Miura K, Ezura H, Nishida K, Ariizumi T, Kondo A (2017) Targeted base editing in rice and tomato using a CRISPR/Cas9 cytidine deaminase fusion. Nat Biotechnol 35:441–443

Shinozaki K, Yamaguchi-Shinozaki K, Seki M (2003) Regulatory network of gene expression in the drought and cold stress responses. Curr Opin Plant Biol 6(5):410–417

Silas S, Mohr G, Sidote DJ, Markham LM, Sanchez-Amat A, Bhaya D, Lambowitz AM, Fire AZ (2016) Direct CRISPR spacer acquisition from RNA by a natural reverse transcriptase–Cas1 fusion protein. Science 351(6276):aad4234

Singh AP, Krishna R, Kumar R, Singh NP (2001a) Assaying divergence in lentil. Crop Res 22(3):469–473

Singh NP, Krishna R, Kumar R (2001b) An assay of effects of different traits on chickpea grain yield. Ann Agric Res New Series 22(4):564–569

Singh NP, Krishana R, Kumar R, Singh AP (2002) Assaying direct selection parameters in chickpea. Prog Agric 2(2):169–170

Singh R, Kumar R, Kumari N (2012) Genetic diversity analysis of chickpea using STMS marker. Prog Agric Int J 12(1):35–40

Singh RK, Singh C, Ambika, Chandana BS, Mahto RK, Patial R, Gupta A, Gahlaut V, Gayacharan, Hamwieh A, Upadhyaya HD, Kumar R (2022) Exploring chickpea germplasm diversity for broadening the genetic base utilizing genomic resources. Front Genet 13:905771. https://doi.org/10.3389/fgene.2022.905771

Singh C, Kumar R, Sehgal H, Bhati S, Singhal T, Nimmy MS, Yadav R, Gupta SK, Abdallah NA, Hamwieh A, Kumar R (2023) Unclasping potentials of genomics and gene editing in chickpea to fight climate change and global hunger threat. Front Genet 14:1085024

Slama I, Abdelly C, Bouchereau A, Flowers T, Savouré A (2015) Diversity, distribution and roles of osmoprotective compounds accumulated in halophytes under abiotic stress. Ann Bot 115(3):433–447

Soi S, Chauhan US, Yadav R, Kumar J, Yadav SS, Yadav HK, Kumar R (2014) STMS based diversity analysis in chickpea (*Cicer arietinum* L.). New Agric 25(2):243–250

Songstad DD, Petolino JF, Voytas DF, Reichert NA (2017) Genome editing of plants. Crit Rev Plant Sci 36:1–23

Soyk S, Müller NA, Park SJ, Schmalenbach I, Jiang K, Hayama R, Zhang L, Van Eck J, Jiménez-Gómez JM, Lippman ZB (2017) Variation in the flowering gene *SELF PRUNING 5G* promotes day-neutrality and early yield in tomato. Nat Genet 49(1):162–168

Srivastava R, Bajaj D, Sayal YK, Meher PK, Upadhyaya HD, Kumar R, Tripathi S, Bharadwaj C, Rao AR, Parida SK (2016) Genome-wide development and deployment of informative intron-spanning and intron-length polymorphism markers for genomics-assisted breeding applications in chickpea. Plant Sci 252(11):374–387. https://doi.org/10.1016/j.plantsci.2016.08.013

Sugano SS, Shirakawa M, Takagi J, Matsuda Y, Shimada T, Hara-Nishimura I, Kohchi T (2014) CRISPR/Cas9-mediated targeted mutagenesis in the liverwort *Marchantia polymorpha* L. Plant Cell Physiol 55(3):475–481

Sun Y, Jiao G, Liu Z, Zhang X, Li J, Guo X, Du W, Du J, Francis F, Zhao Y, Xia L (2017) Generation of high-amylose rice through CRISPR/Cas9-mediated targeted mutagenesis of starch branching enzymes. Front Plant Sci 8:298

Taylor HN, Warner EE, Armbrust MJ, Crowley VM, Olsen KJ, Jackson RN (2019) Structural basis of type IV CRISPR RNA biogenesis by a Cas6 endoribonuclease. RNA Biol 16:1438–1447. https://doi.org/10.1080/15476286.2019.1634965

Tiwari SK, Singh HL, Kumar R, Nigam HK, Singh AP (2001) A postmortem of selection parameters in pea (*Pisum sativum* L.). Res Crops 2(2):237–242

Toda E, Okamoto T (2020) CRISPR/Cas9-based genome editing using rice zygotes. Curr Protoc Plant Biol 5:e20111. https://doi.org/10.1002/cppb.20111

Tripathi JN, Ntui VO, Ron M, Muiruri SK, Britt A, Tripathi L (2019) CRISPR/ Cas9 editing of endogenous banana streak virus in the b genome of Musa spp. overcomes a major challenge in banana breeding. Commun Biol 2:46. https://doi.org/10.1038/s42003-019-0288-7
Viana VE, Pegoraro C, Busanello C, Costa de Oliveira A (2019) Mutagenesis in rice: the basis for breeding a new super plant. Front Plant Sci 10:1326
Wang K (2020) Fixation of hybrid vigor in rice: synthetic apomixis generated by genome editing. aBIOTECH 1(1):15–20
Wang W, Vinocur B, Altman A (2003) Plant responses to drought, salinity and extreme temperatures: towards genetic engineering for stress tolerance. Planta 218:1–14
Wang Y, Cheng X, Shan Q, Zhang Y, Liu J, Gao C, Qiu JL (2014) Simultaneous editing of three homoeoalleles in hexaploid bread wheat confers heritable resistance to powdery mildew. Nat Biotechnol 32(9):947. https://doi.org/10.1038/nbt.2969
Wang S, Zhang S, Wang W, Xiong X, Meng F, Cui X (2015) Efficient targeted mutagenesis in potato by the CRISPR/Cas9 system. Plant Cell Rep 34:1473–1476. https://doi.org/10.1007/s00299-015-1816-7
Wang L, Wang L, Tan Q, Fan Q, Zhu H, Hong Z, Zhang Z, Duanmu D (2016) Efficient inactivation of symbiotic nitrogen fixation related genes in *Lotus japonicus* using CRISPR/Cas9. Front Plant Sci 7:1333
Wang L, Chen L, Li R, Zhao R, Yang M, Sheng J, Shen L (2017) Reduced drought tolerance by CRISPR/Cas9-mediated *SlMAPK3* mutagenesis in tomato plants. J Agric Food Chem 65(39):8674–8682
Wang W, Pan Q, He F, Akhunova A, Chao S, Trick H, Akhunov E (2018a) Transgenerational CRISPR/Cas9 activity facilitates multiplex gene editing in allopolyploid wheat. CRISPR J 1(1):65–74
Wang W, Simmonds J, Pan Q, Davidson D, He F, Battal A, Akhunova A, Trick HN, Uauy C, Akhunov E (2018b) Gene editing and mutagenesis reveal inter-cultivar differences and additivity in the contribution of *TaGW2* homoeologues to grain size and weight in wheat. Theor Appl Genet 131:2463–2475
Wang C, Liu Q, Shen Y, Hua Y, Wang J, Lin J, Wu M, Sun T, Cheng Z, Mercier R, Wang K (2019) Clonal seeds from hybrid rice by simultaneous genome engineering of meiosis and fertilization genes. Nat Biotechnol 37(3):283–286
Wang T, Xun H, Wang W, Ding X, Tian H, Hussain S, Dong Q, Li Y, Cheng Y, Wang C, Lin R (2021) Mutation of *GmAITR* genes by CRISPR/Cas9 genome editing results in enhanced salinity stress tolerance in soybean. Front Plant Sci 12:779598
Wei HH, Yu ST, Wang ZW, Yang Z, Song GS, Wang XZ, Sun XS, Wang CT (2021) In planta genetic transformation to produce CRISPRed high-oleic peanut. Plant Growth Regul 101:443–451. https://doi.org/10.21203/rs.3.rs-109621 l/v2
Wilbie D, Walther J, Mastrobattista E (2019) Delivery aspects of CRISPR/ Cas for in vivo genome editing. Acc Chem Res 52:1555–1564. https://doi.org/10.1021/acs.accounts.9b00106
Wolter F, Puchta H (2018) The CRISPR/Cas revolution reaches the RNA world: Cas13, a new Swiss Army knife for plant biologists. Plant J 94:767–775. https://doi.org/10.1111/tpj.13899
Xie K, Yang Y (2013) RNA-guided genome editing in plants using a CRISPR–Cas system. Mol Plant 6(6):1975–1983
Xu X, Wan T, Xin H, Li D, Pan H, Wu J, Ping Y (2019) Delivery of CRISPR/Cas9 for therapeutic genome editing. J Gene Med 21(7):e3107. https://doi.org/10.1002/jgm.3107
Yadav VK, Kumar R, Ram L (2002) Genetic analysis of malt yield and some of its components in barley. Plant Archiv 2(2):269–273
Yadav JK, Kumar R, Singh HL (2003a) Genetic divergency in chickpea. Adv Plant Sci 16(2):511–514
Yadav JK, Singh HL, Kumar R (2003b) Determining selection components in chickpea (*Cicer arietinum* L.). Plant Archiv 3(1):125–128
Yadav JK, Singh HL, Kumar R (2003c) Perusing selection parameters in chickpea (*Cicer arietinum* L.). New Agriculturist 14(1 & 2):75–80

Yadav R, Aery NC, Kumar R (2006) Effect of nickel and cobalt on the content of carbohydrate and phenols in cowpea (*Vigna unguiculata Var. sinensis L)*. Proc Natl Acad Sci 76(4):368–372
Yadav R, Kumar J, Jain V, Singh IB, Misra JP, Kumar R (2008) Effect of nickel on bio-physio parameters on growth and its accumulation in *Cicer arietinum* L. Prog Agric Int J 8(2):224–230
Yadav R, Jain V, Hegde V, Yadav N, Kumar R (2019) Bio-physico-chemical response of drought tolerant chickpeas to nickel. Legume Res 43(3):345–352. https://doi.org/10.18805/LR-4179
Yadav A et al (2023a) A cellular taxonomy of the adult human spinal cord. Neuron 111:328
Yadav R, Jaiswal S, Singhal T, Mahto RK, Verma SB, Yadav RK, Kumar R (2023b) Potentials of genotypes, morpho-physio-biochemical traits and growing media on shelf life and future prospects of gene editing in tomato. Front Genome Ed 5:1203485. https://doi.org/10.3389/fgeed.2023.1203485
Yang SH, Kim E, Park H, Koo Y (2022) Selection of the high efficient sgRNA for CRISPR/Cas9 to edit herbicide related genes, *PDS*, *ALS*, and *EPSPS* in tomato. Appl Biol Chem 65:13. https://doi.org/10.1186/s13765-022-00679-w
Yeats TH, Rose JK (2013) The formation and function of plant cuticles. Plant Physiol 163(1):5–20
Yen ST, Zhang M, Deng JM, Usman SJ, Smith CN, Parker-Thornburg J, Swinton PG, Martin JF, Behringer RR (2014) Somatic mosaicism and allele complexity induced by CRISPR/Cas9 RNA injections in mouse zygotes. Dev Biol 393(1):3–9
Yu H, Lin T, Meng X, Du H, Zhang J, Liu G, Chen M, Jing Y, Kou L, Li X, Gao Q (2021) A route to *de novo* domestication of wild allotetraploid rice. Cell 184(5):1156–1170
Yuyu C, Aike Z, Pao X, Xiaoxia W, Yongrun C, Beifang W, Yue Z, Liaqat S, Shihua C, Liyong C, Yingxin Z (2020) Effects of *GS3* and *GL3. 1* for grain size editing by CRISPR/Cas9 in rice. Ric Sci 27(5):405–413
Zafar SA, Zaidi SSEA, Gaba Y, Singla-Pareek SL, Dhankher OP, Li X, Mansoor S, Pareek A (2020) Engineering abiotic stress tolerance via CRISPR/Cas-mediated genome editing. J Exp Bot 71(2):470–479
Zeng X, Luo Y, Vu NTQ, Shen S, Xia K, Zhang M (2020a) CRISPR/Cas9-mediated mutation of *OsSWEET14* in rice cv. Zhonghua11 confers resistance to *Xanthomonas oryzae pv. Oryzae* without yield penalty. BMC Plant Biol 20(1):1–11
Zeng Y, Wen J, Zhao W, Wang Q, Huang W (2020b) Rational improvement of rice yield and cold tolerance by editing the three genes *OsPIN5b, GS3*, and *OsMYB30* with the CRISPR–Cas9 system. Front Plant Sci 10:933–946. https://doi.org/10.3389/fpls.2019.01663
Zhang F, Maeder ML, Unger-Wallaced E, Hoshaw JP, Reyon D, Christian M, Li X, Pierick CJ, Dobbs D, Peterson T, Joung JK, Voytas DF (2010) High frequency targeted mutagenesis in *Arabidopsis thaliana* using zinc finger nucleases. Proc Natl Acad Sci U S A 107(26):12028. https://doi.org/10.1073/pnas.0914991107
Zhang Y, Heidrich N, Ampattu BJ, Gunderson CW, Seifert HS, Schoen C, Vogel J, Sontheimer EJ (2013a) Processing-independent CRISPR RNAs limit natural transformation in *Neisseria meningitidis*. Mol Cell 50(4):488–503
Zhang Y, Zhang F, Li X, Baller JA, Qi Y, Starker CG, Bogdanove AJ, Voytas DF (2013b) Transcription activator-like effector nucleases enable efficient plant genome engineering. Plant Physiol 161(1):20–27
Zhang XH, Tee LY, Wang XG, Huang QS, Yang SH (2015) Off-target effects in CRISPR/Cas9-mediated genome engineering. Mol Ther Nucl Acids 4:4
Zhang D, Li Z, Li JF (2016a) Targeted gene manipulation in plants using the CRISPR/Cas technology. J Genet Genomics 43:5. https://doi.org/10.1016/j.jgg.2016.03.001
Zhang Y, Yu H, Yang X, Li Q, Ling J, Wang H, Gu X, Huang S, Jiang W (2016b) *CsWRKY46*, a WRKY transcription factor from cucumber, confers cold resistance in transgenic-plant by regulating a set of cold-stress responsive genes in an ABA-dependent manner. Plant Physiol Biochem 108:478–487
Zhang Y, Bai Y, Wu G, Zou S, Chen Y, Gao C, Tang D (2017) Simultaneous modification of three homoeologs of *ta EDR 1* by genome editing enhances powdery mildew resistance in wheat. Plant J 91(4):714–724

Zhang Y, Li D, Zhang D, Zhao X, Cao X, Dong L, Liu J, Chen K, Zhang H, Gao C, Wang D (2018a) Analysis of the functions of *ta GW 2* homoeologs in wheat grain weight and protein content traits. Plant J 94(5):857–866

Zhang Y, Li D, Zhang D, Zhao X, Cao X, Dong L, Liu J, Chen K, Zhang H, Gao C, Wang D (2018b) Analysis of the functions of *TaGW2* homoeologs in wheat grain weight and protein content traits. Plant J 94(5):857. https://doi.org/10.1111/tpj.13903

Zhang A, Liu Y, Wang F, Li T, Chen Z, Kong D, Bi J, Zhang F, Luo X, Wang J, Tang J (2019) Enhanced rice salinity tolerance via CRISPR/Cas9-targeted mutagenesis of the *OsRR22* gene. Mol Breed 39:1–10

Zheng M, Zhang L, Tang M, Liu J, Liu H, Yang H, Fan S, Terzaghi W, Wang H, Hua W (2020) Knockout of two *Bna MAX 1* homologs by CRISPR/Cas9-targeted mutagenesis improves plant architecture and increases yield in rapeseed (*Brassica napus* L.). Plant Biotechnol J 18(3):644–654

Zhou J, Peng Z, Long J, Sosso D, Liu BO, Eom JS, Huang S, Liu S, Vera Cruz C, Frommer WB, White FF (2015) Gene targeting by the TAL effector PthXo2 reveals cryptic resistance gene for bacterial blight of rice. Plant J 82(4):632–643

Zsögön A, Čermák T, Naves ER, Notini MM, Edel KH, Weinl S, Freschi L, Voytas DF, Kudla J, Peres LEP (2018) *De novo* domestication of wild tomato using genome editing. Nat Biotechnol 36(12):1211–1216

Zuo E, Sun Y, Wei W, Yuan T, Ying W, Sun H, Yuan L, Steinmetz LM, Li Y, Yang H (2019) Cytosine base editor generates substantial off-target single-nucleotide variants in mouse embryos. Science 364(6437):289–292

Plant Genome Editing for Enhanced Biotic Stress Tolerance Using the CRISPR/Cas Technology

Manalisha Saharia, Gargee Dey, Himasri Devi, and Barasha Das

Abstract Plants are continuously attacked by a wide range of disease-causing microorganisms, which damage crops and threaten agricultural sustainability. To deal with this situation, crop varieties that are resistant to biotic stresses have been created by altering their genomes. Plant genome editing with designed nucleases makes it easier to characterize and quantify cells as well as to understand their structural and functional dynamics. The scientific community is well aware of the clustered regularly interspaced short palindromic repeat (CRISPR)/Cas (CRISPR-associated system) as a highly effective tool for genome editing. It is an RNA-guided repetitive DNA spacer sequence that is primarily found in archaea and bacterial cells in the form of adaptive immunity. The physicochemical and functional characterization of these gene-centric CRISPR/Cas genetic scissors, which has revolutionized microarray's limitations, has benefited the production of stress-tolerant plant varieties. The class 2 CRISPR-Cas system's Cas9 and Cas12 proteins cut double-stranded DNA, whereas Cas13 targets mRNA and promises a repair mechanism to accurately manipulate the targeted gene. The interdependent basic links of the redesigned heteroduplex structure along with its stochastic omic studies provide a complete understanding and redefining of the gene ontology (GO) for genetic engineering. This chapter will discuss RNA-guided endonuclease-induced technologies and data-driven from the interactomics of plant cells to manage the complex regulatory and metabolic processes in plants to achieve resistance against biotic stress.

Keywords PAM · CRISPR/Cas technologies · SHERLOCK · NHEJ · HGR · Cas9 · Cas12 · Cas13 · Biotic stress

---

M. Saharia (✉) · G. Dey · H. Devi · B. Das
Post Graduate Department of Botany, Madhab Choudhury College, Barpeta, Assam, India

Gauhati University, Guwahati, Assam, India

J.-T. Chen, S. Ahmar (eds.), *Plant Genome Editing Technologies*, Interdisciplinary Biotechnological Advances,
https://doi.org/10.1007/978-981-99-9338-3_7

1 Introduction

Due to the sessile nature of plants, they are continuously exposed to a wide range of biotic and abiotic factors. In response to such adverse circumstances, plants have developed immense signaling pathways following altered gene expression and transcriptional changes to establish cellular stability and homeostasis, with physiological consequences (Pandey et al. 2015), which is significantly impacting plant growth and global food production as a whole. Biotic stresses in plants are caused by microorganisms, insects, and viral entities and are challenging to control due to the continuous evolution of new pathogen biotypes; high variations in pathogen inoculum, infection, and proliferation rates; and constant interactions and facilitation through abiotic stresses. For instance, pathogen spread has been reported to increase through temperature rise (Suzuki et al. 2014). Certain plants have developed complex gene regulation mechanisms to endure biotic stress. Potato plant has shown biotic defenses through activation/inactivation of nine transcription factors (TFs) from a family of 67TFs, triggering numerous metabolic pathways that allow plants to produce defensive metabolites, proteins, hormones, and/or transcriptional and posttranscriptional modifications. Marker-based breeding strategies have also been used to integrate novel resistance genes into numerous plant species. Resistant cultivars of rice have been produced using a combination of microsatellite markers along with marker-assisted selection (MAS) techniques against rice blast disease (Miah et al. 2013). However, in recent years, resistant crop varieties have been produced using reverse genetics, which offers an understanding of gene function by observing phenotypic changes generated from target modifications in their genomes. Advanced omics studies have been used to perform genome editing involving specific modifications in the genome at target sites to introduce desired alterations to the DNA sequence. The huge breakthrough in genome editing techniques came after the creation of the technology involving site-specific cleaving of DNA (Li et al. 2022).

Site-specific nucleases (SSNs), which are typically created to bind and cleave a particular nucleotide sequence, are used in genome modification to introduce double-stranded breaks (DSBs) at or close to the target location. Zinc finger nucleases (ZFNs), RNA interference (RNAi), transcription activator-like effector nucleases (TALENs), and Cas proteins are the few major groups of SSNs. TALENs and ZFNs were the primary tools used in early genetic modification studies. RNAi technology is used to modify crops to make them resistant to environmental stress. However, rather than inhibiting the targeted genes, this approach often results in their downregulation (Mushtaq et al. 2018). As scientists and researchers looked for approaches that should be effective, precise, stable, and easy to modify as well as have the ability to fix the issues with earlier methods, the CRISPR-Cas strategy came to light. The clustered regularly interspaced short palindromic repeats (CRISPR) along with CRISPR-Associated Protein 9 (Cas9) RNA-guided DNA endonuclease recognizes the process through base pairing between the single guide RNA (sgRNA) and the target DNA sequence and performs a robust targeted double-strand break (DSB) in the aimed DNA sequence with the assistance of a protospacer-adjacent motif (PAM) sequence. The host cell then proceeds to repair these DSBs employing its diverse repair mechanisms that include nonhomologous end-joining (NHEJ), micro-homology-mediated end-joining (MMEJ), or through homology-directed repairs (HDR), causing gene knockout or insertion, deletion, and replacement of DNA fragments (Li et al. 2022). Contemporary genetic engineering research and functional genomics have focused on the improvement of several agronomic traits like biotic stress tolerance, abiotic stress tolerance, herbicide resistance, improved photosynthesis, and improved yield of economically significant plants through saturation mutagenesis by employing CRISPR/Cas9 editor technology. Unlike traditional breeding methods, CRISPR-Cas technology offers an immediate way to produce desired germplasms by deleting unwanted genetic elements responsible for negative traits or establishing gain-of-function mutations by precise genome editing. Cytosine base editing (CBE), adenine base editing, dual base editing, CBE-based precise DNA deletion, and prime editing are some of the precise CRISPR-Cas genome editing technologies (Zhu et al. 2020). Malnoy et al. (2016) used purified CRISPR/Cas9 ribonucleoproteins (RNPs) to perform targeted mutagenesis of MLO-7 (a disease-susceptible gene in grapevine) and DIPM-1, DIPM-2, and DIPM-4 (disease-susceptible genes in apple) to produce resistance toward powdery mildew and fire blight disease, respectively.

CRISPR/Cas9 has emerged as a powerful diagnostic tool with an immense potential to understand host-pathogen interaction, confer disease resistance, and perform targeted modifications in the genome of crop plants, thus revolutionizing the field of biotechnology. In this chapter, we make an effort to discuss the architecture, footprint, and operation of CRISPR/Cas technology and discuss how omics-assisted CRISPR/Cas modification in the plants is helping them to understand the regulatory mechanism behind improved resistance to biotic stressors.

2 CRISPR/Cas Technology

During his studies on the isozyme conversion of alkaline phosphatase in the 1980s, Yoshizumi Ishino sequenced a 1.7 kbp *E. coli* DNA segment traversing the region constituting the iap (isozyme of the alkaline phosphatase) gene. The analysis of this iap gene resulted in the discovery of an unusual but organized repetitive sequence downstream of the translation termination codon of this gene. However, at the time, no similar repeated sequences were found stored in the sequence databases. CRISPR sequence was eventually identified for the first time (Ishino et al. 2018). CRISPR was further identified in 1993 by coworkers while analyzing the sequences of the chromosomal regions of archaeon *Haloferax mediterranei*, allowing these halophilic organisms to survive in the high salt environment (Mojica et al. 1995), and it was concluded that these repeats were present throughout the prokaryotic organism (Mojica et al. 2005). Archaea (both Crenarchaeota and Euryarchaeota), lineages of both Gram-positive and Gram-negative bacteria, and other physiological and phylogenetic groupings of prokaryotes are among those organisms, possessing a substantial presence of CRISPR (Mojica et al. 2000; Jansen et al. 2002). CRISPR comprises a series of 20–50 bp repetitive sequences that are spaced by comparable-sized unique spacer sequences. It is an RNA-mediated nuclease system that has the potential to cleave foreign DNA into fragments. The CRISPR/Cas genome editing system confers adaptive immunity in approximately half of the bacteria and the majority of archaea against mobile genetic elements (MGEs) like bacteriophages and plasmids, first by acquiring the short segment of foreign DNA from the intruding plasmid or virus (called as protospacer) into the leader end of the CRISPR locus. A recognition element present in the leader sequence mediates this integration. The integration machinery might also rely on single-stranded regions of the CRISPR DNA that are made accessible during transcription. Transcription-mediated recombination ensures genome stability, and the following mechanism would connect the process of adaptation to the expression of CRISPR RNA, prioritizing the transcription of the most recent plasmid or virus spacers (Wiedenheft et al. 2012). Short CRISPR RNAs (crRNAs) are produced from precursor CRISPR RNA (pre-crRNA) transcripts that combine with Cas proteins to recognize and destroy the MGEs carrying the complementary sequence (protospacer) through a process that is almost identical to RNA interference in eukaryotes (Westra et al. 2012). Every CRISPR-Cas tools depend on crRNA or, in experimental CRISPR-Cas9 tools, on the guide RNA (gRNA) for guidance and site specificity. By exploiting the unique characteristics of different CRISPR-Cas systems, for example, Protospacer adjacent motif (PAM) specificity, protein content, and nuclease activity, a diverse range of CRISPR-Cas tools can be harnessed for programmable genome editing instead of binding to specific nucleic acid sequences. In addition, the development of mechanisms for detecting both on-target and off-target interactions has revolutionized the site-specificity of the CRISPR-Cas genome editing tools. The ability of these new CRISPR-Cas technologies to impact fundamental biological research by promoting gene regulation and enabling protein engineering besides causing precise nucleotide modifications portends new advancements in agriculture (Fig. 1).

2.1 The Cas Protein

The CRISPR/Cas (clustered regularly interspaced short palindromic repeats/CRISPR-associated proteins) system comprises two integral components: Cas proteins (which have the catalytic function) and the CRISPR locus (which acts as genetic memory) (Westra et al. 2012). The CRISPR-Cas system works by inserting pieces of foreign DNA (referred to as spacers) into CRISPR cassettes and then transcribing the spacer-containing CRISPR arrays and processing them to create a guide crRNA (CRISPR RNA), which is used to selectively target and cleave the corresponding virus or plasmid's genome. The various stages that involve the integration of the new spacers, processing of CRISPR locus transcripts, and cleavage of the target DNA or RNA employ numerous, extremely diverse Cas (CRISPR-associated) proteins. The majority of Cas proteins can be divided into two primary functional

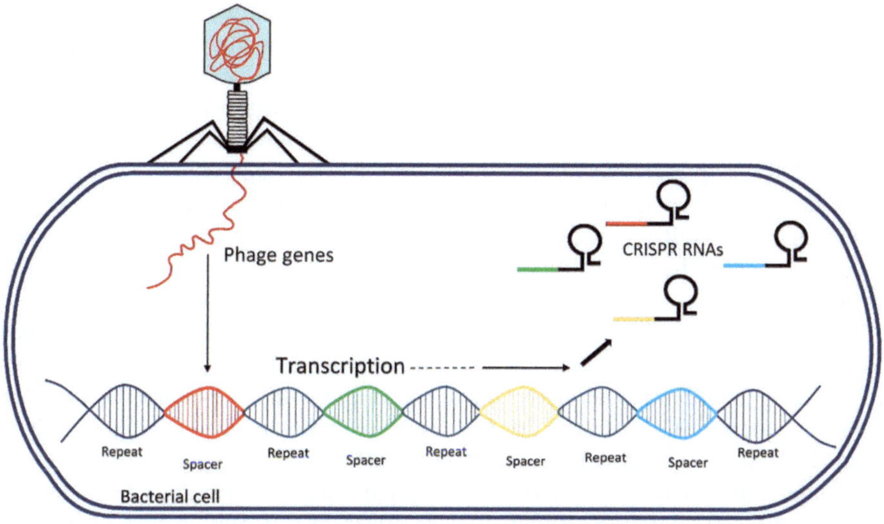

Fig. 1 The microbial defense system CRISPR-Cas9. The microbe's Cas enzymes identify and remove a piece of foreign DNA. The DNA fragment is inserted as a spacer by the Cas enzymes into the bacterial genome's CRISPR region. A Cas9 protein is connected to a spacer sequence after it has been translated. The CRISPR-Cas9 complex may identify the phage DNA sequence during reintroduction by the identical invader and cut it to stop full reinfection

modules: the adaptation module, which inserts genetic material into CRISPR arrays to produce CRISPR RNAs (crRNAs), and the effector module, which when directed by crRNA, targets and destroys invasive nucleic acids. The adaptation modules are made up of the two crucial proteins Cas1 and Cas2, and they are essentially consistent among CRISPR-Cas systems. The effector modules, however, exhibit a great deal of variation. The architecture of the effector modules serves as the basis for the most recent classification of the CRISPR-Cas systems, which separates them into two classes (Shmakov et al. 2015). The CRISPR-Cas genomic loci also exhibit remarkable diversity and complexity. The CRIPSR-Cas systems have been categorized using a multiple criterion technique based on the lineage of Cas1, the signature Cas genes (Cas1 and Cas2), the sequence resemblance of the Cas proteins, as well as the structural arrangement of the complex in the loci. CRISPR-Cas despite being a diverse system in terms of components and cas gene organization, operates in three definite stages: adaptation, expression, and interference.

The CRISPR array is enlarged during adaptation by the inclusion of an MGE-derived spacer genomic sequence, which replicates the repeat sequence. New spacers usually get incorporated at the figurehead-proximal end of the array, which entails duplicating the first repetition of the array. Spacer integration into the CRISPR cassettes is mediated by two proteins, Cas1 and Cas2, that are in the vast majority of the CRISPR-Cas systems (Yosef et al. 2012). These two proteins work together to produce a complex that is necessary for this adaptation process. The endonuclease function of Cas1 is necessary for spacer integration, but Cas2 appears to have a nonenzymatic role. The Cas1-Cas2 complex is the heavily conserved "information processing" component of the CRISPR-Cas that functions independently of the remainder of the system.

During the expression stage, the CRISPR loci are transcribed from an upstream promoter located in the AT-rich region of the leader sequence, and the resulting pre-CRISPR RNA (pre-crRNA) is processed into short crRNA called guide crRNA through double cleavage, one within the spacer and the other within the repeat, by either an exclusive RNA endonuclease complex or through an alternate mechanism that necessitates bacterial RNase III and an added RNA species (Deltcheva et al. 2011).

CrRNAs serve as guides for the Cas proteins during the interference stage as they identify and bind to complementary gene sequences in invading MGEs. Cas proteins under the direction of crRNAs and forming a complex silence the invading foreign sequences. It is believed that Cas9 is the only protein involved in the crRNA-guided silencing of foreign DNA (Jinek et al. 2012).

A highly diverse range of CRISPR-Cas systems, each one having a distinct set of components and operating principles, have arisen in different species of bacteria and archaea due to the co-evolution of the immune systems and the selective pressures imposed by the invading MGEs. CRISPR-Cas system is mainly divided into two main classes (Class 1 and Class 2), six types and loosely split into at least 29 subtypes, constructed based on the diversity of cas genes and the characteristics of the interference complex. Class 1 includes the Types I and III systems which share several defining characters: a large multi-Cas effector protein complex mediates the process of gene silencing, and the rare Type IV has rudimentary CRISPR-Cas loci without the adaptation module, whereas the Class 2 effector modules comprise the Types II, V, and VI systems that constitute a single effector protein, Cas9 (of Type II), Cpf1 (of Type V), and Cas13 (of Type VI) (Jinek et al. 2012). Cas9 requires a transactivating cRNA (tracrRNA) to recognize and cleave complementary nucleic acids through base pairing with the repeats in the pre-crRNA transcript.

Types I and III exhibit complex architectures, with a backbone made up of paralogous RAMPs (Repeat-Associated Mysterious Proteins), like Cas7 and Cas5, which contain the RRM (RNA Recognition Motif) fold and supplementary "large" and "small" subunits. The large subunits comprise Cas8 in Type I and Cas10 in Type III, with a loosely attached Cas6, whereas the small subunit is often present in many copies and engages with the crRNA backbone coupled to Cas7 (Koonin et al. 2017).

Types II, V, and VI of Class 2 CRISPR effectors have their respective CRISPR-Cas loci organized more simply and uniformly compared to those of Class 1. The Cas9 effector protein of the type II CRISPR-Cas system comprises two nuclease domains, an HNH (McrA-like fold) domain and a RuvC-like (RNase H fold) domain for target cleavage of DNA. The multifunctional activity of this protein makes it a significant genome editing tool. The Cpf1 effector protein of type V contains the RuvC domain but not an HNH domain (Schunder et al. 2013; Shmakov et al. 2015) and is represented as an RNA-guided endonuclease which cleaves the target DNA by making a staggered cut.

The Cas proteins are anticipated to play a role in psiRNA biogenesis, psiRNA-mediated resistance to invaders, as well as numerous aspects of CRISPR gene locus maintenance, including the addition of further invader-derived elements in the reaction against infection. Although there is substantial evidence that CRISPR RNAs and Cas proteins work to silence potential invaders in prokaryotes, the CRISPR-Cas pathway's effector complexes and silencing mechanisms are yet unclear (Hale et al. 2009).

2.1.1 Cas9

Cas9 protein is a DNA endonuclease with multiple domains and functions. Base editing precision has consistently been a challenge for genome modification technologies. By integrating catalytically inactive Cas9 variant, dCas9 (dead Cas9), and Cas9 nickase to specialize in deaminase domains and modify particular loci, attempts are being made to increase the accuracy of gene editing tools (Eid et al. 2018). A single DNA endonuclease, Cas9, is used in type II CRISPR systems to recognize double-stranded DNA targets and slice every strand of DNA using a unique nuclease motif (HNH or RuvC) (Garneau et al. 2010; Gasiunas et al. 2012). Cas9 has two lobes: a recognition (REC) lobe and a nuclease (NUC) lobe, as the crystal structure. The REC lobe may be split into three distinct sections: the REC-I domain, the REC-II domain, and a lengthy connecting helix known as the bridge helix. The HNH, RuvC, and PAM-interacting (PI) domains collectively comprise the NUC lobe (Nishimasu et al. 2014; Jiang and Doudna 2017). In an engineered system, a CRISPR locus produces a brief CRISPR RNA (crRNA) that binds to a protospacer-associated motif site (PAM) with an additional sequence on the intended genome (protospacer). In the case of *Streptococcus pyogenes*, the PAM is a traditionally 5′-NGG-3′ nucleotide sequence that needs to be present for Cas9 to specifically recognize and adhere to targeted DNA (Wright et al. 2016) (Fig. 2).

While the RuvC domain cuts the noncomplementary strand, the HNH domain separates the complementary strand. The short-conserved sequence PAM, which comes just after the corresponding sequence to the crRNA at position 3′, is also necessary for Cas9 to function as a double-stranded endonuclease (Nishimasu et al. 2014).

An extra tiny noncoding RNA, known as the trans-activating crRNA (tracrRNA), base pairs along with the repetitive sequence in the crRNA to create a special dual-RNA hybrid structure during this silencing process (Deltcheva et al. 2011). The DNA having an analogous 20-nucleotide (nt) sequence that has its target nearby PAM will be cut by Cas9 under the guidance of this dual-RNA guide (Jinek et al. 2012; Gasiunas et al. 2012). The technique is made simpler while still keeping fully functioning Cas9-mediated sequence-specific DNA cleavage by using a chimeric synthetic single-guide RNA (sgRNA) that merges tracrRNA and the crRNA into a single RNA transcript (Jinek et al. 2012). This streamlined CRISPR-Cas9 system may be intended for targeting almost every DNA structure of interest in the entire genome and further induce a site-specific blunt-ended double-strand break (DSB)

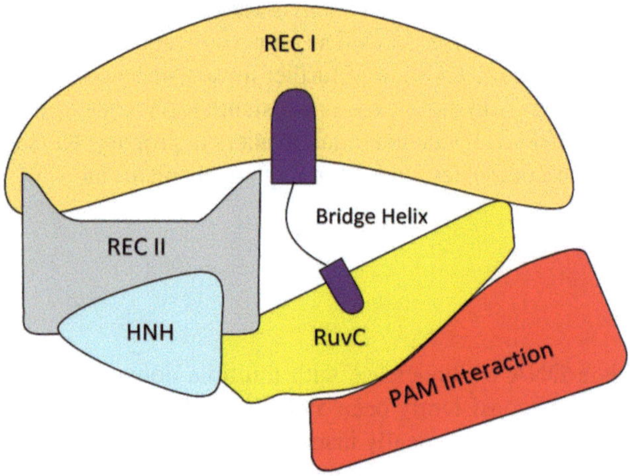

Fig. 2 An overview schematic of the lobes and domains that make up the Cas9 nuclease's structure. REC I is in charge of binding up the guide RNA. Single-stranded DNA is cut by the nucleases HNH and RuvC domains. PAM specificity is conferred by the domain PAM-interacting. Initiation of the cleavage activity after binding of target DNA depends on the bridge helix

by altering the guide RNA sequence (spacer) inside the crRNA (Jinek et al. 2012). The Cas9-induced DSB is then either modified by prone to errors, nonhomologous end joining (NHEJ) (Lieber 2010), which results in minor arbitrary insertions and/or deletions to the point of cleavage, or by wide-fidelity homology-directed repair (HDR) (San Filippo et al. 2008), which modifies the genome precisely at the point of the DSB by employing a homologous repair template.

2.1.2 Cas12

A separate CRISPR/Cas engine that permits the controlled stimulation of DSBs with efficiency similar to SpCas9 was characterized about 2 years ago. This CRISPR/Cas12 system is a class 2, type V CRISPR effector and has been allocated with five more RNA-guided DNA-targeting subgroups (Koonin et al. 2017) that shares certain similar traits with the CRISPR/Cas9 system yet differs significantly, making it a different tool for applications that involves genome editing (Zetsche et al. 2015; Shmakov et al. 2017). Effectors linked with each of these subtypes have been categorized as Cas12 family variations based on similarities in domain organization. The Cas12a (previously known Cpf1) and Cas12b (previously known C2c1) have undergone extensive structural and functional characterization, while the activity of Cas12c (previously known as C2c3) has not been established. Two other Type V effectors, Cas12d (formerly known as CasY) and Cas12e (formerly called CasX), have shown RNA-dependent DNA interference activity through in vivo trials (Dong et al. 2016; Yamano et al. 2016; Gao et al. 2016; Liu et al. 2017a; Shmakov et al. 2017; Swarts et al. 2017). Cas12 attaches to a crRNA and uses a T-rich, up-streamed PAM sequence, i.e., 5′-TTN-3′ (in which N can be adenine, guanine, or cytosine) to split dsDNA targets (Zetsche et al. 2015). With just one RuvC nuclease domain, Cas12a has the capability of cleaving both the target and nontarget strands (Swarts et al. 2017; Swarts and Jinek 2019). Cas12a is frequently employed as an adaptable genome-engineering tool because it demonstrates strong nuclease activity in eukaryotic cells (Zetsche et al. 2015). Due to the absence of an HNH domain, which is comparable to the RuvC domain, the Cas12a properly stands out for having a special fold domain. Change in the catalytic residue in the RuvC domain of Cas12a in the microbe *Acidaminococcus* sp. hindered the cleaving of target and nontarget strands. The positively charged central channel associated with a nuclease (NUC) region may control the cleaving of the target strand (Yamano et al. 2016). Swarts et al. (2017) explained that despite having a NUC domain that controls target strand incision activity and being comparable to Cas12a in that it excludes the HNH domain, the molecular framework of the NUC domain in Cas12b is quite distinct from that of the NUC domain in Cas12a. Both Cas12a and Cas12b break DNA strands using the RuvC-like domain following substantial conformational changes brought on by the initial cleavage of the nontarget DNA strand. *Francisella novicida* mutations that affected the NUC domain led to insufficient deactivation of the desired strand cleaving. This raises the notion that the RuvC-like domain, rather than the NUC domain which had been previously hypothesized, governed DNA cleavage activity. Given that complementary overhangs could change the repair from NHEJ to HR, the CRISPR/Cas12a system may be a crucial tool for plants to encourage the initiation of genomic modifications via HR. Additionally, even when NHEJ-based tiny InDels are created, subsequent cleavage can continue to take place as it happens far away from the PAM and seed area, which may ultimately trigger HR.

2.1.3 Cas13

The RNA-guided endonuclease enzyme known as Cas13 is a component of the CRISPR-Cas immune system found in bacteria. Abudayyeh et al. (2017) at the Broad Institute of MIT and Harvard made the initial discovery of it. The Cas13 evolutionary relationships, traits, and functional characterization are used to additionally categorize this framework into six subtypes: in recent years, VI-X (Cas13X) and VI-Y (Cas13Y) have joined the ranks of VI-A (Cas13a previously known as C2c2), VI-B (Cas13b, C2c4), VI-C (Cas13c, C2c7), and VI-D (Cas13d) (Shmakov et al. 2015, 2017; Smargon et al. 2017; Yan et al. 2018; Xu et al. 2021). The crRNA recognition lobe (REC) along with a nuclease lobe (NUC) make up the bilobed effector polypeptide, cas13, which has distinct biochemical properties from the REC and NUC lobe of cas9 and cas12. The two nucleotide-binding domains of the NUC lobe, namely, higher eukaryotes and prokaryotes nucleotide-binding domains (HEPN domains), contain a Helical-2 domain insertion in tandem with a Linker/ Helical-3 domain. The REC lobe is made up of an N-terminal domain (NTD) and a

Helical-1 domain that play a role in binding as well as in recognizing crRNA (Gupta et al. 2022). Contrary to Cas9, where the DNase cleavage activity was found to be carried out by the RuvC and HNH domains, and Cas12, where the RuvC domain was the sole domain that was found to be involved in cleaving, Cas13 lacks homology with either of these cleavage enzymatic domains and instead has an additional pair of HEPN domains that are solely in charge of implementing the process of RNase cleavage activity (Tang and Fu 2018). Every Cas13 protein comprises two RNase activities that are enzymatically distinct, including the conversion of pre-crRNA into mature and effective crRNA, whose activity is catalyzed by a unique domain called the Helical1 domain, and the second activity is the destruction of target RNA through the HEPN domains (Smargon et al. 2017). The HEPN catalytic site of active Cas13a lies on the outer surface and causes unspecific cleavage of RNAs in addition to the target RNA during the repair process, suggesting that Cas13 causes programmed cell death or systematic inhibition in the natural system (Liu et al. 2017b). Using diverse RNA technologies, including RNA editing, RNA targeting, RNA interference, and RNA detection, Type VI CRISPR/Cas systems offer a variety of applications across different organisms. The most prevalent type of plant viruses are RNA viruses, and Cas13's capability to target RNA is applicable to treat these viruses as well as DNA retroviruses that proliferate via an RNA intermediate during their lives. Cas13 is capable of being utilized in its natural state during its posttranscriptional repression, for cleaving the targeted RNA, especially knocking down the particular transcript. Though DNA-level transcriptional control has little effect on any particular isoform, Cas13 makes it easier to target particular splicing isoforms. In this manner, pathogenic or aberrant splicing isoforms have the potential to be abolished without impacting transcripts that are of the wild type (Mahas et al. 2019). One of Cas13's most intriguing applications is its potential as a diagnostic tool. Researchers developed a technique called SHERLOCK (Specific High-sensitivity Enzymatic Reporter UnLOCKing) that uses Cas13 to pinpoint specific RNA sequences in a sample. By combining isothermal amplification via recombinase polymerase amplification (RPA) with the CRISPR and CRISPR-associated (CRISPR-Cas) RNA-guided endoribonuclease, Cas13 was previously used for an assortment of RNA-targeting purposes metabolically in cells, whereas the newly established nucleic acid detection platform called SHERLOCK offers mobile, programmed, and quick nucleic acid recognition (Abudayyeh et al. 2016). SHERLOCK has shown promise in locating genetic mutations, COVID-19-like viral infections, and cancer biomarkers.

2.2 *Nucleic Acid Repair Mechanism*

Cell survival and genome integrity are gravely impacted by DNA double-strand breaks (DSBs). DSBs are generated from extraneous agents like mutagens, carcinogens, ionizing radiations, or internally during normal cellular metabolism (Uematsu et al. 2007). When the endonuclease-based system of CRISPR-Cas recognizes the specifically targeted locus of the nucleic acid of the eukaryotic organism, the Cas-sgRNA combination causes a DNA double-strand break. The DSBs result in lesions that get counteracted by the cell by activating the cell's repair system. The repair mechanism mostly follows two pathways: homology-directed repair (HDR) and nonhomologous end joining (NHEJ). Homologous recombination (HR) is common in cell stages with double chromosome copies, whereas nonhomologous end joining (NHEJ) involves re-ligating the DSB that frequently results in the loss or addition of a few nucleotides, or "indels" (Yang et al. 2020).

2.2.1 Homology-Directed Repair (HDR)

Homology-directed repair (HDR) is utilized to precisely repair DNA double-strand breaks (DSBs). It depends on the presence of a homologous DNA template to direct the repair process, often a sister chromatid or an exogenously supplied DNA molecule. HDR is thought to be a more precise process for DSB repair. This pathway is utilized for precise genome editing and is primarily active during the late S and G2 phases of the cell cycle. In these stages, the cell is equipped with the parts and machinery needed to effectively use the HDR pathway for repairing DNA double-strand breaks (DSBs) (Puchta and Fauser 2014). Genomic sequence insertions are executed and can be adjusted to accurately establish selection markers, point mutations, fuse affinity tags, remove specified sequences, or introduce heterologous genes. A small percentage of HDR events involve crossovers, which can either stabilize or destabilize the genome. Gene duplications, inversions, translocations, and deletions are often triggered by crossing over between repetitive regions on sister chromatids. Multiple genes have been introduced using this method at specific genomic regions (Van Vu et al. 2019) (Fig. 3).

Resection of the DNA end in a particular direction, which results in the development of a single-stranded DNA (ssDNA) overhang, is a crucial step in the HDR

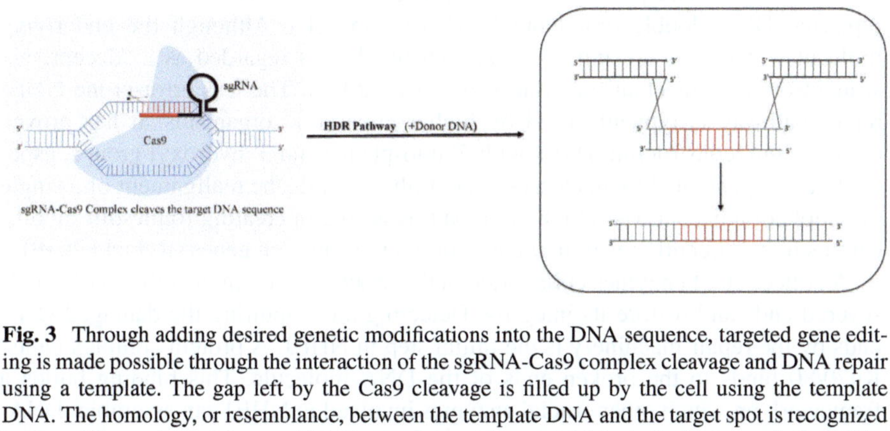

Fig. 3 Through adding desired genetic modifications into the DNA sequence, targeted gene editing is made possible through the interaction of the sgRNA-Cas9 complex cleavage and DNA repair using a template. The gap left by the Cas9 cleavage is filled up by the cell using the template DNA. The homology, or resemblance, between the template DNA and the target spot is recognized by the cell's repair machinery, which then makes precise use of this similarity to repair the DNA

process for fixing DNA double-strand breaks (DSBs) (Symington 2016). The MRN (MRE11-RAD50-NBS1) protein complex, which also serves as a scaffold for enhancing the cellular signaling response to the DSB, is responsible for starting the resection process. The resection process is initiated by MRN, which also recruits the protein CtIP. The NBS1 subunit then produces brief tails of single-stranded DNA (Symington and Gautier 2011). Exo1 and the DNA replication ATP-dependent helicase/nuclease DNA2/bloom syndrome protein (BLM) complex are two enzymes that enter the picture after the initial resection (Garcia et al. 2011). Together, they eliminate DNA over a wider area, which causes a 3′ ssDNA tail to develop. Replication protein A (RPA) quickly binds to and shields the 3′ ssDNA overhang to stop degradation and speed up the repair procedure. The following stages are aided by other proteins referred to as recombination "mediators," such as BRCA1, BRCA2, and partner and localizer of BRCA2 (PALB2). Then, RAD51 homolog 1 (RAD51), a DNA repair protein that creates prolonged nucleoprotein filaments on the ssDNA, takes the place of RPA (Bhat and Cortez 2018). In the following phase of HDR, these RAD51 nucleoprotein filaments are extremely important. They allow for homology searches and make it easier for the ssDNA to invade a section of homologous DNA that matches, resulting in the displacement loop (D-loop) structure. After the development of the D-loop, depending on the details of the repair and the presence of one or two Holliday junctions, at least three possible sub-pathways may develop (West 2009). Finally, a group of enzymes known as resolvases break down the resolution junction, which is created during the repair process. These enzymes complete the repair procedure and return the broken chromosome to its initial, undamaged state. The sister chromatid or foreign DNA that complements the damaged DNA serves as a starting point for the HDR repair process and provides the sequence data required for HDR-based precision editing (Heyer et al. 2010).

2.2.2 Nonhomologous End Joining (NHEJ)

Nonhomologous end joining (NHEJ) represents another remarkable approach for repairing DNA double-strand breaks (DSBs) in cells. Although the underlying molecular processes are intricate and controlled, it is regarded as a "deceptively straightforward" mechanism (Maruyama et al. 2015). The less error-prone NHEJ repair pathway is typically used by higher eukaryotic organisms. It has proved effective in reconstructing DSBs with 5′ phosphates and 3′ hydroxyl groups, especially those generated by nucleases. NHEJ often entails the realignment of a single or couple of adjacent bases to direct repair, resulting in creating frameshift modifications that frequently result in the loss or gain of function genes (Rybicki 2019).

A collection of enzymes collaborate in the process to rejoin the DNA molecule's severed ends and restore its integrity. Detecting and capturing the damaged DNA ends by the repair machinery is the initial step in NHEJ. A protein complex called Ku70/80 binds to the broken ends of the DNA molecule. This binding builds a framework on which other crucial enzymes involved in NHEJ can assemble. DNA-dependent protein kinase catalytic subunit (DNA-PK_{CS}), one of these enzymes, is

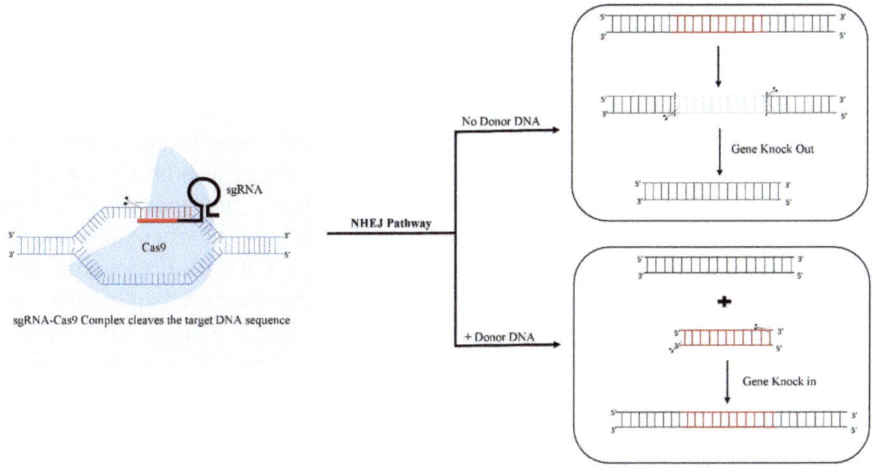

Fig. 4 Nonhomologous end joining (NHEJ) is a mechanism that can occur after the Cas9 enzyme cleaves the target DNA sequence. With or without the involvement of a template, broken DNA ends are joined directly, which might result in minor insertions or deletions in the DNA sequence

drawn to the damaged DNA by the DNA-Ku scaffold. The formation of a complex known as a synaptic complex, which connects the two ends of the DNA, is one of the many activities of DNA-PK_{CS} (Uematsu et al. 2007). Before the DNA double-strand break can be fully repaired, any non-lightable DNA termini (unjoinable ends) must be processed. The Ku and DNA-PK_{CS} complex catch and hold the DNA ends together after they have been cut. Nucleases and polymerases are examples of particular enzymes that can eliminate or fill in single-stranded, incompatible overhangs found at the broken ends. Nucleotides, the building units of DNA, can occasionally be lost during the repair phase of this step in the NHEJ process. The ligation, or joining, of the processed DNA ends is catalyzed by the ligase IV/XRCC4 complex, which is another essential component in NHEJ (Maruyama et al. 2015). A recently identified protein known as XLF/Cernunnos can accelerate this ligation reaction. The DNA molecule is then re-ligated once the ends are close together. To close the gap and restore the original DNA sequence, the repair mechanism orchestrates the joining of the broken ends (Fig. 4).

3 Omics-Based CRISPR/Cas-Mediated Plant Genome Editing

Omics-assisted CRISPR/Cas modification refers to an integrated approach that utilizes omics technologies, such as genomics, transcriptomics, proteomics, and metabolomics, in conjunction with the CRISPR/Cas genome editing system that allows for a comprehensive understanding of the regulatory mechanisms and dynamics of cellular systems which ultimately influence the phylogenetic advancement of an organism, improvement in disease resistance, increase in stress tolerance level, increase in crop production, etc. (Scheben et al. 2017). Researchers can discover a lot about how genetic changes affect a cell's or organism's general functioning by combining CRISPR/Cas-mediated editing systems with omics tools. This conjunction has helped in analyzing regulatory networks, gene expression patterns, protein interactions, and metabolic pathways thoroughly and methodically (Sun et al. 2022). For an organism's phylogenetic development, the regulatory mechanisms and dynamics discovered using omics-assisted CRISPR/Cas modification can be extremely important (Razzaq et al. 2021). Researchers can explore the functions of specific genes or genetic components throughout the evolutionary processes by carefully changing the genomic material. Our comprehension of the processes behind species diversification and adaptability is improved by this information. By identifying key genes or regulatory elements involved in disease susceptibility or resistance, researchers can use CRISPR/Cas-mediated editing to introduce targeted modifications that enhance the organism's immune response or disrupt pathogen interactions. This can lead to the development of crops and livestock with increased resistance to diseases, reducing the need for chemical pesticides or insecticides (Nelson et al. 2018). CRISPR/Cas modification with the use of genomics greatly enhances tolerance to stress in organisms. Researchers can pinpoint the essential genes or molecular networks involved in stress responses by analyzing the modifications in gene expression, protein profiles, and metabolic pathways under various stress situations. The ability of an organism to tolerate environmental conditions like drought, extremely high or low temperatures, or nutrient shortages can be improved through targeted editing utilizing CRISPR/Cas (Shriram et al. 2016). Omics-assisted CRISPR/Cas modification offers a potent toolkit for comprehending the dynamics and regulatory processes of biological systems. Progress in a variety of biological areas, such as phylogenetic understanding, disease resistance, stress tolerance, and crop development, is made possible by this integrated approach and understanding of the complex workings of genetic systems to advance medicine, biotechnology, and agriculture.

3.1 Genomics

Genomics refers to a set of methodologies and analytical methods to study the composition and function of genomes, which includes the complete set of genes. Gene regulation depends on how genetic components work together to produce a dynamic living entity. The term "genomics" was initially put forth by Tom Roderick in 1968 (Ramzi et al. 2020). Functional genomics has undergone technological developments that have made it possible to sequence the full genomes of various plant species and identify the genes that are related to stress in those crops. Genome-wide association studies have successfully identified stress-responsive genes and their accompanying genetic variants to address the complexities of these genes. A key goal of contemporary biotechnology is the creation of innovative crop varieties with high yields, resistance to both biotic and abiotic stressors, and improved nutritional value. Tools for modifying the genome have been used to alter plant immunity at various stages in a variety of crops. By editing the mildew resistance locus O (MLO) gene, for instance, the CRISPR/Cas9 technique has been used to provide resistance to powdery mildew in wheat genotypes (Wang et al. 2014). The potential of the CRISPR/Cas9 technique to confer resistance against geminivirus infection has been investigated through the introduction of sgRNA/Cas9. Successful establishment of geminivirus resistance has been achieved in both *N. benthamiana* and *Arabidopsis* (Ji et al. 2015). In the grape cultivar "Chardonnay," CRISPR-mediated editing targeting the L-donate dehydrogenase (IdnDH) gene resulted in a 100% occurrence rate of mutations. Currently, CRISPR/Cas9 technology is being used to edit the SWEET11, SWEET13, and SWEET14 genes, enhancing the resistance capacity of rice against bacterial blight disease (Oliva et al. 2019). In an experiment, de Toledo Thomazella et al. (2021) demonstrated that the inactivation of the *Downy mildew resistance 6* genes *(DMR6)* artificially resulted in the upregulation of the production of salicylic acid in the tomato plant, offering broad-spectrum resistance against a variety of diseases, including oomycetes and bacteria. He further engineered the tomato plants with small deletions in the SlDMR6–1 gene using the CRISPR-Cas9 technology, which resulted in a frameshift mutation and the premature end of the protein to develop a pathogen-resistant plant. Similarly, the *LATERAL ORGAN BOUNDARIES 1* transcription factor (*CsLOB1*) facilitates the proliferation of *Xanthomonas citri* spp., the causal agent of *Citrus canker* (Giraud et al. 2010). The CRISPR/Cas9 approach, however, is effective in creating resistance to plant viruses including the *Tomato yellow leaf curl virus (TYLCV)* and the *Bean yellow dwarf virus (BeYDV)* in numerous studies (Ji et al. 2015).

3.2 Proteomics

The term proteomics was first coined 20 years ago which is defined as the quantitative and qualitative investigation of total proteins expressed in a cell, tissue, or organism (Luan et al. 2018). Proteins play a crucial role in the plant's response to biotic stress because they are (1) directly involved in the development of new plant phenotypes by regulating physiological traits to adapt to environmental changes and (2) proteins play a crucial role in the execution of cellular functions and are essential for maintaining cellular homeostasis (Liu et al. 2019).

Usually, proteomics can be classified into four classes, viz., "structural proteomics," "functional and interaction proteomics," "sequence proteomics," and "expression proteomics." These classes of proteomics have different functional approaches toward different fields and help in understanding numerous biological processes, disease mechanisms, and medication discovery (Aizat et al. 2018). However, Dolgalev and Poverennaya (2021) explained that the utilization of CRISPR-Cas technology has already had a considerable positive impact on proteomics, particularly in the areas of gene knockout studies, gene tagging and localization, and protein modifications. Regarding gene knockout, CRISPR-Cas allows the targeted disruption or deletion of particular protein-synthesizing genes in a cell or organism. Scientists can evaluate the effects on protein expression, function, and interactions by removing a specific gene, revealing important information about the proteome. With regard to tagging and localization, using CRISPR-Cas, specific labels or tags can be inserted into the genome and coupled with relevant proteins. This method makes it easier to track and observe proteins inside cells or tissues, allowing researchers to investigate their subcellular localization and dynamics. Proteomic research can be conducted with improved accuracy and effectiveness by properly labeling proteins. For protein alterations, using CRISPR-Cas, specific alterations or mutations can be inserted into genes that code for proteins. Researchers are now able to examine how these alterations affect protein structure, function, and relationships. Scientists have learned more about a protein's function in cellular processes and disease pathways by changing particular amino acids or protein domains in that protein.

Biotic stress reactions are heavily reliant on complicated signaling networks and protein interactions. Understanding these intricate systems can be achieved through the combined use of proteomics and CRISPR-Cas technology. The proteins involved in these interactions can be altered or knocked out using CRISPR-Cas technology. This allows them to look into and confirm the functional effects of these relationships in more detail. By utilizing the capabilities of CRISPR-Cas, researchers can uncover important details about the underlying mechanisms of biotic stress and possibly pinpoint targets for the creation of cutting-edge methods to improve plant resilience (Dolgalev and Poverennaya 2021). Exploring protein-protein interactions, researching protein-DNA interactions, and developing biological cellular models are the three primary subcategories of current proteomics and CRISPR-Cas technology applications. Researchers can study the consequences of these alterations on protein structure, function, and relationships by altering certain genes responsible for proteins engaged in stress-related pathways. Protein-protein interactions are key players in the response to biotic stress, and they can be discovered by combining mass spectrometry with methods like co-immunoprecipitation and affinity purification. These methods are now more dependable and effective with the CRISPR-Cas integration (Dolgalev and Poverennaya 2021). Scientists moreover can precisely and methodically investigate protein-protein and protein-DNA interactions with the aid of CRISPR-Cas, to get a better grasp of the intricate mechanisms underlying the biotic stress response and open the door to creating focused strategies to increase plant resilience (Feng et al. 2013). One of the main uses of CRISPR-Cas technology is the construction of cellular models. This entails creating novel approaches for delving into ways genes function inside of cells. These techniques enable researchers to examine gene behavior in a natural setting, providing an improved understanding of the way genes function within pathogens. It has been demonstrated that CRISPR-Cas genome editing is a valuable technique for creating cellular models that enable the evaluation of the consequences of particular genes or mutations. In these study areas, observable traits and potential changes are evaluated using proteomic techniques in combination with other high-throughput techniques like RNA sequencing (Dolgalev and Poverennaya 2021). Loss-of-function studies, which entail deactivating a target gene, are commonly employed as an initial assessment of gene function (Wang et al. 2015). By generating a cell line that lacks that target gene, researchers can meticulously investigate the effects and establish the gene's significance and its relationship to the observed traits. In this process, proteomics plays a pivotal role, working in conjunction with RNA interference, to provide crucial functional insights into numerous protein-coding genes (Brioschi and Banfi 2018). Several examinations have considered the cell wall proteome, proteogenomics, phosphoproteome, and organellar proteome from the perspective of plant stress tolerance (Nakagami et al. 2012). In times of biotic stress, plants execute a complex defense mechanism that involves pattern recognition receptors (PRRs). These receptors are produced in the endoplasmic reticulum and then moved to the plasma membrane. Their function is to detect specific patterns (PAMPs, MAMPs, and DAMPs) found in pathogens, microbes, and molecules associated with virulence. Once these patterns are recognized, the plant initiates pattern-triggered immunity (PTI) as a defense response (Liu et al. 2019). The ability to research protein interactions and gene function has increased as a result of the integration of CRISPR-Cas technology into proteomics. Researchers have enhanced accuracy and dependability by adding CRISPR-Cas into current techniques. Additionally, new opportunities for researching gene activity in the context of cells' natural environments have emerged as a result of the development of novel techniques for creating cellular models.

3.3 Metabolomics

Studies on metabolomics concentrate on examining the net metabolites, which are the byproducts of gene expression and are very important in determining the phenotype of an organism. This is consistent with the "central dogma of molecular biology," which describes how genetic information moves from DNA to RNA to proteins to metabolites. By applying this idea, past research has used metabolomics to look into and comprehend metabolic changes brought on by diverse stresses. Metabolomics studies produce widespread information about organic acids, secondary metabolites, hormones, ketones, aldehydes, lipids, steroids, amino acids, etc. (dos Santos et al. 2017) and therefore have found applications in many fields, especially in stress tolerance activities in plants. Researchers can learn more about how organisms react to biotic stress at the molecular level and pinpoint specific metabolic pathways that can be altered by analyzing the targeted and nontargeted methods for endogenous and exogenous metabolites (Frédérich et al. 2016; Drapal et al. 2023). According to Parida et al. (2018), some metabolomic processes have moved toward advances in liquid chromatography-mass spectrometry, gas chromatography-mass spectrometry, high-performance liquid chromatography, direct injection mass spectrometry, and nuclear magnetic resonance. These methods have further elucidated stress tolerance mechanisms. Stressed environments experience increased levels of metabolites compared to non-stressed environments. A large number of researches have demonstrated the metabolic outline under stress environments in plants (Muthuramalingam et al. 2020). Plants undergo metabolic reprogramming in response to biotic stress to defend themselves and mount a suitable defense (Razzaq et al. 2022). Moreover, researchers can predict changes in metabolite profiles that occur in response to stress by using CRISPR-Cas9 to tweak particular genes involved in the plant's immune system. The Cas9 protein has been designed with codons coding for secondary metabolites such as phytoalexins, substances involved in cellular defense, and signaling molecules that help to understand how specific genes and pathways contribute to the plant's defense mechanisms and acquire insights into the interrelated metabolic networks. These changes were unique to the initial crop of modified plants, indicating that they might be transitory and not stably passed down to the next generations. Extensive whole-genome sequencing would be needed to validate their findings and get a deeper understanding. Researchers can determine changes in metabolite profiles that take place in response to stress, by using CRISPR-Cas9 to tweak particular genes involved in the plant's immune system and then using metabolomics (Razzaq et al. 2022). It further enables the identification of crucial metabolites and defense-related pathways in plants, which aids in the creation of tactics to strengthen plant resistance and raise crop yield. A vast variety of metabolites, including secondary metabolites like phytoalexins, chemicals involved in defense, and signaling molecules, can be detected using metabolomics. It has become possible to isolate particular metabolites that are involved in a plant's defense response by contrasting the metabolomes of genetically modified plants with those of wild-type or non-edited plants.

3.4 Transcriptomics

In simple words, transcriptomics refers to the study of the total RNA profile in a cell. According to Leisner et al. (2017), RNA profiling currently applies microarrays, sequential examination of gene expression, RNA sequencing, and digital profiling which can recognize various stress resistance-related applicant genes, concluding pertinent gene functions. Using CRISPR-Cas9 technology, transcriptomics can be very helpful in examining biotic stress. Researchers can learn more about the changes in gene expression that take place in plants under biotic stress and comprehend the underlying molecular mechanisms by combining transcriptomics with CRISPR-Cas9. For plant stress response, the accessible online directory gives rise to the entire crop genome's extensive transcriptomics (Zhang et al. 2019). A total of 770 unaffected transcripts with 53 divergent specific proteins were recognized as stressed in *Arabidopsis* (Rizhsky et al. 2004). By precisely editing particular genes involved in the plant's immune system and response to invaders, CRISPR-Cas9 enables precise genetic alterations in plants. This strategy is complemented by transcriptomics, which enables researchers to examine the entire set of

RNA molecules generated by the plant in response to biotic stress. Plants employ several defense mechanisms to defend themselves during biotic stress. Specific genes are either activated or suppressed during these defense reactions, changing the entire transcriptome (Jung et al. 2021). Researchers can find and characterize the genes that are differently expressed during biotic stress by using CRISPR-Cas9 to edit target genes. The genes that are up- or downregulated in response to stress can be identified by transcriptomic analysis. It enables scientists to pinpoint important signaling channels and regulators involved in the plant's defense response. Researchers can analyze the effect of the edited genes on gene expression patterns and uncover their involvement in the plant's response to biotic stress by juxtaposing the transcriptomes of genetically modified plants with those of wild-type or non-edited plants. Discovering new genes and regulatory components involved in the plant's pathogen defense can be aided by transcriptomics. Researchers can discover previously undiscovered genes that are active under stressful circumstances by studying the transcriptome data. The intricate molecular networks underpinning plant-pathogen interactions can be better understood with the use of this information.

4 Gene Alteration to Improve Biotic Stress Tolerance Using Cas Protein

Analysis of host-pathogen interactions and the discovery of the virulent gene can both be done using the CRISPR-Cas toolkit. Additionally, precise alterations to the targeted plant sequence, such as knock-ins, knock-outs, or replacements employing Cas endonuclease and a single-guided RNA (sgRNA), aid in the development of disease resistance. Researchers may be able to improve the plant's capacity to recognize and combat pathogens by discovering and altering critical genes linked to biotic stress response pathways. Genes that encode proteins involved in pathogen identification, signal transduction, or the synthesis of defense chemicals may need to be changed to do this. Altering defense-related genes that produce phytoalexins, antimicrobial peptides, or secondary metabolites is also possible with CRISPR-Cas technology. Increasing the production or accumulation of these substances through gene modification may aid in preventing or reducing the growth and spread of pathogens. Increasing the production or accumulation of these substances through gene modification may aid in preventing or reducing the growth and spread of pathogens. Genes that might function as repressive regulators of the plant's defense response can be targeted using CRISPR-Cas. Because CRISPR-Cas editing is so precise, site-specific changes can be made to target genes while minimizing unintended consequences for other genes or physiology disruptions in general. This focused strategy shows promise for creating crop types that are more resilient to biotic stress. However, instead of introducing dominant resistant genes (that leads to the co-evolution of resistance in pathogens), deleting host susceptibility genes using

CRISPR-Cas can be more effective in providing resistance in plants against biotic stress (Zhu et al. 2020).

4.1 Resistance to Bacteria Through CRISPR/CAS-Mediated Editing

CRISPR/Cas9 gene editing is a recently developed method that has raised concerns in most scientific domains with more than 7105 articles so far (Dort et al. 2020). Additionally, the application of this CRISPR/CAS modification approach is motivated by a desire to increase agricultural productivity through resistance to several plant pathogens. Bacteria are generally omnipresent and diverse, and their proliferation capability makes them easy to survive everywhere. Just a handful of all the bacterial species that exist on Earth are responsible for agricultural harm, which frequently manifests various disease signs (Schloss and Handelsman 2004). Some phytopathogenic bacteria pose a significant threat to agriculture, making it difficult to treat diseases like spots and rot and resulting in a significant decrease in crop yield. Controlling phytopathogenic bacteria is challenging, mostly due to undiagnosed infections that are asymptomatic and a lack of effective agrochemicals. Moreover, intricate methods have been developed by pathogens for interacting with hosts and take advantage of their genetic resources. These infections have developed abilities to effectively regulate host plant genes by acquiring specific signals. By doing this, they can successfully infect the plant and get through its defenses. The intricacy makes it more difficult to create effective plans for plant resistance to various diseases (Borrelli et al. 2018). Though plants have a mechanism for a defense that can identify specific signals and react appropriately to eradicate the pathogens, these resistance genes might sometimes fail leading to the successful establishment of infection.

CRISPR-Cas9 genome editing has become a potential strategy for plants to build defenses against bacterial infections. One illustration is the significant threat to economically valuable citrus plants by *Xanthomonas citri*, causing citrus canker. Utilizing CRISPR-Cas9 technology, particular genes in citrus plants are modified to strengthen their built-in defenses, especially by focusing on a specific gene involved in effector-triggered immunity (ETI), known as PthA4. When this effector-binding element (EBE) PthA4, associated with the promoter gene CsLOB1 (lateral organ boundaries 1), is altered, the host plant loses its ability to recognize and respond to bacterial effectors. However, it is crucial to note that this mutation has no discernible impact on the plant's phenotypic characteristics. Instead, the CRISPR/Cas9 modification made the host plant more susceptible to bacterial spread (Jia et al. 2016). In rice plants, the OsSWEET13 gene, which is assigned for transporting sucrose, has been found to have the potential to give resistance to the *Xanthomonas oryzae* pv. oryzae disease. The OsSWEET13 gene has been altered to improve its resistance to infection through the use of the CRISPR/Cas9 gene editing method.

The plant's defense mechanisms hence were bolstered to increase its resistance against *Xanthomonas oryzae* pv. oryzae infection, by introducing specific changes to the gene (Zhou et al. 2015). Currently, CRISPR/Cas9 technology is being used to edit the SWEET11, SWEET13, and SWEET14 genes, enhancing the resistance capacity of rice against bacterial blight disease (Oliva et al. 2019). In tomato plants, the downy mildew resistance 6 (*DMR6*) gene was silenced using the CRISPR/Cas9 technique. In the plant defense mechanism against pathogens, the *DMR6* gene performs as a negative regulator. Researchers were successful in preventing the specific gene from functioning in the tomato plant by targeting it with CRISPR/Cas9. The goal of this change was to strengthen the plant's defenses against the pathogen *Pseudomonas syringae*, which causes infections like downy mildew. The plant defense mechanisms hence were improved by inactivating the *DMR6* gene, possibly resulting in higher resistance to *Pseudomonas syringae* infection. This highlights the capability of CRISPR/Cas9 technology to enhance plant defenses against particular diseases (Langner et al. 2018).

A useful tool for editing the plant genome is CRISPR/Cas12, a protein related to Cas9. Another protein known as CRISPR-Cas13a has been discovered and investigated in addition to Cas9. Because it can precisely cleave single-stranded RNA (ssRNA), CRISPR-Cas13a is special. This characteristic enables it to specifically target and destroy bacterial RNAs, preventing bacterial development. The defense against phages, which are viruses that infect bacteria, is also provided by this mechanism of action. The ssRNA cleavage activity of CRISPR-Cas13a can be used by researchers to improve plant resistance to bacterial infections and possibly mitigate the negative impacts of phages. This increases the CRISPR technology's capabilities in plant biotechnology and creates new opportunities for crop protection and enhancement (Meeske et al. 2019). CRISPR-Cas13a systems come in four different subtypes, according to recent research. These systems were discovered in 11 strains from 6 distinct *Leptotrichia* bacterial species. The RNA-targeting properties of CRISPR-Cas13a, a component of the CRISPR immune system in bacteria, are well recognized (Watanabe et al. 2019). In their 2016 study, Abudayyeh ct al. (2016) concentrated on the CRISPR-Cas13a system from the *Leptotrichia shahii* strain (also known as *LshCaset13a*). It was discovered that this specific CRISPR-Cas13a system significantly inhibited bacterial growth. This finding implies that *LshCaset13a* is capable of efficiently locating and destroying bacterial RNA, which prevents bacterial development. The identification of this robust inhibitory effect emphasizes CRISPR-Cas13a's promise as a tool for preventing bacterial infections and managing bacterial populations. Many crops cause serious bacterial disease due to biotic stress that can directly affect agronomy and also challenge food security in the growing population; therefore, the application of CRISPR/CAS-mediated editing tool endows a new direction to improve the biotic stress like bacterial disease resistance in crop plants (Arora and Narula 2017).

4.2 CRISPR/CAS Mediated Fungal Pathogen Resistivity

Due to the intricate behaviors and genetic adaptability of pathogenic fungi, they are capable of easily damaging host plants. Once inside the host plant, these pathogens produce diseases like rot, smut, mildew, etc., making them a serious threat to crops (Fisher et al. 2012; Das et al. 2019). Mycotoxins, which are released by pathogenic fungi as secondary metabolites, are representative of major health problems that can be fatal to both humans and other animals and also to the growth, development, and general health of plants. To combat mycotoxin contamination in crops and decreased agricultural output, farmers and agricultural professionals have used different methods and techniques. The adoption of disease-resistant crop types, efficient pest and disease control techniques, and postharvest storage procedures are a few of these tactics. They have also included optimal crop rotation (Bebber et al. 2013). Advanced methods for detecting and managing mycotoxins are being developed through research in the fields of plant pathology and agricultural biotechnology by utilizing molecular techniques like DNA-based diagnostics, cutting-edge biocontrol agents, and genetic engineering techniques like CRISPR-Cas genome editing (Tyagi et al. 2018; Das et al. 2019; Dong and Ronald 2019). When a pathogen attacks a plant, it releases chemicals called effectors or avirulence factors. These chemicals act as signals and catalysts for the defense response in plants. R genes, also referred to as resistance genes, are certain genes in plant genomes that are essential for conferring pathogens against the disease. These genes create the R proteins, which play a role in identifying pathogens and reacting to their presence. After identifying the effectors, the R proteins initiate a cascade of signaling events inside the plant that activate the plant's defense mechanisms (Idnurm et al. 2017). However, the species-specific character of infections can limit the ability of R genes to provide long-lasting resistance. Due to the ability of pathogens to genetically mutate and adapt, they can circumvent the plant's defenses and render the R genes ineffective. An evolutionary arms race between plants and diseases may result from their ongoing interaction. Pathogens respond to the development of new defense mechanisms by plants through R genes by altering their genes or acquiring new genetic features that enable them to overcome the plant's defense (Joshi and Nayak 2008; Marraffini 2009).

CRISPR/Cas9 technology is used to target the PMR4 gene ortholog SlPMR4 to increase resistance against *Oidium neolycopersici*, a fungus that causes powdery mildew in tomato plants. By promoting callose deposition, the PMR4 gene is known to contribute to plant defense against powdery mildew. Accurate modification in the SlPMR4 gene in tomato plants using CRISPR/Cas9, and adding particular alterations or mutations, improved the PMR4 enzyme's activity or function (Tyagi et al. 2021). An intriguing result was produced by tandem mutation inactivating the PMR4 gene. It caused a significant increase in salicylic acid levels, which in turn triggered the host plant's hypersensitive response (HR). To confer resistance against powdery mildew disease caused by fungal pathogens, the CRISPR-Cas9 gene editing technique was utilized to modify the MLO (Mildew Resistance Locus O) locus, which is present in the host S gene. Three separate plant species—wheat, tomato, and grapevine—have undergone this modification. The MLO gene modification is intended to interfere with the pathogen and plant interaction, making it more difficult for the fungi to colonize and spread throughout the host plant (Malnoy et al. 2016; Nekrasov et al. 2017). Modification in particular to the MLO genes in various plant species using the CRISPR-Cas9 gene editing technology helped to make resistant cultivars that can self-pollinate and produce CRISPR-Cas9-free crops. Using the ribonucleoprotein (RNP)/CRISPR-Cas9 complex in protoplast culture, the SlMlo1 gene in tomatoes and the VvMLO7 gene in grapevines were both changed. Similar to this, the TaMLO-A1, TaMLO-B1, and TaMLO-D1 MLO genes in wheat were modified using CRISPR-Cas9, resulting in resistance to the pathogenic fungus causing powdery mildew. This strategy enables the emergence of resistant crops without interfering with the growth and development of plants, thus enhancing crop yield (Wang et al. 2016; Malnoy et al. 2016). Watermelons are among the many crop kinds that are significantly threatened by the vascular fungal disease fusarium. This disease has the potential to significantly reduce yields, from 30% to 80% of the total harvest. It interferes with the plants' capabilities for transporting water and nutrients by affecting their vascular system. Wilting, reduced development, and even plant death may result from this. Fusarium must be managed and controlled to reduce the negative economic and agricultural effects it has on the production of watermelons. Fusarium damage is minimized and agricultural yield is maintained using a variety of measures, including cultural practices, resistant cultivars, and disease management techniques (Martyn and Netzer 1991; Zhang et al. 2020). Recent research has shown that the Clpsk1 gene, which generates phytosulfokine (PSK), a particular pentapeptide hormone involved in controlling plant immunity, is crucial in enhancing watermelon's resistance to *Fusarium* wilt. Researchers have reported greater resistance to the disease without any negative impacts on the general health and development of the watermelon plants when the Clpsk1 gene has been knocked out using methods like CRISPR-Cas9. This discovery offers a viable strategy for creating watermelon cultivars that are *Fusarium*-resistant, providing a more sustainable and efficient way for combating this terrible fungal disease (Hammes 2016). Thus, using this CRISPR/CAS technology has significantly increased crop species' yields while reducing fungal disease susceptibility.

4.3 *Viral Resistance Using CRISPR/Cas Technology in Plants*

The threat posed by plant viruses to global agronomy is significant, since, according to research, phytopathogenic viruses are to blame for 10–15% of the decline in crop yield each year. This drop in plant productivity leads to a rise in food security issues. The majority of plant pathogenic viruses have the unique ability to spread from one plant to another through a vector, whose primary function is to survive on the plant and transmit the viruses from one plant to another (Hogenhout et al. 2008). Viruses are divided into two categories based on their genomic characteristic: DNA viruses and RNA viruses. Plant viruses are further divided into six major categories: reverse-transcribing viruses, negative-sense single-stranded RNA (ssRNA−) viruses, positive-sense single-stranded RNA (ssRNA+) viruses, and single-stranded DNA (ssDNA) viruses (Roossinck 2011; Roossinck et al. 2015). In addition to all of these viruses, the Gemini viridian DNA virus family, which carries ssDNA, hijacks plant machinery and replicates using the rolling circle method; as a result, many crops, including tomato, wheat, barley, and others, suffer significant crop yield losses that have an impact on the world economy. Since viruses lack their translational machinery, they take over the plant's system and employ transcription and translation factors to replicate themselves. Targeting the pathogenicity-related genes and altering the host susceptibility genes can be used to develop active resistance against viruses (DNA or RNA). Recently, a ground-breaking method to improve various plants damaged by viruses has emerged: genome engineering. Consequently, the crop species use CRISPR/Cas9 genome editing technology to defend against viruses (Baltes et al. 2015; Ali et al. 2015a). CRISPR/Cas9 is a prokaryotic molecular immune system that can protect against phages and invasive nucleic acids (via horizontal gene transfer), which impact DNA viruses in a different way than RNA viruses (Marraffini and Sontheimer 2008). Ali et al. (2015b) used CRISPR/Cas9 gene editing technology to alter the genome of *Nicotiana benthamiana* plants in their experiment. The goal was to provide resistance against several viral infections, including the *Merremia mosaic virus* (MeMV), *beet curly top virus* (BCTV), and *tomato yellow leaf curl virus* (TYLCV). The CRISPR/Cas9 system was able to make alterations in the plant's genetic makeup by specifically changing a few genes involved in the viral infection process. Improved resistance to the targeted viruses was produced by these genetic alterations. When exposed to the corresponding viruses, the transformed *Nicotiana benthamiana* crops showed a considerable decrease in symptom intensity and viral load (Baltes et al. 2015; Ji et al. 2015). Viral resistance was improved by successfully introducing mutations into the coding and noncoding regions of hypo-pathogenic viruses using CRISPR/Cas9. Tobacco leaf curl Kokhran virus (CLCuKoV) and tomato yellow leaf curl virus (TYLCV) are two specific instances of low pathogenicity viruses that have been addressed. With TYLCV, it has proven possible to lessen the impact of the virus in tomato plants by introducing precise mutations using CRISPR/Cas9. In comparison to unmodified plants, the transformed plants exhibit a lower vulnerability to TYLCV infection, resulting in a higher level of resistance. Furthermore, CRISPR/Cas9-mediated alterations to the viral genome have been introduced in the instance of CLCuKoV, which has reduced viral replication and symptom development in cotton plants. The altered plants exhibit improved resistance to pathogenic CLCuKoV strains, which pose a danger to cotton crops in some areas (Ali et al. 2016). According to a 2019 study by Kis et al. (2019), the use of CRISPR/Cas9 technology has successfully eradicated begomoviral infections that cause cotton plant leaves to curl. A mutated resistance plant against DNA viruses like the wheat dwarf virus (WDV) or the banana streak virus (BSV) was achieved by duplicating the CRISPR/Cas9 system with small guide RNAs (sgRNAs) targeting certain spots within the movement protein (MP) or coat protein (CP) portions of these viruses. The MP or CP of the viruses can be disrupted using CRISPR/Cas technology, which dramatically reduces their capacity to replicate, disseminate, or induce illness symptoms (Tripathi et al. 2019; Kis et al. 2019). After the discovery of RNA targeting and editing capabilities of CRISPR-related proteins, the capacity of molecules like *Francisella novicida*'s FnCas9 variant and *Leptotrichia wadei*'s LwaCas13a to target and alter viral RNA implies a potential approach for combating RNA viruses. RNA can be targeted by the CRISPR-associated protein LwaCas13a. It can be used to specifically target and break viral RNA sequences, preventing RNA viruses from replicating and propagating. Another CRISPR-related protein called FnCas9 was discovered to have RNA-targeting properties. Due to the ability of this protein variation to recognize and attach to viral RNA molecules, viral replication processes may be manipulated or interfered with. The identification of RNA-targeting CRISPR-related proteins, such as LwaCas13a and FnCas9, opens up new possibilities for the defense against RNA viruses (Zhang et al. 2018a; Aman et al. 2018).

When RNA viruses or double-stranded DNA viruses (dsDNA) exploit multiplex sites, these genes are used as multipurpose targets (Aman et al. 2020). The tobacco mosaic virus (TMV) and turnip mosaic virus (TuMV) have both been successfully targeted by the CRISPR/Cas13 system, offering protection to the respective hosts against these viruses. Specific viral genome areas are identified and cleaved by using Cas13's RNA-targeting activity, which destroys viral RNA and the prevention of viral replication. Cas13 recognizes and binds to viral RNAs when the CRISPR/Cas13 system is programmed with guide RNAs that match the target sequences in the viral genomes of TMV and TuMV. Once bound, they go through a conformational change that causes the viral RNA to be cleaved. The CRISPR/Cas13 system offers a potential remedy for guarding plants against viral infections by specifically targeting RNA viruses like TMV and TuMV (Aman et al. 2018). CRISPR/CAS13 modification also helped to provide resistance to potatoes from potato virus Y (PVY) (Zhan et al. 2019). CRISPR/Cas12 targets primarily dsDNA and ssDNA viruses and is used in plant viral resistance (Aman et al. 2020).

4.4 Resistance to Insect Infection by CRISPR/Cas System in Plants

The primary vector for the spread of many diseases is insects. Around 80% of the 1480 plant viruses and many other microbial pathogens, such as fungi and bacteria, are transmitted utilizing insect vectors that account for the total loss of crop yield (Perilla-Henao and Casteel 2016; Eigenbrode et al. 2018; Tomkins et al. 2018). Due to either a lack of adequate resistance genes in most cultivars or the use of synthetic pesticides for providing resistance to diseases in crop fields, which also hurts the surroundings, over 300 crop plants have been impacted by insect-transmitted microbes that produce huge damage to the crops. Owing to the lack of embryonic microinjection technologies and a precise understanding of insect genomic sequences, it is challenging to increase plant resistance to insects using CRISPR/CAS9 technological advancements. However, given the lack of effective genes that are resistant to insect-borne pathogens, the main goal of the technology for genome editing is targeting pathogenic organisms or susceptible genes in the host. It can be challenging to pinpoint and target certain genes in plants involved with insect resistance, without a thorough grasp of the insect genome. CRISPR/Cas modification represents the current incorporating technique for the resistance against insect-borne infectious agents. The CRISPR/Cas9 genome modification technique has successfully discovered developed resistance against the insect-borne bean yellow dwarf virus (BeYDV), TuMV, rice stripe mosaic virus, Southern rice black-streaked dwarf virus, and CMV (Zhang et al. 2018b, 2019b; Mahas et al. 2019). *Arabidopsis* plants have been genetically altered to integrate the potato protease inhibitor II (pinII) using the CRISPR/Cas9 technology. The pinII gene was introduced into *Arabidopsis* using CRISPR/Cas9, enhancing the plant's inherent defenses against Lepidopteran insects. The PinII protein functions by preventing the activity of proteases, which are enzymes necessary for the digestion of proteins. This PinII protein, which serves as a molecular barrier against insect feeding, is produced as a result of this genetic alteration. The PinII protein interferes with Lepidopteran insects' capacity to digest the modified *Arabidopsis* plants, which reduces their ability to absorb nutrients from the plant and ultimately hinders their ability to grow and develop (Bu et al. 2006). However, the availability of a well-annotated and characterized genomic sequence of the target insect species is essential for the effective deployment of CRISPR/Cas9. Hence, developing insect resistance in plants through the CRISPR/Cas9 technique can be challenging due to certain limitations. Additionally, there are legal and moral issues raised by the use of genetic editing techniques on insects to increase plant resistance. The discharge of GM organisms, especially insects, into the environment, is subject to stringent laws.

4.5 CRISPR/Cas-Based Modification to Increase Nematode Resistance

Nematodes are parasitic, self-sustaining saprophytes. They can be found throughout the world, from the north and south poles to the ocean's depths (Amjad Ali et al. 2015). Only 7% of phyto-parasitic nematodes (PPNs), which are responsible for major illnesses in various economically significant crops, remain extant on Earth. Nematodes are obligatory as well as mutual parasites that provide a barrier to prevent water from moving from the roots to other areas of the plant (Webster 1969). Nutrient transport and redistribution in plants are eventually blocked from moving from leaves to roots. Cereal cyst nematodes (CCNs), which are among the most deadly species of plant parasitic nematodes are Heteroderidae members that severely damage agricultural plants like wheat, barley, and others. As a result, the yield and productivity of these cereal crops get constrained (Smiley and Nicol 2009; Nicol et al. 2011). Different genome editing techniques have been used to improve resistance against various nematode diseases in numerous crop plants, but the outdated breeding methods and their labor-intensive processes have proved unsustainable. CRISPR/Cas9 system has shown some potential for nematode resistance in plants, though there aren't many reports available. In recent years, genomic alterations via the CRISPR/Cas9 system have been established in the nematode *Caenorhabditis elegans*. *Meloidogyne graminicola*, a significant pest of rice, is a root-knot nematode (RKN) that poses several serious threats to rice crops. Significant yield losses from RKN infection in rice might range from 17% to 32%. Researchers have used the CRISPR/Cas system to specifically target the rice susceptibility (S) gene OsHPP04 in an effort to solve this issue. They have succeeded in producing rice plants that are resistant to *M. graminicola* by introducing mutations that lead to the loss of function of OsHPP04. When compared to wild-type rice plants, these genetically altered rice plants with the mutant OsHPP04 gene display increased resistance to RKN infection. As a result of this loss-of-function mutation in the S gene, the rice plants displayed immunity against *M. graminicola* (Kyndt et al. 2014; Mantelin et al. 2017).

4.6 Weed Resistance by CRISPR/Cas-Mediated Gene Editing

Weeds are a major factor in harming croplands (Oerke 2006). Due to the existence of weeds between crops, they take up the most space; compete for sunshine, water, and nutrients; and either directly or through indirect means transmit pests and illnesses to survive (Délye et al. 2013; Quareshy et al. 2018). Consequently, weed species adversely affect crop quality and the rate of crop growth (Rao et al. 2007; Akbar et al. 2011). Some hazardous weeds infect the host crop and spread disease to both the aerial and root systems. Some examples of weed parasites include *Arceuthobium* sp., *Cuscuta* sp., *Orobanche* sp., and *Phelipanche aegyptiaca*. One of the best ways to eliminate weeds and boost agricultural productivity is to employ crops that are resistant to herbicides. The genome editing tool CRISPR/Cas for its exceptional specificity and adaptability can be deployed to manage weeds. A point mutation in the rice acetolactate synthase (ALS) gene has led to the development of resistance in rice to the herbicide bispyribac-sodium. Particularly, the ALS protein's 548th and 627th amino acid positions experience this mutation. Herbicides that impede the ALS enzyme's activity are effective at killing plants that are vulnerable to them. However, some plants can acquire mutations in the ALS gene that make the enzyme less susceptible to the herbicidal effects, leading to resistance to these herbicides. The precise point mutations using CRISPR/Cas technology in the ALS gene at the 548th and 627th amino acid positions in the case of rice and bispyribac-sodium resistance change the structure and activity of the ALS enzyme (Li et al. 2015; Svitashev et al. 2015). The obligatory parasitic weeds *Orobanche* and *Phelipanche* spread diseases to the roots of practically all commercially significant plants in the Brassicaceae, Solanaceae, Fabaceae, Apiaceae, etc. families (Joel

2007; Westwood et al. 2010). A cabbage variety's upstream open reading frame (uORF) of the LsGGP2 gene was modified to give paraquat tolerance. This variety's uORF mutation via CRISPR/Cas resulted in the production of a lot of ascorbic acids, which functions as an antioxidant to fend off oxidative damage. As a result, the paraquat herbicide has become significantly resistant to the cabbage cultivar. The increased ascorbic acid production in the paraquat-resistant cabbage variety with the mutated LsGGP2 uORF offers improved defense against oxidative damage. By scavenging free radicals, neutralizing reactive oxygen species, and reducing oxidative stress, ascorbic acid works as an antioxidant. This characteristic offers the cabbage plants a selection advantage in agricultural contexts by enabling them to live and flourish even when exposed to paraquat (Zhang et al. 2019a). The purpose of employing the CRISPR/Cas9 technology to modify strigolactones (SLs), a plant hormone, in tomato crops was to stop the germination and proliferation of parasite plant seeds. Strigolactones are a group of plant hormones that are involved in several developmental processes, such as the establishment of symbiotic relationships with helpful fungi and the branching of shoots. Host plants are necessary for the survival and growth of parasitic plants like *Phelipanche aegyptiaca*. They recognize and react to strigolactones that host plant release, which causes their seeds to germinate and subsequently connect to the host's root system. The generation of strigolactones was probably inhibited or decreased by CRISPR/Cas9-mediated modification of the tomato plant's CCD8 gene (Carotenoid Cleavage Dioxygenase 8). A key enzyme in the manufacture of strigolactones is called CCD8. In tomato plants, altering this gene with CRISPR/Cas9 may modify the amount or makeup of strigolactones. The capacity of parasitic plant seeds, like *Phelipanche aegyptiaca*, to detect and respond to these hormones may be hampered by lowering the amounts or changing the structure of strigolactones. The altered CCD8 gene expression in the transgenic tomato plant boosts resistance to *Phelipanche aegyptiaca* infestation or decreases susceptibility to it. The parasitic weed may have less of an effect on crop productivity if its seeds do not germinate or grow poorly on the roots of the modified tomato plants. This use of CRISPR/Cas9 technology shows the possibility to improve resistance to particular weeds and parasite plants by changing the CCD8 gene in tomato plants (Bari et al. 2019).

5 Future Perspective

CRISPR/Cas offers a powerful set of molecular instruments that allow us to accurately modify the desired genomic sequence, specific to an organism. The most well-researched and often applied CRISPR mechanisms in plants are CRISPR-Cas9, Cas12, and Cas13 (Jinek et al. 2012; Zetsche et al. 2015). In plant cells, CRISPR reagents are introduced as DNA, RNA, or protein-RNA, where they are put together to form a functional site-directed nuclease (SDN) that can cleave specific DNA sequences resulting in DSBs. CRISPR technology is one of the major emerging tools for developing stress-resistant crops in modern agriculture. It is preferable to see the use of CRISPR in crop production as simply "a novel breeding method," one that can get the same results as present procedures in a way that is noticeably more established, quicker, and even less expensive. Crop species including wheat, maize, and tomatoes have already demonstrated potential uses for CRISPR techniques in agriculture. The use of CRISPR/Cas technology has the potential to completely alter plant pathogen biocontrol tactics. It is now possible to create long-lasting disease management strategies by using CRISPR/Cas for biocontrol. Targeting critical genes in pathogens can stop them from developing resistance, making disease control in crops more enduring and long lasting. The emphasis on crop sequence editing for improved defense response is part of ongoing attempts to update the better recognition of PAMP/MAMP. Scientists can prevent pathogens from causing diseases in plants by altering and targeting key genes in the pathogens. However, new scientific understanding is necessary for the CRISPR/Cas approach to be successful in plant protection over the long run.

The CRISPR/Cas technology has shown a lot of promise, but it also has significant drawbacks that scientists are striving to resolve. DNA sequences other than the target site may occasionally undergo unintentional alterations as a result of CRISPR/Cas systems. These unintended consequences have the potential to impair other genes' ability to operate and have unknown results. To reduce off-target effects, scientists are constantly enhancing the specificity of CRISPR/Cas systems. At present, researchers are looking at how to modify multiple genes in plants to make them resistant to a variety of diseases using the CRISPR/Cas9 technique. Blocking of disease susceptible genes in plants has been made using the CRISPR/Cas9 technique to make them disease-resistant. But there are a lot of clarifications to be made because these studies are still in its infancy. Several of these genes may play additional functions in the plant's growth and development aside from defending against infections. For instance, in an experiment using EMS-induced triple mutations targeted at a particular gene named TaMLO, it was aimed to insert mutations into the gene to provide resistance against powdery mildew disease. Triple knockouts were used to disable the TaMLO gene in wheat, which caused chlorosis, or yellowing of the leaves (Wang et al. 2014). Disabling the TaMLO gene has pleiotropic properties, which means it affected several traits or characteristics of the plant, along with the unanticipated effect on leaf color. This demonstrates that altering genes can have a variety of consequences on plants, some of which may be unintended side effects (Acevedo-Garcia et al. 2017). Pathogens can rapidly evolve, creating techniques to get beyond plant defenses. The genetic alterations brought about by CRISPR/Cas-mediated gene editing may also allow them to adapt. This creates a barrier to the robustness and long-term efficacy of CRISPR-based approaches to biotic stress management. Despite these drawbacks, ongoing studies and improvements in CRISPR/Cas technology are aimed at overcoming them and enhancing the efficiency and dependability of CRISPR-based methods for controlling biotic stress in plants.

6 Conclusion

Synthetic biology is positioned to play a crucial role in solving the rising global food demand in light of population growth and expected climate change. We must approach convergent multidisciplinary resources to accomplish sustained agricultural development. While CRISPR offers a practical and target-specific gene alteration to address the probing plant sensors, omics examines the molecular components of the cellular metabolic process to determine the regulatory mechanisms. Without the need for tissue culture, this efficient modification of the plant's genome can be made, and it can be used to make plants more resilient to biotic stress.

References

Abudayyeh OO, Gootenberg JS, Konermann S et al (2016) C2c2 is a single-component programmable RNA-guided RNA-targeting CRISPR effector. Science 353:aaf5573. https://doi.org/10.1126/science.aaf5573

Abudayyeh OO, Gootenberg JS, Essletzbichler P et al (2017) RNA targeting with CRISPR–Cas13. Nature 550:280–284. https://doi.org/10.1038/nature24049

Acevedo-Garcia J, Spencer D, Thieron H et al (2017) *Mlo*-based powdery mildew resistance in hexaploid bread wheat generated by a non-transgenic TILLING approach. Plant Biotechnol J 15:367–378. https://doi.org/10.1111/pbi.12631

Aizat WM, Ismail I, Noor NM (2018) Recent development in omics studies. Adv Exp Med Biol 1102:1–9

Akbar N, Ehsanullah, Jabran K, Ali MA (2011) Weed management improves yield and quality of direct seeded rice. Aust J Crop Sci 5(6):688–694

Ali MA, Azeem F, Abbas A et al (2015a) Transgenic strategies for enhancement of nematode resistance in plants. Front Plant Sci 8:750. https://doi.org/10.3389/fpls.2017.00750

Ali Z, Abulfaraj A, Idris A et al (2015b) CRISPR/Cas9-mediated viral interference in plants. Genome Biol 16:238. https://doi.org/10.1186/s13059-015-0799-6

Ali Z, Ali S, Tashkandi M et al (2016) CRISPR/Cas9-mediated immunity to Geminiviruses: differential interference and evasion. Sci Rep 6:26912. https://doi.org/10.1038/srep26912

Aman R, Ali Z, Butt H et al (2018) RNA virus interference via CRISPR/Cas13a system in plants. Genome Biol 19:1. https://doi.org/10.1186/s13059-017-1381-1

Aman R, Mahas A, Mahfouz M (2020) Nucleic acid detection using CRISPR/Cas biosensing technologies. ACS Synth Biol 9:1226–1233. https://doi.org/10.1021/acssynbio.9b00507

Amjad Ali M, Abbas A, Azeem F et al (2015) Plant-nematode interactions: from genomics to metabolomics. Int J Agric Biol 17:1071–1082. https://doi.org/10.17957/IJAB/15.0037

Arora L, Narula A (2017) Gene editing and crop improvement using CRISPR-Cas9 system. Front Plant Sci 8:1932. https://doi.org/10.3389/fpls.2017.01932

Baltes NJ, Hummel AW, Konecna E et al (2015) Conferring resistance to geminiviruses with the CRISPR–Cas prokaryotic immune system. Nat Plants 1:15145. https://doi.org/10.1038/nplants.2015.145

Bari VK, Nassar JA, Kheredin SM et al (2019) CRISPR/Cas9-mediated mutagenesis of CAROTENOID CLEAVAGE DIOXYGENASE 8 in tomato provides resistance against the parasitic weed Phelipanche aegyptiaca. Sci Rep 9:11438. https://doi.org/10.1038/s41598-019-47893-z

Bebber DP, Ramotowski MAT, Gurr SJ (2013) Crop pests and pathogens move polewards in a warming world. Nat Clim Chang 3:985–988. https://doi.org/10.1038/nclimate1990

Bhat KP, Cortez D (2018) RPA and RAD51: fork reversal, fork protection, and genome stability. Nat Struct Mol Biol 25:446–453. https://doi.org/10.1038/s41594-018-0075-z

Borrelli VMG, Brambilla V, Rogowsky P et al (2018) The enhancement of plant disease resistance using CRISPR/Cas9 technology. Front Plant Sci 9:1245. https://doi.org/10.3389/fpls.2018.01245

Brioschi M, Banfi C (2018) The application of gene silencing in proteomics: from laboratory to clinic. Expert Rev Proteomics 15:717–732. https://doi.org/10.1080/14789450.2018.1521275

Bu Q-Y, Wu L, Yang S-H, Wan J-M (2006) Cloning of a potato proteinase inhibitor gene PINII-2x from diploid potato (*Solanum phurejia* L.) and transgenic investigation of its potential to confer insect resistance in rice. J Integr Plant Biol 48:732–739. https://doi.org/10.1111/j.1744-7909.2006.00258.x

Das A, Sharma N, Prasad M (2019) CRISPR/Cas9: a novel weapon in the arsenal to combat plant diseases. Front Plant Sci 9:2008. https://doi.org/10.3389/fpls.2018.02008

de Toledo Thomazella DP, Seong K, Mackelprang R et al (2021) Loss of function of a DMR6 ortholog in tomato confers broad-spectrum disease resistance. Proc Natl Acad Sci 118:e2026152118. https://doi.org/10.1073/pnas.2026152118

Deltcheva E, Chylinski K, Sharma CM et al (2011) CRISPR RNA maturation by trans-encoded small RNA and host factor RNase III. Nature 471:602–607. https://doi.org/10.1038/nature09886

Délye C, Jasieniuk M, Le Corre V (2013) Deciphering the evolution of herbicide resistance in weeds. Trends Genet 29:649–658. https://doi.org/10.1016/j.tig.2013.06.001

Dolgalev G, Poverennaya E (2021) Applications of CRISPR-Cas technologies to proteomics. Genes (Basel) 12:1790. https://doi.org/10.3390/genes12111790

Dong OX, Ronald PC (2019) Genetic engineering for disease resistance in plants: recent Progress and future perspectives. Plant Physiol 180:26–38. https://doi.org/10.1104/pp.18.01224

Dong D, Ren K, Qiu X et al (2016) The crystal structure of Cpf1 in complex with CRISPR RNA. Nature 532:522–526. https://doi.org/10.1038/nature17944

Dort EN, Tanguay P, Hamelin RC (2020) CRISPR/Cas9 gene editing: an unexplored frontier for forest pathology. Front Plant Sci 11:1126. https://doi.org/10.3389/fpls.2020.01126

dos Santos VS, Macedo FA, do Vale JS et al (2017) Metabolomics as a tool for understanding the evolution of Tabebuia sensu lato. Metabolomics 13:72. https://doi.org/10.1007/s11306-017-1209-8

Drapal M, Enfissi EMA, Almeida J et al (2023) The potential of metabolomics in assessing global compositional changes resulting from the application of CRISPR/Cas9 technologies. Transgenic Res 32:265. https://doi.org/10.1007/s11248-023-00347-9

Eid A, Alshareef S, Mahfouz MM (2018) CRISPR base editors: genome editing without double-stranded breaks. Biochem J 475:1955–1964. https://doi.org/10.1042/BCJ20170793

Eigenbrode SD, Bosque-Pérez NA, Davis TS (2018) Insect-borne plant pathogens and their vectors: ecology, evolution, and complex interactions. Annu Rev Entomol 63:169–191. https://doi.org/10.1146/annurev-ento-020117-043119

Feng Z, Zhang B, Ding W et al (2013) Efficient genome editing in plants using a CRISPR/Cas system. Cell Res 23:1229–1232. https://doi.org/10.1038/cr.2013.114

Fisher MC, DanielA H, Briggs CJ et al (2012) Emerging fungal threats to animal, plant and ecosystem health. Nature 484:186–194. https://doi.org/10.1038/nature10947

Frédérich M, Pirotte B, Fillet M, de Tullio P (2016) Metabolomics as a challenging approach for medicinal chemistry and personalized medicine. J Med Chem 59:8649–8666. https://doi.org/10.1021/acs.jmedchem.5b01335

Gao P, Yang H, Rajashankar KR et al (2016) Type V CRISPR-Cas Cpf1 endonuclease employs a unique mechanism for crRNA-mediated target DNA recognition. Cell Res 26:901–913. https://doi.org/10.1038/cr.2016.88

Garcia V, Phelps SEL, Gray S, Neale MJ (2011) Bidirectional resection of DNA double-strand breaks by Mre11 and Exo1. Nature 479:241–244. https://doi.org/10.1038/nature10515

Garneau JE, Dupuis M-È, Villion M et al (2010) The CRISPR/Cas bacterial immune system cleaves bacteriophage and plasmid DNA. Nature 468:67–71. https://doi.org/10.1038/nature09523

Gasiunas G, Barrangou R, Horvath P, Siksnys V (2012) Cas9–crRNA ribonucleoprotein complex mediates specific DNA cleavage for adaptive immunity in bacteria. Proc Natl Acad Sci 109:E2579. https://doi.org/10.1073/pnas.1208507109

Giraud T, Gladieux P, Gavrilets S (2010) Linking the emergence of fungal plant diseases with ecological speciation. Trends Ecol Evol 25:387–395. https://doi.org/10.1016/j.tree.2010.03.006

Gupta R, Ghosh A, Chakravarti R et al (2022) Cas13d: a new molecular scissor for transcriptome engineering. Front Cell Dev Biol 10:866800. https://doi.org/10.3389/fcell.2022.866800

Hale CR, Zhao P, Olson S et al (2009) RNA-guided RNA cleavage by a CRISPR RNA-Cas protein complex. Cell 139:945–956. https://doi.org/10.1016/j.cell.2009.07.040

Hammes UZ (2016) Novel roles for phytosulfokine signalling in plant–pathogen interactions. Plant Cell Environ 39:1393–1395. https://doi.org/10.1111/pce.12679

Heyer W-D, Ehmsen KT, Liu J (2010) Regulation of homologous recombination in eukaryotes. Annu Rev Genet 44:113–139. https://doi.org/10.1146/annurev-genet-051710-150955

Hogenhout SA, Ammar E-D, Whitfield AE, Redinbaugh MG (2008) Insect vector interactions with persistently transmitted viruses. Annu Rev Phytopathol 46:327–359. https://doi.org/10.1146/annurev.phyto.022508.092135

Idnurm A, Urquhart AS, Vummadi DR et al (2017) Spontaneous and CRISPR/Cas9-induced mutation of the osmosensor histidine kinase of the canola pathogen Leptosphaeria maculans. Fungal Biol Biotechnol 4:12. https://doi.org/10.1186/s40694-017-0043-0

Ishino Y, Krupovic M, Forterre P (2018) History of CRISPR-Cas from encounter with a mysterious repeated sequence to genome editing technology. J Bacteriol 200(7):e00580–e00517. https://doi.org/10.1128/JB.00580-17

Jansen R, Embden JD, Gaastra W, Schouls LM (2002) Identification of genes that are associated with DNA repeats in prokaryotes. Mol Microbiol 43:1565–1575. https://doi.org/10.1046/j.1365-2958.2002.02839.x

Ji X, Zhang H, Zhang Y et al (2015) Establishing a CRISPR–Cas-like immune system conferring DNA virus resistance in plants. Nat Plants 1:15144. https://doi.org/10.1038/nplants.2015.144

Jia H, Orbovic V, Jones JB, Wang N (2016) Modification of the PthA4 effector binding elements in type I CsLOB1 promoter using Cas9/sgRNA to produce transgenic Duncan grapefruit alleviating XccΔpthA4:dCsLOB1.3 infection. Plant Biotechnol J 14:1291–1301. https://doi.org/10.1111/pbi.12495

Jiang F, Doudna JA (2017) CRISPR–Cas9 structures and mechanisms. Annu Rev Biophys 46:505–529. https://doi.org/10.1146/annurev-biophys-062215-010822

Jinek M, Chylinski K, Fonfara I et al (2012) A programmable dual-RNA–guided DNA endonuclease in adaptive bacterial immunity. Science 337:816–821. https://doi.org/10.1126/science.1225829

Joel DM (2007) Direct infection of potato tubers by the root parasite Orobanche aegyptiaca. Weed Res 47:276–279. https://doi.org/10.1111/j.1365-3180.2007.00570.x

Joshi RK, Nayak S (2008) Gene pyramiding—a broad spectrum technique for developing durable stress resistance in crops. Biotechnol Mol Biol Rev 5(3):51–60

Jung YJ, Lee HJ, Yu J et al (2021) Transcriptomic and physiological analysis of OsCAO1 knockout lines using the CRISPR/Cas9 system in rice. Plant Cell Rep 40:1013–1024. https://doi.org/10.1007/s00299-020-02607-y

Kis A, Hamar É, Tholt G et al (2019) Creating highly efficient resistance against wheat dwarf virus in barley by employing CRISPR Cas9 system. Plant Biotechnol J 17:1004–1006. https://doi.org/10.1111/pbi.13077

Koonin EV, Makarova KS, Zhang F (2017) Diversity, classification and evolution of CRISPR-Cas systems. Curr Opin Microbiol 37:67–78. https://doi.org/10.1016/j.mib.2017.05.008

Kyndt T, Fernandez D, Gheysen G (2014) Plant-parasitic nematode infections in rice: molecular and cellular insights. Annu Rev Phytopathol 52:135–153. https://doi.org/10.1146/annurev-phyto-102313-050111

Langner T, Kamoun S, Belhaj K (2018) CRISPR crops: plant genome editing toward disease resistance. Annu Rev Phytopathol 56:479–512. https://doi.org/10.1146/annurev-phyto-080417-050158

Leisner CP, Yendrek CR, Ainsworth EA (2017) Physiological and transcriptomic responses in the seed coat of field-grown soybean (*Glycine max* L. Merr.) to abiotic stress. BMC Plant Biol 17:242. https://doi.org/10.1186/s12870-017-1188-y

Li Z, Liu Z-B, Xing A et al (2015) Cas9-guide RNA directed genome editing in soybean. Plant Physiol 169:960–970. https://doi.org/10.1104/pp.15.00783

Li Y, Wu X, Zhang Y, Zhang Q (2022) CRISPR/Cas genome editing improves abiotic and biotic stress tolerance of crops. Front Genom Ed 4:987817. https://doi.org/10.3389/fgeed.2022.987817

Lieber MR (2010) The mechanism of double-Strand DNA break repair by the nonhomologous DNA end-joining pathway. Annu Rev Biochem 79:181–211. https://doi.org/10.1146/annurev.biochem.052308.093131

Liu L, Chen P, Wang M et al (2017a) C2c1-sgRNA complex structure reveals RNA-guided DNA cleavage mechanism. Mol Cell 65:310–322. https://doi.org/10.1016/j.molcel.2016.11.040

Liu L, Li X, Ma J et al (2017b) The molecular architecture for RNA-guided RNA cleavage by Cas13a. Cell 170:714–726.e10. https://doi.org/10.1016/j.cell.2017.06.050

Liu Y, Lu S, Liu K, Wang S, Huang L, Guo L (2019) Proteomics: a powerful tool to study plant responses to biotic stress. Plant Methods 15:1–20

Luan H, Shen H, Pan Y et al (2018) Elucidating the hypoxic stress response in barley (*Hordeum vulgare* L.) during waterlogging: a proteomics approach. Sci Rep 8:9655. https://doi.org/10.1038/s41598-018-27726-1

Mahas A, Aman R, Mahfouz M (2019) CRISPR-Cas13d mediates robust RNA virus interference in plants. Genome Biol 20:263. https://doi.org/10.1186/s13059-019-1881-2

Malnoy M, Viola R, Jung M-H et al (2016) DNA-free genetically edited grapevine and apple protoplast using CRISPR/Cas9 ribonucleoproteins. Front Plant Sci 7:1904. https://doi.org/10.3389/fpls.2016.01904

Mantelin S, Bellafiore S, Kyndt T (2017) *Meloidogyne graminicola*: a major threat to rice agriculture. Mol Plant Pathol 18:3–15. https://doi.org/10.1111/mpp.12394

Marraffini LA, Sontheimer EJ (2008) CRISPR interference limits horizontal gene transfer in staphylococci by targeting DNA. Science 322:1843–1845. https://doi.org/10.1126/science.1165771

Martyn RD, Netzer D (1991) Resistance to races 0, 1, and 2 of fusarium wilt of watermelon in Citrullus sp. PI-296341 -FR. Hort Sci 26:429–432. https://doi.org/10.21273/HORTSCI.26.4.429

Maruyama T, Dougan SK, Truttmann MC et al (2015) Increasing the efficiency of precise genome editing with CRISPR-Cas9 by inhibition of nonhomologous end joining. Nat Biotechnol 33:538–542. https://doi.org/10.1038/nbt.3190

Meeske AJ, Nakandakari-Higa S, Marraffini LA (2019) Cas13-induced cellular dormancy prevents the rise of CRISPR-resistant bacteriophage. Nature 570:241–245. https://doi.org/10.1038/s41586-019-1257-5

Miah G, Rafii MY, Ismail MR, Puteh AB, Rahim HA, Islam KN, Latif MA (2013) A review of microsatellite markers and their applications in rice breeding programs to improve blast disease resistance. Int J Mol Sci 14(11):22499–22528. https://doi.org/10.3390/ijms141122499

Mojica FJM, Ferrer C, Juez G, Rodríguez-Valera F (1995) Long stretches of short tandem repeats are present in the largest replicons of the Archaea Haloferax mediterranei and Haloferax volcanii and could be involved in replicon partitioning. Mol Microbiol 17:85–93. https://doi.org/10.1111/j.1365-2958.1995.mmi_17010085.x

Mojica FJM, Diez-Villasenor C, Soria E, Juez G (2000) Biological significance of a family of regularly spaced repeats in the genomes of archaea, bacteria and mitochondria. Mol Microbiol 36:244–246. https://doi.org/10.1046/j.1365-2958.2000.01838.x

Mojica FJ, Díez-Villaseñor C, García-Martínez J, Soria E (2005) Intervening sequences of regularly spaced prokaryotic repeats derive from foreign genetic elements. J Mol Evol 60:174–182. https://doi.org/10.1007/s00239-004-0046-3

Mushtaq M, Bhat JA, Mir ZA et al (2018) CRISPR/Cas approach: a new way of looking at plant-abiotic interactions. J Plant Physiol 224–225:156–162. https://doi.org/10.1016/j.jplph.2018.04.001

Muthuramalingam P, Jeyasri R, Selvaraj A et al (2020) Integrated transcriptomic and metabolomic analyses of glutamine metabolism genes unveil key players in *Oryza sativa* (L.) to ameliorate the unique and combined abiotic stress tolerance. Int J Biol Macromol 164:222–231. https://doi.org/10.1016/j.ijbiomac.2020.07.143

Nakagami H, Sugiyama N, Ishihama Y, Shirasu K (2012) Shotguns in the front line: phosphoproteomics in plants. Plant Cell Physiol 53:118–124. https://doi.org/10.1093/pcp/pcr148

Nekrasov V, Wang C, Win J et al (2017) Rapid generation of a transgene-free powdery mildew resistant tomato by genome deletion. Sci Rep 7:482. https://doi.org/10.1038/s41598-017-00578-x

Nelson R, Wiesner-Hanks T, Wisser R, Balint-Kurti P (2018) Navigating complexity to breed disease-resistant crops. Nat Rev Genet 19:21–33. https://doi.org/10.1038/nrg.2017.82

Nicol JM, Turner SJ, Coyne DL et al (2011) Current nematode threats to world agriculture. In: Genomics and molecular genetics of plant-nematode interactions. Springer, Dordrecht, pp 21–43

Nishimasu H, Ran FA, Hsu PD et al (2014) Crystal structure of Cas9 in complex with guide RNA and target DNA. Cell 156:935–949. https://doi.org/10.1016/j.cell.2014.02.001

Oerke EC (2006) Crop losses to pests. J Agric Sci 144:31–43. https://doi.org/10.1017/S0021859605005708

Oliva R, Ji C, Atienza-Grande G et al (2019) Broad-spectrum resistance to bacterial blight in rice using genome editing. Nat Biotechnol 37:1344–1350. https://doi.org/10.1038/s41587-019-0267-z

Pandey P, Ramegowda V, Senthil-Kumar M (2015) Shared and unique responses of plants to multiple individual stresses and stress combinations: physiological and molecular mechanisms. Front Plant Sci 6. https://doi.org/10.3389/fpls.2015.00723

Parida AK, Panda A, Rangani J (2018) Metabolomics-guided elucidation of abiotic stress tolerance mechanisms in plants. In: Plant metabolites and regulation under environmental stress. Elsevier, Amsterdam, pp 89–131

Perilla-Henao LM, Casteel CL (2016) Vector-borne bacterial plant pathogens: interactions with hemipteran insects and plants. Front Plant Sci 7:1163. https://doi.org/10.3389/fpls.2016.01163

Puchta H, Fauser F (2014) Synthetic nucleases for genome engineering in plants: prospects for a bright future. Plant J 78:727–741. https://doi.org/10.1111/tpj.12338

Quareshy M, Prusinska J, Li J, Napier R (2018) A cheminformatics review of auxins as herbicides. J Exp Bot 69:265–275. https://doi.org/10.1093/jxb/erx258

Ramzi AB, Baharum SN, Bunawan H, Scrutton NS (2020) Streamlining natural products biomanufacturing with omics and machine learning driven microbial engineering. Front Bioeng Biotechnol 8:608918. https://doi.org/10.3389/fbioe.2020.608918

Rao AN, Johnson DE, Sivaprasad B et al (2007) Weed management in direct-seeded rice, pp 153–255

Razzaq MK, Aleem M, Mansoor S et al (2021) Omics and CRISPR-Cas9 approaches for molecular insight, functional gene analysis, and stress tolerance development in crops. Int J Mol Sci 22:1292. https://doi.org/10.3390/ijms22031292

Razzaq A, Wishart DS, Wani SH et al (2022) Advances in metabolomics-driven diagnostic breeding and crop improvement. Metabolites 12:511. https://doi.org/10.3390/metabo12060511

Rizhsky L, Liang H, Shuman J et al (2004) When defense pathways collide. The response of Arabidopsis to a combination of drought and heat stress. Plant Physiol 134:1683–1696. https://doi.org/10.1104/pp.103.033431

Roossinck MJ (2011) The big unknown: plant virus biodiversity. Curr Opin Virol 1:63–67. https://doi.org/10.1016/j.coviro.2011.05.022

Roossinck MJ, Martin DP, Roumagnac P (2015) Plant virus metagenomics: advances in virus discovery. Phytopathology 105:716–727. https://doi.org/10.1094/PHYTO-12-14-0356-RVW

Rybicki EP (2019) CRISPR–Cas9 strikes out in cassava. Nat Biotechnol 37:727–728. https://doi.org/10.1038/s41587-019-0169-0

San Filippo J, Sung P, Klein H (2008) Mechanism of eukaryotic homologous recombination. Annu Rev Biochem 77:229–257. https://doi.org/10.1146/annurev.biochem.77.061306.125255

Scheben A, Wolter F, Batley J et al (2017) Towards CRISPR Cas crops—bringing together genomics and genome editing. New Phytol 216:682–698. https://doi.org/10.1111/nph.14702

Schloss PD, Handelsman J (2004) Status of the microbial census. Microbiol Mol Biol Rev 68:686–691. https://doi.org/10.1128/MMBR.68.4.686-691.2004

Schunder E, Rydzewski K, Grunow R, Heuner K (2013) First indication for a functional CRISPR/Cas system in *Francisella tularensis*. Int J Med Microbiol 303:51–60. https://doi.org/10.1016/j.ijmm.2012.11.004

Shmakov S, Abudayyeh OO, Makarova KS et al (2015) Discovery and functional characterization of diverse class 2 CRISPR-Cas systems. Mol Cell 60:385–397. https://doi.org/10.1016/j.molcel.2015.10.008

Shmakov S, Smargon A, Scott D et al (2017) Diversity and evolution of class 2 CRISPR–Cas systems. Nat Rev Microbiol 15:169–182. https://doi.org/10.1038/nrmicro.2016.184

Shriram V, Kumar V, Devarumath RM et al (2016) MicroRNAs as potential targets for abiotic stress tolerance in plants. Front Plant Sci 7:817. https://doi.org/10.3389/fpls.2016.00817

Smargon AA, Cox DBT, Pyzocha NK et al (2017) Cas13b is a type VI-B CRISPR-associated RNA-guided RNase differentially regulated by accessory proteins Csx27 and Csx28. Mol Cell 65:618–630.e7. https://doi.org/10.1016/j.molcel.2016.12.023

Smiley RW, Nicol JM (2009) Nematodes which challenge global wheat production. Wiley-Blackwell, Ames, IA

Sun Y, Li J, Zhu L, Jiang L (2022) Cooperation and competition between CRISPR- and omics-based technologies in foodborne pathogens detection: a state of the art review. Curr Opin Food Sci 44:100813. https://doi.org/10.1016/j.cofs.2022.10081

Suzuki N, Rivero RM, Shulaev V, Blumwald E, Mittler R (2014) Abiotic and biotic stress combinations. New Phytol 203(1):32–43. https://doi.org/10.1111/nph.12797

Svitashev S, Young JK, Schwartz C et al (2015) Targeted mutagenesis, precise gene editing, and site-specific gene insertion in maize using Cas9 and guide RNA. Plant Physiol 169:931–945. https://doi.org/10.1104/pp.15.00793

Swarts DC, Jinek M (2019) Mechanistic insights into the cis- and trans-acting DNase activities of Cas12a. Mol Cell 73:589–600.e4. https://doi.org/10.1016/j.molcel.2018.11.021

Swarts DC, van der Oost J, Jinek M (2017) Structural basis for guide RNA processing and seed-dependent DNA targeting by CRISPR-Cas12a. Mol Cell 66:221–233.e4. https://doi.org/10.1016/j.molcel.2017.03.016

Symington LS (2016) Mechanism and regulation of DNA end resection in eukaryotes. Crit Rev Biochem Mol Biol 51:195–212. https://doi.org/10.3109/10409238.2016.1172552

Symington LS, Gautier J (2011) Double-Strand break end resection and repair pathway choice. Annu Rev Genet 45:247–271. https://doi.org/10.1146/annurev-genet-110410-132435

Tang Y, Fu Y (2018) Class 2 CRISPR/Cas: an expanding biotechnology toolbox for and beyond genome editing. Cell Biosci 8:59. https://doi.org/10.1186/s13578-018-0255-x

Tomkins M, Kliot A, Marée AF, Hogenhout SA (2018) A multi-layered mechanistic modelling approach to understand how effector genes extend beyond phytoplasma to modulate plant hosts, insect vectors and the environment. Curr Opin Plant Biol 44:39–48. https://doi.org/10.1016/j.pbi.2018.02.002

Tripathi JN, Ntui VO, Ron M et al (2019) CRISPR/Cas9 editing of endogenous banana streak virus in the B genome of Musa spp. overcomes a major challenge in banana breeding. Commun Biol 2:46. https://doi.org/10.1038/s42003-019-0288-7

Tyagi S, Mulla SI, Lee K-J et al (2018) VOCs-mediated hormonal signaling and crosstalk with plant growth promoting microbes. Crit Rev Biotechnol 38:1277–1296. https://doi.org/10.1080/07388551.2018.1472551

Tyagi S, Kumar R, Kumar V et al (2021) Engineering disease resistant plants through CRISPR-Cas9 technology. GM Crops Food 12:125–144. https://doi.org/10.1080/21645698.2020.1831729

Uematsu N, Weterings E, Yano KI, Morotomi-Yano K, Jakob B, Taucher-Scholz G et al (2007) Autophosphorylation of DNA-PKCS regulates its dynamics at DNA double-strand breaks. J Cell Biol 177(2):219–229. https://doi.org/10.1083/jcb.200608077

Van Vu T, Sung YW, Kim J et al (2019) Challenges and perspectives in homology-directed gene targeting in monocot plants. Rice 12:95. https://doi.org/10.1186/s12284-019-0355-1

Wang Y, Cheng X, Shan Q et al (2014) Simultaneous editing of three homoeoalleles in hexaploid bread wheat confers heritable resistance to powdery mildew. Nat Biotechnol 32:947–951. https://doi.org/10.1038/nbt.2969

Wang T, Birsoy K, Hughes NW et al (2015) Identification and characterization of essential genes in the human genome. Science 350:1096–1101. https://doi.org/10.1126/science.aac7041

Wang Y-S, Yao H-Y, Xue H-W (2016) Lipidomic profiling analysis reveals the dynamics of phospholipid molecules in *Arabidopsis thaliana* seedling growth. J Integr Plant Biol 58:890–902. https://doi.org/10.1111/jipb.12481

Watanabe S, Cui B, Kiga K et al (2019) Composition and diversity of CRISPR-Cas13a systems in the Genus *Leptotrichia*. Front Microbiol 10:2838. https://doi.org/10.3389/fmicb.2019.02838

Webster JM (1969) The host-parasite relationships of plant-parasitic nematodes. Adv Parasitol 7:1–40

West SC (2009) The search for a human Holliday junction resolvase. Biochem Soc Trans 37:519–526. https://doi.org/10.1042/BST0370519

Westra ER, van Erp PBG, Künne T et al (2012) CRISPR immunity relies on the consecutive binding and degradation of negatively supercoiled invader DNA by Cascade and Cas3. Mol Cell 46:595–605. https://doi.org/10.1016/j.molcel.2012.03.018

Westwood JH, Yoder JI, Timko MP, dePamphilis CW (2010) The evolution of parasitism in plants. Trends Plant Sci 15:227–235. https://doi.org/10.1016/j.tplants.2010.01.004

Wiedenheft B, Sternberg SH, Doudna JA (2012) RNA-guided genetic silencing systems in bacteria and archaea. Nature 482:331–338. https://doi.org/10.1038/nature10886

Wright AV, Nuñez JK, Doudna JA (2016) Biology and applications of CRISPR systems: harnessing nature's toolbox for genome engineering. Cell 164:29–44. https://doi.org/10.1016/j.cell.2015.12.035

Xu C, Zhou Y, Xiao Q et al (2021) Programmable RNA editing with compact CRISPR–Cas13 systems from uncultivated microbes. Nat Methods 18:499–506. https://doi.org/10.1038/s41592-021-01124-4

Yamano T, Nishimasu H, Zetsche B et al (2016) Crystal structure of Cpf1 in complex with guide RNA and target DNA. Cell 165:949–962. https://doi.org/10.1016/j.cell.2016.04.003

Yan WX, Chong S, Zhang H et al (2018) Cas13d is a compact RNA-targeting type VI CRISPR effector positively modulated by a WYL-domain-containing accessory protein. Mol Cell 70:327–339.e5. https://doi.org/10.1016/j.molcel.2018.02.028

Yang H, Ren S, Yu S et al (2020) Methods favoring homology-directed repair choice in response to CRISPR/Cas9 induced-double Strand breaks. Int J Mol Sci 21:6461. https://doi.org/10.3390/ijms21186461

Yosef I, Goren MG, Qimron U (2012) Proteins and DNA elements essential for the CRISPR adaptation process in *Escherichia coli*. Nucleic Acids Res 40:5569–5576. https://doi.org/10.1093/nar/gks216

Zetsche B, Gootenberg JS, Abudayyeh OO et al (2015) Cpf1 is a single RNA-guided endonuclease of a class 2 CRISPR-Cas system. Cell 163:759–771. https://doi.org/10.1016/j.cell.2015.09.038

Zhan X, Zhang F, Zhong Z et al (2019) Generation of virus-resistant potato plants by RNA genome targeting. Plant Biotechnol J 17:1814–1822. https://doi.org/10.1111/pbi.13102

Zhang H, Si X, Ji X et al (2018a) Genome editing of upstream open reading frames enables translational control in plants. Nat Biotechnol 36:894–898. https://doi.org/10.1038/nbt.4202

Zhang T, Zheng Q, Yi X et al (2018b) Establishing RNA virus resistance in plants by harnessing CRISPR immune system. Plant Biotechnol J 16:1415–1423. https://doi.org/10.1111/pbi.12881

Zhang H-X, Zhang Y, Yin H (2019a) Genome editing with mRNA encoding ZFN, TALEN, and Cas9. Mol Ther 27:735–746. https://doi.org/10.1016/j.ymthe.2019.01.014

Zhang T, Zhao Y, Ye J et al (2019b) Establishing CRISPR Cas13a immune system conferring RNA virus resistance in both dicot and monocot plants. Plant Biotechnol J 17:1185–1187. https://doi.org/10.1111/pbi.13095

Zhang M, Liu Q, Yang X et al (2020) CRISPR/Cas9-mediated mutagenesis of Clpsk1 in watermelon to confer resistance to *Fusarium oxysporum* f. sp. niveum. Plant Cell Rep 39:589–595. https://doi.org/10.1007/s00299-020-02516-0

Zhou J, Peng Z, Long J et al (2015) Gene targeting by the TAL effector PthXo2 reveals cryptic resistance gene for bacterial blight of rice. Plant J 82:632–643. https://doi.org/10.1111/tpj.12838

Zhu H, Li C, Gao C (2020) Applications of CRISPR–Cas in agriculture and plant biotechnology. Nat Rev Mol Cell Biol 21(11):661–677

The Application of Genome Editing Technologies in Soybean (*Glycine max* L.) for Abiotic Stress Tolerance

Xuanbo Zhong, Longlong Hu, and Guixiang Tang

Abstract Soybean (*Glycine max* L.) is one of the five major cultivated crops that are important in the world, and it is also a major source of oil and protein for humans and livestock. But its seed yield is relatively low compared with rice, wheat, and other food crops. Soybean is subject to many biotic and abiotic stresses throughout its whole growth stages, and different growth habits have an impact on yield. Based on the rapid development of whole-genome sequencing and genome editing technology, precise improvement of soybean cultivars gains an unprecedented possibility. The clustered regularly interspaced short palindromic repeats/CRISPR-associated protein 9 (CRISPR/Cas9) system is widely utilized for genome editing following zinc-finger nucleases (ZFNs) and transcription activator-like effector nucleases (TALENs). This system provides scientists with the capability to efficiently, precisely, and simply modify target genes. In this chapter, we provide a comprehensive review of three primary genome editing techniques (ZFNs, TALENs, and CRISPR/Cas9) in the application of soybean research. Numerous scholarly works have thoroughly recorded CRISPR/Cas9's ability to alter soybean genes responsible for controlling agronomic and quality traits. However, limited attention has been given to the editing of genes associated with abiotic stress. Nonetheless, a substantial number of soybean genes have been meticulously analyzed and examined in this regard. Due to the pronounced impact of abiotic stress on soybean seed yield, this chapter primarily consolidated and examined the editing occurrences of abiotic stress-associated genes utilizing CRISPR/Cas9 technology. Specifically, it addressed drought stress, salt stress, and heavy metal stress. It appraised the potential of novel genome editing technologies such as CRISPR/Cas9-based base editor, prime editor, and dual-sgRNA/Cas9 system in relation to soybean research. This updated chapter was intended to furnish comprehensive theoretical guidance for the investigation of genetic mechanisms underlying abiotic stress tolerance and the corresponding breeding efforts in soybeans.

X. Zhong · L. Hu · G. Tang (✉)
Key Laboratory of Crop Genetic Resources, Institute of Crop Science, Zhejiang University, Hangzhou, Zhejiang, China
e-mail: tanggx@zju.edu.cn

J.-T. Chen, S. Ahmar (eds.), *Plant Genome Editing Technologies*, Interdisciplinary Biotechnological Advances,
https://doi.org/10.1007/978-981-99-9338-3_8

Keywords Crop breeding · CRISPR/Cas9 · Drought stress · Heavy metal stress · Salt stress · Seed yield · TALENs · ZFNs

1 Introduction

Soybean (*Glycine max* [L.] Merr.) is a significant staple and oilseed crop, serving as a major source of protein and oil worldwide. It has extensive wide applications in the food, feed, and industrial sectors. Originating from China, soybeans play a vital role in the development of ancient Chinese agricultural civilization. Over time, soybean cultivation gradually expands to different regions such as Asia, the Americas, and other regions. At present, soybeans are extensively grown and utilized worldwide, being an essential agricultural commodity. Their cultivation area and yield have shown continuous growth. According to statistical data from FAOSTAT (2022), the global soybean cultivation area reached 129.2 million hectares in 2022, with a total production of 353 million metric tons. The widespread cultivation of soybeans can mainly be attributed to their strong stress tolerance, wide adaptability, and positive impact on soil improvement through nitrogen fixation (Roy et al. 2020). However, in recent years, abnormal climate patterns and extreme weather events worldwide have become more frequent. As a result, adverse conditions such as imbalanced water resource distribution, and abnormal temperature fluctuations have risen. In addition, improper human practices have led to issues such as soil degradation, salinization, and heavy metal pollution further exacerbating the challenges to soybean productivity. These factors have become the primary obstacles to increasing soybean production. It was observed that each year, millions of acres of soybean crops are lost due to various abiotic factors. To maintain stable yield under multiple abiotic stresses, it is necessary to improve soybean varieties to achieve climate resilience and sustainable production. These improved varieties should also possess higher yield potential and nutritional value. Enhancing biotic stress tolerance in soybean through traditional breeding methods such as backcrossing, pedigree breeding, and population breeding is greatly limited due to the complex genetic control and substantial environmental influence on plant resilience. Nevertheless, recent advancements in the complete sequencing of the soybean reference genome, functional genomics research, and the progress made in technologies like DNA recombinant, genetic transformation, and genome editing have rapidly accumulated valuable genomic data and provided significant tools to improve soybean varieties. Specifically, the utilization of genome editing technology has greatly expedited the development of soybean varieties that are tolerant to abiotic stress.

Genome editing technology, which is a novel developed biotechnology that emerged in the early twenty-first century, utilizes sequence-specific nucleases (SSNs) to introduce double-strand breaks (DSBs) at targeted genomic loci. The repair of DSBs is mediated through either nonhomologous end-joining (NHEJ) or homology-directed repair (HDR) mechanisms to generate targeted mutations in the genes of interest by inserting, deleting, or replacing nucleotides (Symington et al. 2011; Zhu et al. 2020). At present, three major types of SSNs are used: zinc-finger nucleases (ZFNs), transcription activator-like effector nucleases (TALENs), and clustered regularly interspaced short palindromic repeat-associated protein (CRISPR/Cas) systems. ZFNs and TALENs, like earlier generations of genome editing systems, suffer from complex design and cumbersome operation, ultimately resulting in their limited applicability (Boch et al. 2009; Christian et al. 2010). The third-generation system, known as CRISPR/Cas, offers a simplified design for guide RNA and provides a high level of targeting efficiency. It requires only one sgRNA to guide Cas endonucleases for sequence cleavage. Particularly, the genome editing technology based on CRISPR/Cas9, renowned for its broad range of applications, has been extensively utilized in diverse crops. With the continuous advancement of technology, researchers have developed base editing and prime editing systems based on the CRISPR/Cas system (Komor et al. 2016; Gaudelli et al. 2017). Base editing involves the fusion of a nucleotide deaminase with Cas9-D10A nickase (nCas9). This fusion enables precise mutations at targeted genomic loci guided by sgRNA. The process is controlled by cytidine base editors (CBEs) and adenine base editors (ABEs), which allows for accurate substitutions between C-T, G-A, A-G, and T-C single nucleotides. Prime editing, considered a highly precise and targeted method for editing the genome, facilitates the accurate insertion, deletion, and arbitrary nucleotide substitutions of small fragments located at genomic target sites. This technique holds unparalleled significance and potential for enhancing key characteristics in agricultural produce, allowing for precise trait advancements in crops.

Despite the fact that the application of genome editing technology in soybean is not as advanced as in plants like *Arabidopsis* and rice, it has started to show promise in gene functional identification and improvement of crucial agronomic traits. Starting from the initial mutagenesis achieved through the utilization of the CRISPR/Cas9 system in hairy roots via *Agrobacterium*-mediated transformation in 2015 (Cai et al. 2015), numerous studies have since been conducted. This chapter has furnished extensive documentation and reviews concerning the modification of soybean quality and agronomic traits with the aid of this technique. Expanding on this basis, this chapter provides a comprehensive overview of the present applications of genome editing technology in enhancing resistance to abiotic stress and aiding in gene functional identification in soybean. The objective of this review is to offer valuable perspectives on incorporating genome editing technology to expedite the process of genetic enhancement in soybeans.

2 CRISPR-Cas9 Role in Salinity Stress

Soil salinization constitutes a significant factor in contributing to the degradation of the global environment. It is estimated that around 434 million hectares of land are currently experiencing issues related to salt stress (Raza et al. 2022). It is projected that soil salinization will continue to escalate in the future, with regional disparities becoming more apparent. Among the various abiotic stresses, salt stress has a substantial impact on the growth and productivity of crops in global agricultural production. Enhancing the tolerance of crops to salt is not only an effective strategy for mitigating yield losses caused by soil salinization but also a fundamental requirement for the reclamation of saline lands, ensuring future food production security.

Soybean is classified as a crop with moderate tolerance to salt, with a salinity level threshold of 5.0 dS/m in the soil (Ashraf and Wu 1994). Mild salt stress negatively affects the germination of seeds and vegetative growth in soybean, leading to substantial reductions in plant height, biomass, pod number, internode number, branch number, and hundred-seed weight (Wang and Shannon 1999). Severe salt stress can cause direct harm to the entire life cycle of the soybeans, severely impacting their yield (Papiernik et al. 2005). Additionally, salt stress observed during the nodulation stage significantly diminishes the efficiency of biological nitrogen fixation in soybean by causing a sharp decline in the number and biomass of root nodules (Duzan et al. 2004).

Plant response to salt stress encompasses a variety of biological processes, including ion absorption, transportation, and homeostasis, as well as osmotic regulation, scavenging of reactive oxygen species, and regulation of transcription factors. The main factor contributing to ion toxicity in plants is the excessive accumulation of Na^+. The variations in salt tolerance observed among soybean germplasms were primarily linked to the balance of ion homeostasis (Chen and Yu 2007). Similar to other higher plants, soybean employs mechanisms such as sodium/hydrogen exchangers (NHX) to expel or store Na^+ in vacuoles, and cation/H^+ exchangers (CHX) to remove Na^+ from the xylem to the phloem, thereby reducing leaf Na^+ accumulation (van Zelm et al. 2020). In *Arabidopsis*, *AtNHX1* and *AtNHX7* (also known as *AtSOS1*) play crucial roles in maintaining Na^+ homeostasis and enhancing tolerance to salt stress (Li et al. 2010; Shi et al. 2003). In soybean, the homologous *GmNHX1* is localized in the vacuolar membrane and functions as a transporter for cytoplasmic Na^+ into vacuoles. The overexpression of *GmNHX1* in soybean and its heterologous expression in *Arabidopsis* both resulted in enhanced salt tolerance in transgenic plants, as demonstrated by Sun et al. (2019). Similarly, *GmNHX2*, which was localized in the plasma membrane, improved salt tolerance in transgenic plants when heterologously expressed in *Arabidopsis*, as indicated by Joshi et al. (2021). Sun et al. (2021) conducted a CRISPR/Cas9-mediated knockout of *GmNHX5*, which was localized in the Golgi apparatus of young leaves and vascular cells, and found that the mutant had a notable decrease in fresh weight and root length when subjected to salt treatment. This evidence suggested an elevated susceptibility to salt-induced stress. Zhang et al. (2022b) generated three *gmsos1* mutants in soybean through the utilization of the CRISPR-Cas9 system, with one and three base insertions introduced at the target site, respectively. The mutant plants exhibited the compromised ability to remove Na^+ from the roots, consequently leading to a significant Na^+ accumulation within the cells and enhanced efflux of K^+ ions. Furthermore, Nie et al. (2015) determined that the introduction of *GmSOS1* into the *Arabidopsis sos1–1* mutant restored salt tolerance to a level comparable to that of the wild type, suggesting that the existence of analogous functions in *GmSOS1* and *AtSOS1*, which involved both the expulsion of Na^+ ions from the roots and the regulation of long-distance transportation of Na^+ ions from the roots to the shoots. The CRISPR-Cas9 technology has demonstrated significant advances in the disruption of genes within wild soybean. Niu et al. (2020) generated successful mutations of *GsSOS1* and the *GsNSCC* which is responsible for nonselective cation channels, in the hairy roots of wild soybean. This resulted in an editing efficiency of up to 28.5%. These findings provide a solid foundation for the research and functional identification of salt tolerance genes in wild soybean. Knocking out genes encoding cation diffusion facilitators can also be a strategy to improve salt tolerance in soybean. Genome-wide association studies (GWAS) conducted by Zhang et al. (2019) identified several single nucleotide polymorphisms (SNPs) significantly associated with salt tolerance within the same genetic region on chromosome 8. The further analysis pinpointed the candidate gene *GmCDF1* (Glyma.08g102000), which played a negative role in regulating salt tolerance by maintaining K^+-Na^+ homeostasis in soybean. By employing genome editing techniques, *GmCDF1* was manipulated in hairy roots, resulting in enhanced salt tolerance.

Transcription factors are of significant importance as they play a crucial role in governing the expression of stress-responsive genes. Numerous transcription factor families, such as MYB, bZIP, WRKY, and NAC, have been reported to be connected with salt tolerance in soybean. Studies have demonstrated that *GmbZIP132* and *GmWRKY54* enhanced salt tolerance in *Arabidopsis* by upregulating the expression of *DREB2A* and other genes. The overexpression of *GmMYB118* in soybean hairy roots had been shown to enhance salt tolerance, as described by Du et al. (2018). On the other hand, the knockout of *GmMYB118* in hairy roots using the CRISPR/Cas9-mediated method had been found to increase the plant's sensitivity to salt stress. This resulted in a higher accumulation of reactive oxygen species and malondialdehyde content compared to control and overexpression plants. Furthermore, there was a significant decrease in proline and chlorophyll content in the knockout plants. Li et al. (2021a) similarly conducted a knocked out of the NAC transcription factor gene, *GmNAC06* (Glyma06g21020.1), in soybean hairy roots. This led to a decrease level of proline and betaine, as well as an increased Na^+/K^+ ratio, indicating increased sensitivity to salt stress in the mutants. Conversely, negative regulators of salt stress, such as *GmMYB3a* and *GmNAC2*, can be targeted for gene editing and directly applied to improve salt-tolerant soybean germplasms. Wang et al. (2021b) utilized CRISPR/Cas9 to simultaneously target six *GmAITRs* genes, resulting in the generation of a *gmaitr36* double mutant and a *gmaitr23456* quintuple mutant without Cas9 protein. These *gmaitr* mutants exhibited enhanced salt tolerance and ABA sensitivity during seed germination and seedling growth stages.

Long noncoding RNAs (lncRNAs) and microRNAs (miRNAs) are regarded as significant regulators of abiotic stress responses. Niu et al. (2021) employed the double sgRNA/Cas9 design for the first time in soybean to induce a substantial deletion of a DNA fragment in lncRNA77580, resulting in the alteration of expression of multiple adjacent salt stress-responsive genes. Overexpression of *gmamiR172c* and *miR172c* knockout in soybean demonstrated a significant increase and decrease in salt stress sensitivity in the roots, respectively (Sahito et al. 2017).

3 CRISPR-Cas9 Role in Drought Stress

Drought is a worldwide issue, as arid and semiarid regions encompass approximately 34.9% of the overall land area across more than 50 countries and regions. Nearly 40% of cultivated land receives less than 500 mm of precipitation. The growth of soybean requires substantial water, and water scarcity poses a bottleneck for increasing seed yield and quality (Cotrim et al. 2021). About 40% of the global reduction in soybean yield is attributed to water scarcity (Wei et al. 2018). Additionally, future forecasts of climate change are more frequent, and further water scarcity aggravated yield losses.

Soybean yield losses can exceed 80% based on the severity and stage of drought stress. Drought stress results in a notable decrease in plant height during the vegetative stage. While water deficiency during the flowering and pod-filling stages can directly lead to a substantial flower and pod abscission, ultimately resulting in a decrease in the number of seeds and seed yield (Frederick et al. 2001). Drought could disrupt the activity of nitrogenase, which affected soybean's ability to fix nitrogen through symbiotic interactions, ultimately hindering growth and reducing yield (Kunert et al. 2016). Soybean implements various strategies such as the regulation of roots and stomata, increased solute concentration in intracellular through metabolic activities, and the maintenance of membrane stability by scavenging reactive oxygen species in order to adapt to drought environments and mitigate damage caused by drought. The emergence of high-throughput sequencing technology has enabled the identification of transcription factor families in soybean, such as NAC, MYB, WRKY, AREB, and DREB, which play a role in regulating drought responses by modulating the synthesis of drought-responsive hormones like ABA and ethylene, as well as other signaling compounds. These gene data not only contribute to our understanding of the mechanisms involved in responding to drought stress but also provide valuable targets for precisely breeding drought-tolerance soybean varieties.

Recently, there have been notable advancements in the field of gene editing technology with regard to the precise modification of the soybean genome. The application of CRISPR/Cas9 has proven to be successfully employed for the knockout of drought-responsive genes within the soybean genome. For instance, Zhong et al.

(2022) conducted the manipulation of the *GmHdz4* transcription factor through knockout and overexpression techniques in soybean hairy roots. The *gmhdz4* mutants exhibited more complex root system architecture and showed enhanced capability in osmotic regulation and scavenging of reactive oxygen species when subjected to PEG-simulated drought stress. As a result, these mutants displayed improved tolerance to drought stress. In contrast, when compared to the wild type, the *GmNAC8* knockout lines exhibited significantly diminished superoxide dismutase (SOD) activity and proline content, leading to a decrease in drought tolerance (Yang et al. 2020). Yang et al. (2022b) also generated *gmnac12* mutant lines using the CRISPR/Cas9 system and found that these knockout lines experienced a minimum decrease of 46% in survival rate during the period of drought stress. The gene *GmNF-YC14*, which encodes the NF-YC transcription factor, functions as a positive regulator of drought stress by activating the ABA signaling pathway. *GmNF-YC14* mutants exhibited a decreased stem basal circumference when observed in the field drought conditions, indicating a reduced ability to adapt to drought stress (Yu et al. 2021). A study conducted by Osakabe et al. (2016) employed an improved CRISPR/Cas9 system that integrated tru-sgRNA and Cas9 to edit a plasma membrane H^+-ATPase gene OST2, which is essential for the stomatal response. Consequently, when measuring the stomatal response induced by ABA, it was observed that the *ost2* mutants improved drought tolerance by promoting stomatal closure and decreasing water loss in comparison to the wild type. Zhang et al. (2022c) conducted targeted genetic modifications to knock out three ABA receptor genes in soybean, *GmPYL17*, *GmPYL18*, and *GmPYL19*, resulting in the creation of heritable double and triple mutant genotypes known to exhibit decreased sensitivity to ABA. CRISPR/Cas9-mediated mutation of the soybean circadian clock gene *GmLCL* also led to reducing water loss under drought stress. This mutant upregulated the expression of *GmABI2* and *GmSnRK2s* (Yuan et al. 2021). The aforementioned findings suggested that *GmLCL* acted as a suppressor of leaf ABA signaling in response to drought. Another rhythm gene called *GmLHYs* had also been identified as a negative regulator of drought stress in soybean (Wang et al. 2021a). Two CRISPR/Cas9-generated homologs, *LHY1a* and *LHY2a*, had revealed that *GmLHYs* played a crucial role in maintaining cellular homeostasis by modulation of the abscisic acid signaling pathway.

The small ubiquitin-like protein modifier (SUMO) has emerged as a focus of investigation in the research field of abiotic stress signal transduction. However, its biological function in soybean has yet to be fully elucidated. To address this knowledge gap, the CRISPR-Cas9 gene editing technique was employed to functionally evaluate to *GmOTSb*, a SUMO gene in soybean. It revealed that *GmOTSb* hairy root mutant decreased levels of malondialdehyde (MDA) and impaired stomatal closure, thereby indicating a decrease in drought tolerance (Guo et al. 2023). The T2 generation of *GmHSP26* homozygous mutant lines showed a considerable decrease in yields when subjected to drought stress in comparison to the overexpression lines (Liu et al. 2022a). The phospholipases A (PLAs) plays a crucial role in plant growth, development, and stress response. In order to examine the role of PLAs under drought and phosphorus and iron deficiencies, two paralogous genes, *GmpPLA-IIε* and *GmpPLA-IIζ*, were knocked out to conduct the experiment (Xiao et al. 2021). Under drought conditions, certain mutant lines (*ppla-IIε-1*, *ppla-IIε-2*, and *ppla-IIε/ppla-IIζ-3*) demonstrated enhanced survival rate, increased plant height, and improved freshness compared to the wild type. Additionally, all mutants performed superior performance under iron deficiency. Furthermore, in comparison to overexpression lines, soybean mutants lacking miR398c showed the augmented capacity for scavenging reactive oxygen species and reduced electrolyte leakage under drought stress (Zhou et al. 2020). These findings present a promising prospect for precise gene editing utilizing CRISPR/Cas9 technology.

4 CRISPR-Cas9 Role in Heat Stress

Temperature is a highly crucial environmental factor that greatly affects plant growth, development, yield, and quality. In the past century, there has been an approximate increase of 0.74 °C in the global average surface temperature, and it is anticipated that this will further rise by approximately 1.5 °C within the next two decades (IPCC 2018). The impact of high temperatures on plant growth, development, yield, and quality is considerable. Soybean, being a major staple and oil crop worldwide, exhibits high sensitivity to heat stress. For every 1 °C rise in temperature, there is an estimated decrease of approximately 17% in soybean yield (Jianing et al. 2022). Heat stress profoundly influences seed germination, resulting in inadequate germination and increased vulnerability to pathogen attacks in soybean. The flowering stage of soybean is particularly vulnerable to high-temperature stress, causing flower and young pod abscission and leaf senescence, ultimately diminishing the number and weight of pods and influencing yield at last (Krishnan et al. 2020).

The development of heat-tolerant soybean germplasms is a matter of great urgency. The soybean heat tolerance is often constrained due to the complexity of its genome (Yang et al. 2018). Novel gene editing tools such as CRISPR/Cas9 play a crucial role in overcoming biological barriers and creating new heat-tolerant soybean germplasms. However, the application of gene editing technology in improving soybean heat tolerance is currently limited. Many proteins, known as heat shock proteins (HSPs), are upregulated under heat stress and play a significant role in protecting the plant. Out of these, Hsp90 is one of the most conserved and abundant molecular chaperones and is essential for mounting protective responses against stress. Xu et al. (2013) cloned a total of 12 *GmHsp90* genes and found that these genes could be induced by high temperatures, potentially influencing proline synthesis and stress response mechanisms. Among them, *GmHsp90A1* and *GmHsp90A2* responded swiftly to heat stress. *GmHsp90A2*, *GmHsp90A4*, and *GmHsp90C1.1* could enhance resistance to high-temperature stress by preventing levels of chlorophyll and lipid peroxidation. The *GmHsp90A2* mutant lines generated by the CRISPR/Cas9 gene editing technology have demonstrated the responsiveness to heat stress and its active involvement in regulating heat tolerance by the formation of complexes with *GmHsp90A1* in soybean (Huang et al. 2019). Li et al. (2021b) showed that simultaneous mutations in *FT2a* and *FT5a* resulted in an enhanced long juvenile phenotype, leading to an increase in soybean seed yield in tropical environments.

The *GmCDPKSK5* gene, responsible for encoding proteins located in cytoplasmic and membrane-associated proteins, interacted with *GmTCTP*, and this interaction contributed to soybean seeds' response to heat stress (Wang et al. 2017). The promoter regions of Hsfs harbored cis-elements that had the potential to contribute to response against drought and heat stress in soybean. Overexpression of the *GmHsf-34* gene ameliorated drought tolerance in rice and enhanced resistance to heat stress in *Arabidopsis* (Li et al. 2014). Additionally, *GmRAR1*, *GmSGT1*, *GmSBH1P*, and *GmSBH1* were also important partners in the soybean's defense response against stress (Chen et al. 2019). The AQP gene *GmTIP2;6* is also associated with heat stress in soybean. Overexpression of this gene served to improve plant growth when subjected to heat stress (Feng et al. 2019). The homologous gene of *GmGBP1* had a positive role in enhancing heat tolerance in both soybeans and tobacco. Its main effect was observed to enhance seed germination during heat stress (Zhao et al. 2013). Furthermore, the soybean gene *Gmdnjl* contributed to the enhancement of heat tolerance by searching for misfolded proteins and promoting refolding to maintain cellular functions (Zhao et al. 2013). The aforementioned genes had a significant role in bestowing high-temperature resilience in soybean. Overexpression of these genes can significantly enhance their expression levels; however, it may be a better strategy to finely tune their expression levels using precise gene editing techniques. Screening lines with exogenous DNA-free in homozygous progeny would enable plants to enhance the expression of heat tolerance genes while avoiding the safety issues linked to transgenic plants.

5 CRISPR-Cas9 Role in Heavy Metal Stress

The accumulation of heavy metals in soil has been accelerated by industrialization and technological advancement, which has disrupted normal geochemical cycles and presented a serious threat to ecosystems and agricultural production (Luo et al. 2015). The primary pollutants comprise cadmium (Cd), nickel (Ni), mercury (Hg), arsenic (As), lead (Pb), zinc (Zn), chromium (Cr), and copper (Cu), with Cd contamination exceeding one-fourth of the total contamination. Heavy metal pollution acts as a significant factor limiting crop growth and productivity in agricultural fields. Soybean, being widely utilized in animal feed and human consumption, is prone to heavy metal accumulation. This accumulation poses a potential risk of transferring into the human food chain and thereby increases the risk of various diseases (Huang et al. 2008). Exploring the soybean's response mechanisms to heavy metals, along with the implementation of genome editing techniques for enhancing tolerance and reducing the uptake and translocation of heavy metals in plants, holds significant importance in safeguarding yield, quality, and food safety in soybean.

Plants possess a comprehensive regulatory system for maintaining the metal ions within their tissues. HIPP (heavy metal-associated isoprenylated plant protein) is a kind of metal-related plant protein that is involved in balancing metal concentrations and detoxifying in plants. The CRISPR/Cas9 system was employed to investigate the function of the *GmHIPP26* gene. Mutants treated with $CdCl_2$ exhibited significant reductions in plant height, aboveground and underground biomass, as well as impaired root development compared to the control group. Moreover, the mutant exhibited a fourfold elevation in Cd^+ ion accumulation in the underground parts, indicating the influence of *GmHIPP26* on the transportation and uptake of cadmium in soybeans, consequently exerting a positive regulatory effect on their cadmium stress tolerance (Cui 2021).

Currently, there is a lack of reports regarding the application of gene editing to improve aluminum stress tolerance in soybean. Therefore, conducting a study on the genotypes of soybeans subjected to aluminum stress may provide potential targets for future genetic engineering investigation. Wang et al. (2019) identified five quantitative trait loci (QTLs) by analyzing and detecting recombinant inbred lines in soybeans. One of these QTLs is the *Glyma.04g218700* gene showed a considerable upregulation in expression by several hundred times following treatment with Al^{3+}, indicating its potential significance as a gene crucially involved in the response to aluminum-induced stress. Research indicated that the involvement of *GmGSTU9* and *GmPrx145* played a crucial role in enhancing plant antioxidant activity under aluminum stress (Cai et al. 2019). Moreover, the potential of *GmWRKY81* in enhancing soybean's ability to withstand aluminum stress had been discovered by its regulation of downstream components involved in Al^{3+} transportation. Plants overexpressing *GmWRKY81* exhibited significant changes in the secretion of organic acid and the reduction of antioxidants in comparison to the control group. Researchers had successfully obtained soybean plants that overexpressed *GmMATE75*, *GmMATE79*, and *GmMATE87*. These plants showed an elevated secretion of citrate in roots when subjected to aluminum stress treatment. As a result, the aluminum content in the roots decreased. This effectively alleviated the negative impact of aluminum stress on the growth of soybean roots. The findings highlighted the potential role of *MATE* genes in improving soybean's tolerance to aluminum stress as well as their ability to regulate other abiotic stress responses (Rasheed et al. 2023).

6 Other Abiotic Stress

Furthermore, aside from the aforementioned stress conditions, soybeans can also encounter a range of abiotic stresses including low nitrogen, low phosphorus, herbicides, and flooding during their complete growth cycle. The ongoing advancements in genomics have led to the discovery and understanding of a growing multitude of stress-related gene functions. For instance, the utilization of CRISPR/Cas9 gene editing to target the genes *DD20* and *DD43* has demonstrated improved resistance to the herbicide chlorsulfuron in mutant soybeans (Li et al. 2015). Furthermore, mutants of

GmpPLA exhibited superior growth performance when subjected to phosphorus-deficient conditions. Yang et al. (2022b) generated *GmHSP17.9* mutants by CRISPR/Cas9 gene editing techniques. The results shed light on the significant role of small heat shock proteins (sHSPs) in the process of soybean root nodule development and nitrogen fixation. These genes exhibited substantial promise in enhancing the quality of soybean, regulating resistance, and maintaining yield stability. Therefore, researchers must prioritize the mutation or knockout of stress-sensitive genes.

7 Predicament and Solution of Genome Editing Technology in Soybean

The advent of genome editing technology has brought about significant advancements in molecular breeding techniques for soybeans, offering novel biotechnological tools. Genome editing's ability to perform precise modifications at specific sites is gradually supplanting conventional mutagenesis breeding approaches. However, in comparison to crops such as rice (*Oryza sativa* L.) and tomato (*Solanum lycopersicum*), the implementation of genome editing technology in soybeans is presently at a comparatively slower rate. The utilization of CRISPR/Cas9 gene editing technology in soybeans encounters various obstacles, which can be distilled as follows.

First, the availability of target genes for genome editing in soybeans is inherently limited. This is primarily attributed to the complexity of the soybean genome, which has resulted in the identification of only a small number of functionally characterized genes associated with desirable traits (Schmutz et al. 2010). Furthermore, the CRISPR/Cas9 gene editing technology has been restricted to identifying known gene functions. A detailed compilation of the genes explored in current research on abiotic stress in soybeans through genome editing techniques can be found in Table 1. Second, the progress in genome editing systems appropriate for soybeans has been relatively sluggish. The predominant focus in soybean studies has been on the CRISPR/Cas9 system, while the exploration of alternative editing systems has remained constrained. Conversely, in rice, a diverse array of Cas variants, including xCas9, SpCas9-NG, LbCas12a-RR, and LbCas12a-RVR, have been developed to broaden the scope for genome editing. In contrast, soybean has only explored xCas9, SpCas9-NG, and XNG-Cas9 variants, with limited practical applications being reported (He et al. 2022; Zhang et al. 2022a; Carrijo et al. 2021). Additionally, with regard to accurate genome editing, proficient mature editors like CBE, ABE, and prime editors have been effectively developed for rice, thereby providing powerful technical assistance for genetic improvement. However, in the case of soybeans, only Cai et al. (2020) have reported the existence of tools for editing single bases, which have successfully transitioned from random mutations to precise single-base substitutions. Nevertheless, there still exist significant challenges that need to be addressed in

Table 1 Application of genome editing in soybean genes in relation to abiotic stress

Stress	Genes	Function	Reference
Salinity	*GmNHX5*	Sodium/hydrogen exchanger	Sun et al. (2019)
	GmSOS1	Sodium/hydrogen exchanger	Zhang et al. (2022b) and Nie et al. (2015)
	GmCDF1	Cation diffusion facilitators	Zhang et al. (2019)
	GmMYB118	REDOX defense, synthesis of proline and chlorophyll	Du et al. (2018)
	GmNAC06	Regulating Na^{+}/K^{+} ratio	Li et al. (2021a)
	GmAITR	Salt stress and abscisic acid resistance	Wang et al. (2021b)
	lncRNA77580	Regulating stress-related genes expression	Niu et al. (2021)
	miR172c	Salt stress resistance	Sahito et al. (2017)
Drought	*GmHdz4*	Root structure, osmotic regulation, ROS scavenging	Zhong et al. (2022)
	GmNAC8	SOD activity, proline synthesis	Yang et al. (2020)
	GmNF-YC14	Salt stress resistance and stem growth	Yu et al. (2021)
	OST2	Stomata regulating	Osakabe et al. (2016)
	GmPYL17, GmPYL18, GmPYL19	Abscisic acid receptors	Zhang et al. (2022c)
	GmLCL	Negative regulation of ABA signaling	Yuan et al. (2021)
	GmLHYs	Regulation of ABA signaling	Wang et al. (2021a)
	GmOTSb	The regulation of stomata and MDA	Guo et al. (2023)
	GmpPLA-Iε, GmpPLA-IIζ	Resistance to drought, phosphorus deficiency, and iron deficiency	Xiao et al. (2021)
	miR398c	Clearance of ROS and relieve leakage of electrolytes	Zhou et al. (2020)
Heat	*GmHsp90A2*	Heat stress resistance	Huang et al. (2019)
	FT2a, FT5a	Long juvenile	Li et al. (2021b)
Herbicide	*DD20, DD43*	Herbicide resistance	Li et al. (2015)
Phosphorus deficiency	*GmpPLA*	Phospholipases A	Li et al. (2015)
Nitrogen deficiency	*GmHSP17.9*	Nitrogen fixation	Yang et al. (2022a)

order to enhance replacement efficiency, minimize off-target effects, and improve overall stability. It is worth noting that an important breakthrough has recently been made by Huang et al. (2023), as they have managed to overcome the application bottleneck of Cas deaminase and developed a single-stranded base editing system based on Sdd3/Sdd7 deaminases, as well as a double-stranded base editing system based on Ddd1/Ddd9 deaminases. This advancement has addressed the limitation of conventional editors when it comes to editing GC sequences. The novel Sdd7-CBE system had demonstrated outstanding editing efficiency in soybeans, reaching up to 22.1%, and holds promising potential for broad and diverse applications.

Acknowledgments This work was supported by the Key Research Foundation of Science and Technology Department of Zhejiang Province (2021C02064-5-5), Hainan Provincial Joint Project of Sanya Yazhou Bay Science and Technology City (320LH033), and Key R&D Projects of Zhejiang Province (2021C02057). We thank Ms. Xiaoxiao Wang from Changxing Experimental Station of Zhejiang University, Changxing, Zhejiang, to collect and organize the references.

References

Ashraf M, Wu L (1994) Breeding for salinity tolerance in plants. Crit Rev Plant Sci 13(1):17–42. https://doi.org/10.1080/07352689409701906

Boch J, Scholze H, Schornack S, Landgraf A, Hahn S, Kay S et al (2009) Breaking the code of DNA binding specificity of TAL-type III effectors. Science 326(5959):1509–1512. https://doi.org/10.1126/science.1178811

Cai Y, Chen L, Liu X, Sun S, Wu C, Jiang B et al (2015) CRISPR/Cas9-mediated genome editing in soybean hairy roots. PLoS One 10(8):e136064. https://doi.org/10.1371/journal.pone.0136064

Cai Z, Cheng Y, Xian P, Lin R, Xia Q, Liang Q et al (2019) Fine-mapping QTLs and the validation of candidate genes for aluminum tolerance using a high-density genetic map. Plant and Soil 444:119–137. https://doi.org/10.1007/s11104-019-04261-0

Cai Y, Chen L, Zhang Y, Yuan S, Su Q, Sun S et al (2020) Target base editing in soybean using a modified CRISPR/Cas9 system. Plant Biotechnol J 18(10):1996–1998. https://doi.org/10.1111/pbi.13386

Carrijo J, Illa-Berenguer E, LaFayette P, Torres N, Aragao FJL, Parrott W et al (2021) Two efficient CRISPR/Cas9 systems for gene editing in soybean. Transgenic Res 30(3):11. https://doi.org/10.1007/s11248-021-00246-x

Chen XQ, Yu BJ (2007) Ionic effects of Na+ and Cl- on photosynthesis in *Glycine max* seedlings under isoosmotic salt stress. Zhi Wu Sheng Li Yu Fen Zi Sheng Wu Xue Xue Bao 33(4):294–300

Chen S, Guo Y, Sirault X, Stefanova K, Saradadevi R, Turner NC et al (2019) Nondestructive phenomic tools for the prediction of heat and drought tolerance at anthesis in brassica species. Plant Phenom 9:989. https://doi.org/10.34133/2019/3264872

Christian M, Cermak T, Doyle EL, Schmidt C, Zhang F, Hummel A et al (2010) Targeting DNA double-strand breaks with TAL effector nucleases. Genetics 186(2):757–761. https://doi.org/10.1534/genetics.110.120717

Cotrim MF, Gava R, Campos CNS, de David CHO, de Assis Reis I, Teodoro LPR et al (2021) Physiological performance of soybean genotypes grown under irrigated and rainfed conditions. J Agron Crop Sci 207(1):10. https://doi.org/10.1111/jac.12448

Cui N (2021) Editing GmHIPP26 gene by CRISPR/Cas9 and its function under cadmium stress in soybean (*Glycine max* L.). Zhejiang University, Hangzhou. https://doi.org/10.27461/d.cnki.gzjdx.2021.001954

Du YT, Zhao MJ, Wang CT, Gao Y, Wang YX, Liu YW et al (2018) Identification and characterization of GmMYB118 responses to drought and salt stress. BMC Plant Biol 18(1):320. https://doi.org/10.1186/s12870-018-1551-7

Duzan HM, Zhou X, Souleimanov A, Smith DL (2004) Perception of Bradyrhizobium japonicum nod factor by soybean [*Glycine max* (L.) Merr.] root hairs under abiotic stress conditions. J Exp Bot 55(408):6. https://doi.org/10.1093/jxb/erh265

FAOSTAT (2022) Food and Agriculture Organization of the United Nations. http://www.fao.org/faostat/en/

Feng ZJ, Liu N, Zhang GW, Niu FG, Xu SC, Gong YM (2019) Investigation of the AQP family in soybean and the promoter activity of TIP2;6 in heat stress and hormone responses. Int J Mol Sci 20(2):262. https://doi.org/10.3390/ijms20020262

Frederick JR, Camp CR, Bauer PJ (2001) Drought-stress effects on branch and mainstem seed yield and yield components of determinate soybean. Crop Sci 41(3):5. https://doi.org/10.2135/cropsci2001.413759x

Gaudelli NM, Komor AC, Rees HA, Packer MS, Badran AH, Bryson DI et al (2017) Programmable base editing of a.T to G.C in genomic DNA without DNA cleavage. Nature 551(7681):21. https://doi.org/10.1038/nature24644

Guo J, Wang S, Chen R, Guo Y, Han J, Li G et al (2023) SUMO protease GmOTSa positively regulates drought tolerance in transgenic tobacco and soybean hairy roots. Environ Exp Bot 210:16. https://doi.org/10.1016/j.envexpbot.2023.105329

He R, Zhang P, Yan Y, Yu C, Jiang L, Zhu Y et al (2022) Expanding the range of CRISPR/Cas9-directed genome editing in soybean. Abiotech 3(2):89–98. https://doi.org/10.1007/s42994-021-00051-4

Huang M, Zhou S, Sun B, Zhao Q (2008) Heavy metals in wheat grain: assessment of potential health risk for inhabitants in Kunshan, China. Sci Total Environ 405(1–3):54–61. https://doi.org/10.1016/j.scitotenv.2008.07.004

Huang Y, Xuan H, Yang C, Guo N, Wang H, Zhao J et al (2019) GmHsp90A2 is involved in soybean heat stress as a positive regulator. Plant Sci 285:26–33. https://doi.org/10.1016/j.plantsci.2019.04.016

Huang JY, Lin QP, Fei HY et al (2023) Discovery of deaminase functions by structure-based protein clustering. Cell 186:3182. https://doi.org/10.1016/j.cell.2023.05.041

IPCC (2018) Fifth assessment report. Cambridge University Press, New York

Jianing G, Yuhong G, Yijun G, Rasheed A, Qian Z, Zhiming X et al (2022) Improvement of heat stress tolerance in soybean (*Glycine max* L), by using conventional and molecular tools. Front Plant Sci 13:993189. https://doi.org/10.3389/fpls.2022.993189

Joshi S, Kaur K, Khare T, Srivastava AK, Suprasanna P, Kumar V (2021) Genome-wide identification, characterization and transcriptional profiling of NHX-type (Na(+)/H(+)) antiporters under salinity stress in soybean. 3 Biotech 11(1):16. https://doi.org/10.1007/s13205-020-02555-0

Komor AC, Kim YB, Packer MS, Zuris JA, Liu DR (2016) Programmable editing of a target base in genomic DNA without double-stranded DNA cleavage. Nature 533(7603):420–424. https://doi.org/10.1038/nature17946

Krishnan HB, Kim WS, Oehrle NW, Smith JR, Gillman JD (2020) Effect of heat stress on seed protein composition and ultrastructure of protein storage vacuoles in the cotyledonary parenchyma cells of soybean genotypes that are either tolerant or sensitive to elevated temperatures. Int J Mol Sci 21(13):4775. https://doi.org/10.3390/ijms21134775

Kunert KJ, Vorster BJ, Fenta BA, Kibido T, Dionisio G, Foyer CH (2016) Drought stress responses in soybean roots and nodules. Front Plant Sci 7:1015. https://doi.org/10.3389/fpls.2016.01015

Li T, Zhang Y, Liu H et al (2010) Stable expression of Arabidopsis vacuolar Na^+/H^+ antiporter gene AtNHX1, and salt tolerance in transgenic soybean for over six generations. Chin Sci Bull 55(12):1127. https://doi.org/10.1007/s11434-010-0092-8

Li PS, Yu TF, He GH, Chen M, Zhou YB, Chai SC et al (2014) Genome-wide analysis of the Hsf family in soybean and functional identification of GmHsf-34 involvement in drought and heat stresses. BMC Genomics 15(1):1009. https://doi.org/10.1186/1471-2164-15-1009

Li Z, Liu ZB, Xing A, Moon BP, Koellhoffer JP, Huang L et al (2015) Cas9-guide RNA directed genome editing in soybean. Plant Physiol 169(2):960–970. https://doi.org/10.1104/pp.15.00783

Li M, Chen R, Jiang Q, Sun X, Zhang H, Hu Z (2021a) GmNAC06, a NAC domain transcription factor enhances salt stress tolerance in soybean. Plant Mol Biol 105(3):333–345. https://doi.org/10.1007/s11103-020-01091-y

Li X, Fang C, Yang Y, Lv T, Su T, Chen L et al (2021b) Overcoming the genetic compensation response of soybean florigens to improve adaptation and yield at low latitudes. Curr Biol 31(17):3755–3767. https://doi.org/10.1016/j.cub.2021.06.037

Liu S, Liu J, Zhang Y, Jiang Y, Hu S, Shi A et al (2022a) Cloning of the soybean sHSP26 gene and Analysis of its drought resistance. Phyton Int J Exp Bot 91(7):18. https://doi.org/10.32604/phyton.2022.018836

Liu Y, Ren L, Zhao J, Xia Y, Zhang Z, Guan X et al (2022b) Ergosterol production at elevated temperatures by Upc2-overexpressing Kluyveromyces marxianus using Jerusalem artichoke tubers as feedstock. Bioresour Technol 362:127878. https://doi.org/10.1016/j.biortech.2022.127878

Luo XS, Xue Y, Wang YL, Cang L, Xu B, Ding J (2015) Source identification and apportionment of heavy metals in urban soil profiles. Chemosphere 127:152–157. https://doi.org/10.1016/j.chemosphere.2015.01.048

Nie WX, Xu L, Yu BJ (2015) A putative soybean GmsSOS1 confers enhanced salt tolerance to transgenic Arabidopsis sos1-1 mutant. Protoplasma 252(1):127–134. https://doi.org/10.1007/s00709-014-0663-7

Niu FJ, Jiang QY, Chen R et al (2020) CRISPR/Cas9-mediated targeted mutagenesis of wild soybean (Glycine soja) hairy roots altered the transcription profile of the mutant. J Agric Sci 12(9):14. https://doi.org/10.5539/jas.v12n9p14

Niu F, Jiang Q, Sun X, Hu Z, Wang L, Zhang H (2021) Large DNA fragment deletion in lncRNA77580 regulates neighboring gene expression in soybean (Glycine max). Funct Plant Biol 48(11):1139–1147. https://doi.org/10.1071/FP20400

Osakabe Y, Watanabe T, Sugano SS, Ueta R, Ishihara R, Shinozaki K et al (2016) Optimization of CRISPR/Cas9 genome editing to modify abiotic stress responses in plants. Sci Rep 6:26685. https://doi.org/10.1038/srep26685

Papiernik SK, Grieve CM, Lesch SM, Yates SR (2005) Effects of salinity, imazethapyr, and chlorimuron application on soybean growth and yield. Commun Soil Sci Plant Anal 36(7–8):17. https://doi.org/10.1081/CSS-200050280

Rasheed A, Khan AA, Nawaz M, Mahmood A, Arif U, Hassan MU et al (2023) Development of Aluminium (Al)-tolerant soybean using molecular tools: limitations and future directions. J Plant Growth Regul 42:7403. https://doi.org/10.1007/s00344-023-11051-7

Raza A, Tabassum J, Fakhar AZ, Sharif R, Chen H, Zhang C et al (2022) Smart reprograming of plants against salinity stress using modern biotechnological tools. Crit Rev Biotechnol 43:1035–1062. https://doi.org/10.1080/07388551.2022.2093695

Roy S, Liu W, Nandety RS, Crook A, Mysore KS, Pislariu CI et al (2020) Celebrating 20 years of genetic discoveries in legume nodulation and symbiotic nitrogen fixation. Plant Cell 32(1):15–41. https://doi.org/10.1105/tpc.19.00279

Sahito, Z.A., Wang, L.X., Sun, Z.X., et al. (2017) The miR172c- NNC1 module modulates root plastic development in response to salt in soybean. BMC Plant Biol 17(1):1–12. doi:https://doi.org/10.1186/s12870-017-1161-9

Schmutz J, Cannon SB, Schlueter J, Ma J, Mitros T, Nelson W et al (2010) Genome sequence of the palaeopolyploid soybean. Nature 463(7278):178–183. https://doi.org/10.1038/nature08670

Shi H, Lee BH, Wu SJ, Zhu JK (2003) Overexpression of a plasma membrane Na+/H+ antiporter gene improves salt tolerance in *Arabidopsis thaliana*. Nat Biotechnol 21(1):81–85. https://doi.org/10.1038/nbt766

Sun, T.J., Fan, L., Yang, J., et al. (2019) A Glycine max sodium/hydrogen exchanger enhances salt tolerance through maintaining higher Na^+ efflux rate and K^+/Na^+ ratio in Arabidopsis. BMC Plant Biol, 19(1):1–10. doi:https://doi.org/10.1186/s12870-019-2084-4

Sun T, Ma N, Wang C, Fan H, Wang M, Zhang J et al (2021) A golgi-localized sodium/hydrogen exchanger positively regulates salt tolerance by maintaining higher K(+)/Na(+) ratio in soybean. Front Plant Sci 12:638340. https://doi.org/10.3389/fpls.2021.638340

Symington LS, Gautier J (2011) Double-strand break end resection and repair pathway choice. Annu Rev Genet 45:247–271. http://doi:10.1146/annurev-genet-110410-132435

van Zelm E, Zhang Y, Testerink C (2020) Salt tolerance mechanisms of plants. Annu Rev Plant Biol 71:403–433. https://doi.org/10.1146/annurev-arplant-050718-100005

Wang D, Shannon MC (1999) Emergence and seedling growth of soybean cultivars and maturity groups under salinity. Plant and Soil 214(1–2):8. https://doi.org/10.1023/A:1004719420806

Wang S, Tao Y, Zhou Y, Niu J, Shu Y, Yu X et al (2017) Translationally controlled tumor protein GmTCTP interacts with GmCDPKSK5 in response to high temperature and humidity stress during soybean seed development. Plant Growth Regul 82:187. https://doi.org/10.1007/s10725-017-0250-y

Wang X, Cheng Y, Yang C, Yang C, Mu Y, Xia Q et al (2019) QTL mapping for aluminum tolerance in RIL population of soybean (*Glycine max* L.) by RAD sequencing. PLoS One 14(10):e223674. https://doi.org/10.1371/journal.pone.0223674

Wang K, Bu T, Cheng Q, Dong L, Su T, Chen Z et al (2021a) Two homologous LHY pairs negatively control soybean drought tolerance by repressing the abscisic acid responses. New Phytol 229(5):2660–2675. https://doi.org/10.1111/nph.17019

Wang T, Xun H, Wang W, Ding X, Tian H, Hussain S et al (2021b) Mutation of GmAITR genes by CRISPR/Cas9 genome editing results in enhanced salinity stress tolerance in soybean. Front Plant Sci 12:779598. https://doi.org/10.3389/fpls.2021.779598

Wei Y, Jin J, Jiang S, Ning S, Liu L (2018) Quantitative response of soybean development and yield to drought stress during different growth stages in the Huaibei plain, China. Agronomy 8(7):16. https://doi.org/10.3390/agronomy8070097

Xiao Y, Karikari B, Wang L, Chang F, Zhao T (2021) Structure characterization and potential role of soybean phospholipases a multigene family in response to multiple abiotic stress uncovered by CRISPR/Cas9 technology. Environ Exp Bot 188:14. https://doi.org/10.1016/j.envexpbot.2021.104521

Xu J, Xue C, Xue D, Zhao J, Gai J, Guo N et al (2013) Overexpression of GmHsp90s, a heat shock protein 90 (Hsp90) gene family cloning from soybean, decrease damage of abiotic stresses in Arabidopsis thaliana. PLoS One 8(7):e69810. https://doi.org/10.1371/journal.pone.0069810

Yang H, Wu JJ, Tang T, Liu KD, Dai C (2018) Author correction: CRISPR/Cas9-mediated genome editing efficiently creates specific mutations at multiple loci using one sgRNA in *Brassica napus*. Sci Rep 8(1):4877. https://doi.org/10.1038/s41598-018-23161-4

Yang C, Huang Y, Lv W, Zhang Y, Bhat JA, Kong J et al (2020) GmNAC8 acts as a positive regulator in soybean drought stress. Plant Sci 293:110442. https://doi.org/10.1016/j.plantsci.2020.110442

Yang Z, Du H, Xing X, Li W, Kong Y, Li X et al (2022a) A small heat shock protein, GmHSP17.9, from nodule confers symbiotic nitrogen fixation and seed yield in soybean. Plant Biotechnol J 20(1):103–115. https://doi.org/10.1111/pbi.13698

Yang C, Huang Y, Lv P, Antwi-Boasiako A, Begum N, Zhao T et al (2022b) NAC transcription factor GmNAC12 improved drought stress tolerance in soybean. Int J Mol Sci 23(19):12029. https://doi.org/10.3390/ijms231912029

Yu TF, Liu Y, Fu JD, Ma J, Fang ZW, Chen J et al (2021) The NF-Y-PYR module integrates the abscisic acid signal pathway to regulate plant stress tolerance. Plant Biotechnol J 19(12):2589–2605. https://doi.org/10.1111/pbi.13684

Yuan L, Xie GZ, Zhang S, Li B, Wang X, Li Y et al (2021) GmLCLs negatively regulate ABA perception and signaling genes in soybean leaf dehydration response. Plant Cell Environ 44(2):412–424. https://doi.org/10.1111/pce.13931

Zhang W, Liao X, Cui Y, Ma W, Zhang X, Du H et al (2019) A cation diffusion facilitator, GmCDF1, negatively regulates salt tolerance in soybean. PLoS Genet 15(1):e1007798. https://doi.org/10.1371/journal.pgen.1007798

Zhang Y, Cai Y, Sun S, Han T, Chen L, Hou W (2022a) Using *Staphylococcus aureus* Cas9 to expand the scope of potential gene targets for genome editing in soybean. Int J Mol Sci 23(21):12789. https://doi.org/10.3390/ijms232112789

Zhang M, Cao J, Zhang T, Xu T, Yang L, Li X et al (2022b) A putative plasma membrane Na(+)/H(+) antiporter GmSOS1 is critical for salt stress tolerance in *Glycine max*. Front Plant Sci 13:870695. https://doi.org/10.3389/fpls.2022.870695

Zhang Z, Wang W, Ali S, Luo X, Xie L (2022c) CRISPR/Cas9-mediated multiple knockouts in abscisic acid receptor genes reduced the sensitivity to ABA during soybean seed germination. Int J Mol Sci 23(24):16173. https://doi.org/10.3390/ijms232416173

Zhao L, Wang Z, Lu Q, Wang P, Li Y, Lv Q et al (2013) Overexpression of a GmGBP1 ortholog of soybean enhances the responses to flowering, stem elongation and heat tolerance in transgenic tobaccos. Plant Mol Biol 82(3):279–299. https://doi.org/10.1007/s11103-013-0062-z

Zhong X, Hong W, Shu Y, Li J, Liu L, Chen X et al (2022) CRISPR / Cas9 mediated gene-editing of GmHdz4 transcription factor enhances drought tolerance in soybean (*Glycine max* [L.] Merr.). Front Plant Sci 13:15. https://doi.org/10.3389/fpls.2022.988505

Zhou Y, Liu W, Li X, Sun D, Xu K, Feng C et al (2020) Integration of sRNA, degradome, transcriptome analysis and functional investigation reveals gma-miR398c negatively regulates drought tolerance via GmCSDs and GmCCS in transgenic Arabidopsis and soybean. BMC Plant Biol 20(1):190. https://doi.org/10.1186/s12870-020-02370-y

Zhu H, Li C, Gao C (2020) Applications of CRISPR-Cas in agriculture and plant biotechnology. Nat Rev Mol Cell Biol 21(11):661–677. https://doi.org/10.1038/s41580-020-00288-9

Improving Qualities of Horticultural Crops Using Various CRISPR Delivery Methods

Chetan Kaur and Geung-Joo Lee

Abstract The increase in the global population has severely impacted the climate. A wide range of issues such as resilient plant pathogens and harsh weather conditions have resulted in food shortages. Scientists all over the world are working tirelessly to forecast the ever-changing climatic scenarios and the resulting social and economic challenges. These issues necessitate a rapid response from the scientific community in order to incorporate emerging techniques to reduce the breeding time for producing robust crops. Horticultural crops provide a diverse pool of nutrients and medicinal alternatives including functional food with stress-relieving properties. The advancement of CRISPR/Cas9-mediated gene editing has given a much-needed boost to improve the horticulture industry. The popularity of CRISPR/Cas9 technology is owed to its cost-effectiveness associated with speed breeding and versatility. In this chapter, we discuss the latest advances in CRISPR/Cas9 technology and its potential applications to enhance the future of plant genetics. The CRISPR delivery methodologies in plants and the experimental procedures to improve the gene editing efficiencies in plants like tomato, petunia, and tobacco have been briefly discussed. This chapter focuses on the use of gene editing to address the issue of allergens in common horticultural crops, develop more resilient crops, and enhance traits like color, nutritional value, and yield. This chapter was put together to aid in establishing a thorough understanding of the CRISPR/Cas9 systems and serve as a resource for furthering the development of this technology to successfully control agricultural plant features.

C. Kaur
Department of Horticulture, Chungnam National University, Daejeon, South Korea

G.-J. Lee (✉)
Department of Horticulture, Chungnam National University, Daejeon, South Korea

Department of Smart Agriculture Systems, Chungnam National University, Daejeon, South Korea
e-mail: gjlee@cnu.ac.kr

J.-T. Chen, S. Ahmar (eds.), *Plant Genome Editing Technologies*, Interdisciplinary Biotechnological Advances,
https://doi.org/10.1007/978-981-99-9338-3_9

Keywords CRISPR/Cas delivery · Horticulture · Genome editing · Plant development · Vegetable and fruit crops · Floriculture

1 Introduction

Horticulture plays a vital role in agriculture, encompassing a diverse range of crops such as fruits, flowers, vegetables, spices, medicinal plants, and aromatic crops. These crops are not only a significant source of essential nutrients for a balanced diet but also hold economic importance. By cultivating high-value horticultural crops, farmers can increase their income, thereby contributing to the overall economic growth of the country.

In today's world, the most pressing global challenges revolve around food security and climate change. The horticultural industry faces significant losses due to adverse climatic conditions and the sudden emergence of resilient pests and diseases. As a result, the fundamental goal of genetic breeding programs includes developing cultivars with increased resistance to biotic and abiotic stressors. Moreover, the objectives of plant breeding have expanded to include increasing shelf life, reducing post-harvest losses, and improving fruit quality traits that are directly linked to consumer preferences, such as flavor and nutraceutical components (Sharangi and Datta 2015). Over the years, conventional plant breeding has proven successful in tackling numerous pivotal challenges. Nevertheless, conventional techniques faced certain limitations primarily stemming from the absence of target alleles within a species' germplasm and the inherent crossing incompatibility among species and genera. This led to a shift toward transgenic breeding, which enabled the introduction of specific genes that would otherwise be unattainable from alternate species or organisms (Holme et al. 2013). Many transgenic crops with improved traits were popularized and commercialized leading to a boom in biotechnological crop production to date (Brookes and Barfoot 2018). However, despite the initial excitement surrounding transgenic crops, their usage encountered numerous challenges leading to their decreased popularity. Recently, a more advanced and precise approach has piqued the interest of breeders: genome editing using clustered regularly interspaced short palindromic repeats (CRISPR)-associated proteins (CRISPR/Cas) system, which allows deleting, replacing, or inserting specific sequences in a targeted location of the genome to generate new valuable traits with the potential to bring innovative solutions to agriculture (Kato-Inui et al. 2018).

The process involves using sequence-specific nucleases (SSNs) to identify and introduce double-stranded breaks (DSBs) at specific locations in the plant's DNA. Natural DNA repair processes in plants then take over, either through nonhomologous end joining (NHEJ) or homologous recombination (HR). HR is a precise approach that uses a donor template to introduce desired gene alterations, whereas NHEJ is error-prone and frequently results in gene knockouts (Sun et al. 2016). Compared to genetically modified (GM) crops, genome editing techniques typically involve making small changes to a few nucleotides, without significantly altering the overall genetic makeup of a species. Furthermore, the modified genes in offspring plants follow a natural Mendelian segregation pattern, making them indistinguishable from naturally occurring mutations (Songstad et al. 2017). These genome editing tools have been successfully applied in a wide range of plant species for various purposes, such as improving grain yield, enhancing stress tolerance, and facilitating crop domestication. The use of genome editing in current plant breeding programs holds immense potential for accelerating crop improvement.

2 Advanced Genome Editing by CRISPR Systems and Current Delivery Methods

The field of gene editing is continuously advancing, with notable developments in various Cas effectors. These advancements have expanded the range of target sites, improved efficacy, and enhanced the ability to achieve precise modifications in DNA through techniques like base editing and prime editing. The availability of diverse CRISPR reagents provides researchers with a toolbox for performing gene knockouts, targeted gene insertions, precise base substitutions, and multiplexing of genetic modifications. A variety of Cas nucleases, including their variants, are now available, offering advanced and precise genome editing capabilities. The choice of Cas nuclease depends on the specific target site modification required. Additionally, the delivery of gene editing reagents into plant cells is a crucial factor to consider. Significant progress has been made in the field of CRISPR reagent delivery, opening up possibilities for expanding the range of plant species amenable to gene editing techniques. The different Cas effectors and current delivery methods are discussed below.

2.1 CRISPR-Cas Nucleases and Their Variants

The CRISPR-Cas9 nuclease is classified as a class 2, type-II CRISPR system, is guided by a single-guide RNA (sgRNA), and requires a specific "NGG" protospacer adjacent motif (PAM) sequence in the target DNA (Doudna and Charpentier 2014). It has HNH and RuvC-like domains that cleave complementary and noncomplementary DNA strands, resulting in a double-stranded break (DSB) (Fagerlund et al. 2015). The limitations of the CRISPR/Cas9 system include the restricted number of target recognition sites due to the "NGG" PAM requirement, reducing system versatility, and the potential for off-target cleavage, which compromises editing precision and specificity by causing unintended mutations in nontarget genomic regions.

Cas12, a type V CRISPR nuclease, lacks the HNH domain but possesses the RuvC-like domain, generating a staggered cut with a 4–5 nt overhang. LbCas12a is widely used in plant gene editing and recognizes the T-rich PAM sequence "TTTV"

(Gleditzsch et al. 2019). Additionally, engineered Cas12a variants with improved activities and expanded target ranges have been developed (Kleinstiver et al. 2019; Ooi et al. 2021). Cas13, a type VI CRISPR nuclease, targets RNA using crRNA. It activates endoribonuclease activity, degrading the target RNA and silencing genes (O'Connell 2019). Cas13 is mainly used in RNA manipulation and transcriptome engineering, and Cas13 offers possibilities for RNA manipulation and transcriptome engineering.

Base editors are a precise tool that can convert specific DNA nucleotides using either a catalytically impaired dead Cas9 (dCas9) or a nickase Cas9 (nCas9), resulting in minimal undesired byproducts compared to traditional gene-editing methods (Molla and Yang 2019; Monsur et al. 2020). Cytosine base editors (CBEs) utilize cytosine deaminase to catalyze C-to-T conversions, while adenine base editors (ABEs) use a modified deoxyadenosine deaminase to enable A-to-G conversions. By exploiting the single-stranded nature of RNA-DNA hybrids formed by the guide RNA (sgRNA), the target DNA strand is exposed, allowing the base conversion within a defined "activity window."

The newly developed prime editing (PE) method allows precise genetic modifications like targeted insertions, deletions, and specific mutations without the need for double-stranded breaks or donor DNA templates. It utilizes a prime editor protein, combining nCas9 (H840A) with an engineered reverse transcriptase (RT) enzyme, along with a prime editing guide RNA (pegRNA) that encodes the RNA template for reverse transcription. This mechanism copies the desired edits from the pegRNA into the target DNA site (Huang and Puchta 2021). Prime editing offers advantages such as reduced bystander mutations and greater flexibility in target site selection compared to other methods like base editing or HDR (Table 1).

2.2 *Delivery of CRISPR-Cas Components into Plant Cells*

2.2.1 Protoplast Transformation Using PEG-Mediated CRISPR/Cas9 Delivery

Cell walls pose a major barrier to CRISPR/Cas9 delivery. The cell wall can be chemically or mechanically removed, leaving behind the protoplast cells which allows for a much easier manipulation for further study (Bekkaoui et al. 1987; Cocking 1974). Plant protoplast transformation is widely accomplished in a variety of ways, including polyethylene glycol or PEG-mediated transformation, electroporation-mediated transformation, and microinjection-based transformation (Krens et al. 1982; Hauptmann et al. 1987; Crossway et al. 1986). PEG-mediated transformation is a common choice among these procedures because of its ease of use, low cost, lack of specialist equipment, and ability to produce steady and repeatable results. The process typically involves the isolation of protoplasts, followed by PEG-mediated transfection wherein PEG is added to the protoplast-CRISPR/Cas9 mixture (Wu et al. 2020; Shen et al. 2014). PEG serves as a transfection agent,

Table 1 Comparison of different CRISPR gene editing systems

	Cas9	Cas12	Cas13	Base editors	Prime editors
Nuclease activity	RNA-guided DNA endonuclease	RNA-guided DNA endonuclease	RNA-guided RNA endonuclease	Catalytic conversion of nucleotides	RNA-guided DNA endonuclease and reverse transcription (RT)
Guide RNA type	Single-guide RNA (sgRNA)	CRISPR RNA (crRNA)	CRISPR RNA (crRNA)	Single guide RNA (sgRNA)	Prime editing guide RNA (pegRNA)
Target molecule	DNA	DNA	RNA	DNA	DNA
Cleavage mechanism	Double-strand DNA cleavage	Double-strand DNA cleavage	Single-stranded RNA cleavage	Deaminates DNA bases	Nicking and reverse transcription
PAM requirement	PAM sequence required for target recognition (e.g., "NGG" for SpCas9)	PAM sequence required for target recognition (e.g., "TTTV" for LbCas12a)	PAM sequence not required	PAM sequence not required	Required for Cas9 component
Repair pathway	Nonhomologous end joining (NHEJ) or homology-directed repair (HDR)	Nonhomologous end joining (NHEJ) or homology-directed repair (HDR)	N/A (RNA degradation)	Base excision repair (BER)	DNA repair machinery
Editing type	Indels, insertions, deletions	Indels, insertions, deletions	RNA interference, RNA cleavage	Single nucleotide conversions	Indels, insertions, deletions, base conversions

facilitating the entry of CRISPR/Cas9 components into the protoplasts. After transfection, the protoplasts are cultured to promote regeneration into whole plants (Fig. 1). The PEG-mediated method allows for DNA-free genome editing.

In a study reported in watermelon, the researchers were able to successfully produce an albino phenotype by targeting the *ClPDS* (*Citrullus lanatus* phytoene desaturase gene) using PEG-mediated CRISPR/cas9 transfection of the protoplast cells. They also reported the likelihood of no off-target mutation on examination of the sequences homologous to the sgRNA sequence. Similarly, the *F3H* genes in *Petunia* were completely knocked, leading to a significant decrease in the anthocyanin content in the flower petals of the mutant plants (Yu et al. 2021). Unlike conventional genetic engineering methods, which involve introducing foreign DNA into the plant genome, this approach involves the transient delivery of CRISPR/Cas9 components to the cells. By avoiding the integration of external DNA into the plant genome, it addresses concerns associated with genetically modified organisms (GMOs) and

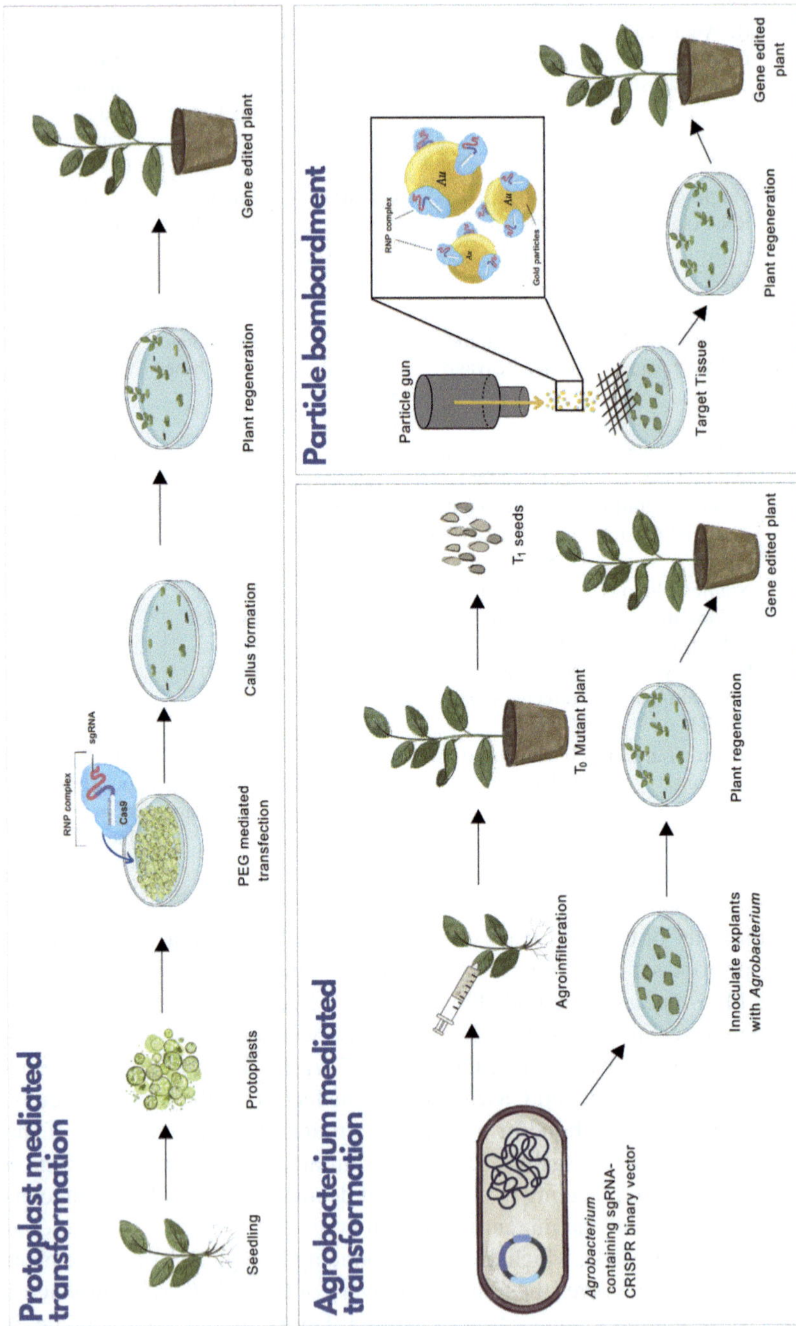

Fig. 1 Diverse approaches for delivering CRISPR/Cas in plant systems

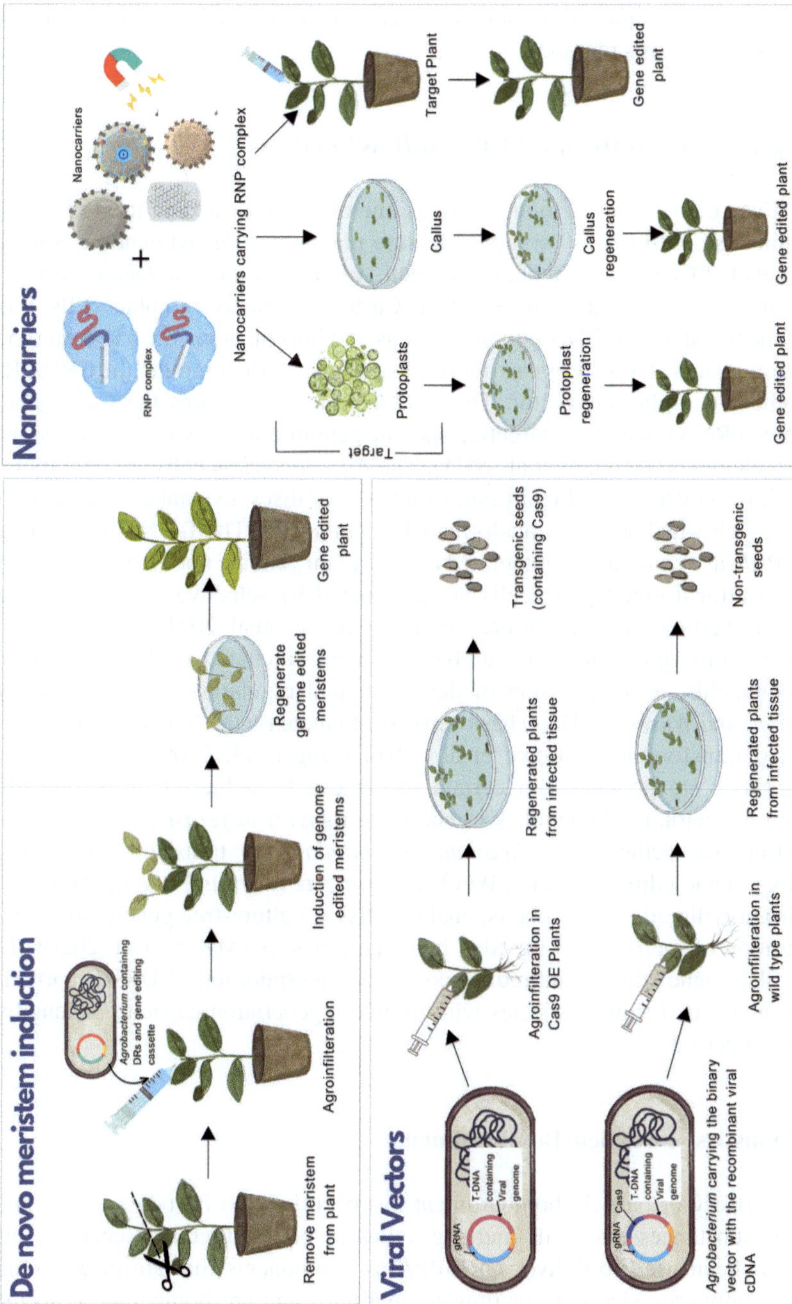

Fig. 1 (continued)

potential regulatory challenges in certain countries (Lazzeri 1995). One of the limiting factors of PEG-mediated protoplast transformation is that many plants have a limited regeneration capacity. Achieving a high transformation rate can be difficult in some cases using this method.

2.2.2 *Agrobacterium*-Mediated CRISPR/Cas9 Delivery

CRISPR/Cas9 delivery by *Agrobacterium* is one of the most widely used methods for introducing CRISPR/Cas9 components into plants for targeted genome editing (Sandhya et al. 2020). A natural plant pathogen, *Agrobacterium tumefaciens*, is used as a delivery vehicle to transfer the CRISPR/Cas9 components into plant cells. It is based on the transfer of T-DNA into the nucleus of a host plant via the bacterial type IV secretion system. The process involves constructing and assembling the CRISPR/Cas system into a plasmid vector. This vector typically contains the Cas9 gene, guide RNA (gRNA) sequences targeting specific genomic sites, as well as necessary regulatory elements (Ghogare et al. 2021). The *Agrobacterium* cells are then transformed with CRISPR/Cas9. Plant tissues such as leaf discs, explants, or callus cultures are co-cultivated with the transformed *Agrobacterium*. The T-DNA (containing the CRISPR/Cas9 construct) is integrated into the plant genome upon *Agrobacterium* infection. Untransformed plant cells are eliminated by selective screening using antibiotics or herbicides in the culture medium (Ghogare et al. 2021). Several plants are recalcitrant to *Agrobacterium*, thereby limiting this method to only a few plants. To overcome this, overexpression of developmental regulators (DRs) like *Baby Boom (Bbm)* and *Wushel2 (Wus2)* has been done to increase the frequency of regeneration and transformation (Lowe et al. 2016; Deng et al. 2009). Additionally, inducing shoot organogenesis in certain dicots can be achieved by ectopically expressing the cytokinin biosynthesis gene *Isopentenyl transferase (Ipt)*. Heritable gene editing was achieved in *Nicotiana benthamiana* by transiently delivering CRISPR/Cas gene-editing cassette, *Wus2*, and *Isopentenyl transferase (Ipt)* through agroinfiltration directly to the plants, enabling tissue culture-free gene editing and the generation of meristems with edits in somatic tissues (Maher et al. 2020). To expedite plant gene editing and reduce costs, the incorporation of DRs is particularly advantageous for plant species with limited regeneration capacity or lengthy regeneration periods.

2.2.3 Biolistics or Particle Bombardment

Biolistic delivery or particle bombardment has been used to introduce genes and exogenous substances into cells and organelles of plants directly. Recently, this method has been used to deliver CRISPR/Cas components that are coated onto micron or submicron-size gold or tungsten particles and are bombarded onto the plant tissues. On entering the plant tissue, the metal and the RNP complex dissociate, and the RNP is available for transient expression (Hamada et al. 2018). Particle bombardment or biolistics is advantageous because it allows direct delivery of DNA/RNA, bypassing the need for bacterial or viral vectors. This method also has minimal size limitations. Particle bombardment has been employed to deliver the CRISPR/Cas9 cassette into the immature maize and wheat embryos; however, the resultant plants were chimeric (Hamada et al. 2017). This further necessitates the selection of plants in the subsequent progenies. In another study a transient transformation assay was performed on the explants obtained from the cotyledons of tomato, using the particle gun method to deliver the CRISPR vectors and sgRNAs (García-Murillo et al. 2023). Biolistic transformation, while known for its speed and reliability, is not without limitations. This technique presents several drawbacks, including a low copy number of introduced DNA and integration into multiple sites within the plant genome. Additionally, its efficiency is comparatively lower than that of other delivery systems, making it a costly option (Lacroix and Citovsky 2020).

2.2.4 Viral Vectors

Viral vectors derived from plant viruses offer advantages for transient gene expression of T-DNA constructs by enhancing the gene expression through viral genome amplification and facilitate the systemic spread of genes throughout the entire plant. The use of viruses for gene delivery in plants provides several advantages, including transient and systemic gene expression without transgenics, high levels of gene expression, and the ability to engineer a variety of viruses for infecting different plant species (Kujur et al. 2021). Virus-induced gene editing (VIGE) vectors can be categorized based on their cargo capacity and the delivered reagents. In the first category, VIGE vectors express a single guide RNA (sgRNA), infect plants with stable Cas9 expression, and produce gene-edited Cas9 transgenic seeds (Ali et al. 2015; Luo et al. 2021). The second category involves VIGE vectors that deliver both Cas9 and sgRNA, spreading systematically in plants (Ariga et al. 2020; Zhang et al. 2022). Upon infection the viral particles enter the cells and release the CRISPR/Cas machinery, allowing it to recognize and bind to the target DNA sequences within the plant genome. Several plant viruses have been utilized as delivery vectors for CRISPR/Cas components. In *Nicotiana benthamiana*, a dual RNA virus system was used for carrying out DNA-free genome editing in plants using potyvirus tobacco etch virus (TEV) and potato virus X (PVX) for expressing the Cas nuclease and guide RNA respectively within a cell (Uranga et al. 2021).

The size constraint of DNA delivery is a significant limitation of viral vector-based methods, as viruses have a limited capacity to carry genetic material. This restriction poses a challenge when attempting to deliver larger DNA constructs. Additionally, the use of viral vectors for CRISPR/Cas delivery raises concerns about potential off-target effects. The CRISPR/Cas9 components, when delivered via viral vectors, may inadvertently introduce unintended mutations in the plant genome, posing a risk to the desired genetic modifications.

2.2.5 Nanocarriers

Nanoparticles (NPs), characterized by their small size (typically below 500 nm), have demonstrated the ability to traverse the plant cell wall and cellular membranes, enabling the delivery of genetic cargo and detection of biomolecules. Multiple studies have shown that the use of nanoparticles can enhance drought stress tolerance, improve nutrient uptake, increase photosynthesis, and increase yield (Rasheed et al. 2022; Ahmed et al. 2021; Maity et al. 2018). Nanocarriers have gained attention as potential delivery systems for CRISPR/Cas components delivery in plants due to their rapid uptake. One of the primary challenges in genome editing using CRISPR/Cas in plants is the inability to regenerate many plant species, coupled with low editing efficiency. This limitation results in a labor-intensive process that hampers the widespread application of CRISPR/Cas in plant genetic engineering. Nanoparticles may offer potential solutions to overcome these challenges faced. Several types of nanoparticles have been used in plants such as mesoporous silica nanoparticles, clay nanosheets, carbon nanotubes, peptides, quantum dots, nanostructured lipid carriers, and DNA nanoparticles. Most of the reported studies in genome editing using nanoparticle delivery methods have been reported in tobacco and *Arabidopsis*. In an initial study, researchers developed a honeycomb mesoporous silica nanoparticle (MSN) capable of delivering DNA and chemicals into the protoplasts of *Nicotiana tabacum*, enabling controlled-release gene expression (Torney et al. 2007). Many studies have used nanocarriers for gene silencing using RNA interference (RNAi). In a study using the intact leaves of tobacco, single-walled carbon nanotubes (SWNTs) were used to deliver siRNA. The siRNA-SWNT infiltration resulted in up to 95% miRNA knockdown efficiency (Demirer et al. 2020). Nevertheless, the delivery of CRISPR/Cas9 using nanoparticles presents challenges, particularly due to the larger size of CRISPR/DNA plasmids compared to the optimal DNA size for nanoparticle loading. This limit is likely to depend on the type of nanoparticle and its surface chemistry. Further research is needed to establish the compatibility of nanoparticles with plant tissue culture and regeneration protocols, especially in cases where germline transformation is not feasible (Demirer et al. 2021).

To date, the delivery of CRISPR-Cas components using nanoparticles has shown success in animal models, particularly in the context of gene therapy (Taha et al. 2022). This delivery process is independent of plant species, with transgenic-free results, and target specificity. However, the technique also presents challenges related to NP-plant interactions, carrier design, and protocol optimization. There are ongoing efforts to overcome the challenges and limitations associated with nanoparticle-based delivery for CRISPR-Cas systems (Demirer et al. 2021).

3 Genome Editing in Vegetables and Fruits

The global rise in population has led to a sudden emergence of abiotic and biotic stresses, causing a massive decline in vegetable crop production (Komarnytsky et al. 2022). Moreover, there is a rise in numerous disease occurrences in humans. These issues are currently being addressed by breeding resilient and fortified fruits and vegetables.

The quality of a vegetable is greatly influenced by its flavor, fruit size, color, and presence of nutrient-rich and health-promoting components. Several genes have been identified to improve the nutritional value of fruits and vegetables. Tomato is a model plant species and is highly consumed all around the world. Many varieties of tomato have been produced with increased nutrients, long shelf life, and resistance to biotic and abiotic stresses. Tomatoes have high levels of γ-aminobutyric acid (GABA). GABA has been associated with potential benefits in humans such as improving sleep and enhancing mood (Lee et al. 2022). In a study, researchers were able to increase the production of GABA in the fruits by deleting the autoinhibitory domains of the key inhibitory enzymes: glutamate decarboxylase (*SlGAD2* and *SlGAD3*) using CRISPR/Cas9-agroinfiltration (Nonaka et al. 2017). Plants with a truncated C-terminal in *SlGAD2* had increased GABA accumulation but exhibited smaller plant size, reduced flowering, and less fruit yield, whereas plants with a truncated C-terminal in *SlGAD3* showed increased GABA accumulation in fruits without any impact on plant size or flowering (Nonaka et al. 2017).

The rise in food allergies in recent years is a multifaceted issue influenced by various factors like dietary changes, undeveloped immune systems in some individuals, and various epigenetic factors (Benedé et al. 2016). Peanut allergy is responsible for the majority of severe food-related allergic reactions. Even small amounts of peanuts can trigger an allergic response in individuals who are highly sensitized to them. Peanut allergy is primarily caused by a group of proteins found in peanuts, namely, *Ara h 1*, *Ara h 2*, *Ara h 3,* and *Ara h 6* (Porterfield et al. 2009; Pandey et al. 2019). *Ara h 2*, in particular, is recognized as the most potent allergen and a significant contributor to peanut allergies. Using CRISPR/Cas9 and PEG-mediated protoplast transformation, researchers obtained mutant peanut plants with different modifications in the major allergen gene in peanut, *Ara h 2A,* resulting in the loss of *Ara h 2A* protein expression and the production of a non-allergenic *Ara h 2A* protein (Biswas et al. 2022). Other allergens found in fruits include profilins, lipid transfer proteins (LTPs), Bet v 1 homologs, and seed storage proteins (Sánchez-Monge et al. 1999; Santos and Van Ree 2011; Marzban et al. 2005; Vicths 2020). Silencing the allergen-producing genes can help in providing safer food options for people with severe allergies.

The ability of fruits to maintain a long shelf life is highly important for ensuring their quality, and it is a significant focus in breeding programs due to its impact on the marketability of the fruit from both the perspectives of farmers and consumers. The long shelf life traits in tomato (Yu et al. 2017), banana (Hu et al. 2013), and melon (Nonaka et al. 2023) have already been harnessed using CRISPR/Cas technology. *BRASSINAZOLE RESISTANT 1 (BZR1)* is known to regulate brassinosteroid-induced growth. Many studies targeting abiotic and biotic stress-related genes in fruits have been published. Through CRISPR/Cas9 gene editing, *BZR1* in tomato was mutated which led to a decrease in the activation of *respiratory burst oxidase homolog1 (RBOH1)* and subsequent hydrogen peroxide (H_2O_2) production in tomato. The suppression of *BZR1* resulted in improved tolerance to heat stress (Yin et al. 2018). These examples represent a subset of numerous studies conducted on fruits and vegetables aimed at enhancing their quality, yield, and resistance to biotic and abiotic stresses. For a comprehensive overview of CRISPR/Cas-mediated gene editing in fruits and vegetables, refer to Table 2.

Table 2 Recent progress in CRISPR/Cas genome editing in vegetables and fruits

Plant	CRISPR/Cas delivery method	Target gene	Plant phenotype/target trait	References
Actinidia chinensis	*Agrobacterium*-mediated transformation	*Actinidia BFT2 (AcBFT2)*	Ever-growing habit with reduced chilling requirement for spring budbreak	Herath et al. (2022)
Arachis hypogaea L.	PEG-mediated protoplast transformation	*Ara h 2A*	Mutant plants with different modifications in the *Ara h 2A* gene, resulting in the loss of Ara h 2A protein expression	Biswas et al. (2022)
Brassica napus	*Agrobacterium*-mediated transformation	*BnS6-Smi2*	Self-incompatible plants	Dou et al. (2021)
Brassica napus	PEG-mediated protoplast transformation	*FRIGIDA (FRI) and phytoene desaturase (PDS)*	Mutated protoplasts were not regenerated into plants	Murovec et al. (2018)
Brassica oleracea	*Agrobacterium*-mediated transformation	*Gibberellin biosynthesis (BolC. GA4.a)*	Dwarf phenotype	Lawrenson et al. (2015)
Brassica oleracea	PEG-mediated protoplast transformation	*FRIGIDA (FRI) and phytoene desaturase (PDS)*	The mutated protoplasts were not regenerated into plants	Murovec et al. (2018)
Brassica rapa	PEG-mediated protoplast transformation	*FRIGIDA (FRI) and phytoene desaturase (PDS)*	The mutated protoplasts were not regenerated into plants	Murovec et al. (2018)
Capsicum annuum L.	*Agrobacterium*-mediated transformation	*CaERF28*	Increased resistance to anthracnose infection	Mishra et al. (2021)
Citrus lanatus	PEG-mediated protoplast transformation	*Citrus lanatus phytoene desaturase (ClPDS)*	Albino phenotype	Tian et al. (2017)
Cucumis melo var. reticulatus	*Agrobacterium*-mediated transformation	*Melon ACC oxidase 1 (CmACO1)*	Reduction in ethylene levels, long shelf life and firmer texture	Nonaka et al. (2023)
Cucumis sativus	*Agrobacterium*-mediated transformation	*Eukaryotic translation initiation factor 4E (eIF4E)*	Cucumber vein yellowing virus, zucchini yellow mosaic virus, and papaya ring spot mosaic virus W	Chandrasekaran et al. (2016)
Cucumis sativus	*Agrobacterium*-mediated transformation	*Csa2G264590 (encodes an auxin transport protein)*	Increase in the number of fruit spines	Liu et al. (2022)

(continued)

Table 2 (continued)

Plant	CRISPR/Cas delivery method	Target gene	Plant phenotype/target trait	References
Daucus carota L. var. sativus Hoffm.	*Agrobacterium*-mediated callus transformation	*Flavanone-3-hydroxylase (DcF3H)*	Inhibition of anthocyanin production in purple carrot	Klimek-Chodacka et al. (2018)
Malus domestica	PEG-mediated protoplast transformation	*DIPM-1,2*	No significant changes in growth and development	Malnoy et al. (2016)
Musa accuminata	*Agrobacterium*-mediated transformation	*Ethylene biosynthesis (MaACO1)*	Prolonged shelf life and decreased ethylene synthesis	Hu et al. (2013)
Nicotiana benthamiana	Viral vector	*Nicotiana benthamiana ARGONAUTE (NbAGO1)*	Disruption of the coding region in the gene paralogs *NbAGO1-H* and *NbAGO1-L*	Cody et al. (2017)
Nicotiana benthamiana	Viral vector	*NbZIP65*	Loss-of-function mutation in the *NbZIP65* gene	Ariga et al. (2020)
Solanum lycopersicum	*Agrobacterium*-mediated transformation	*ARGONAUTE7 (SlAGO7)*	The first leaves have leaflets that are sessile, while the later-formed leaves lack laminae	Brooks et al. (2014)
Solanum lycopersicum	*Agrobacterium*-mediated transformation	*Ripening inhibitor (RIN)*	Fruits have a delayed ripening process and have less red pigment than normal	Ito et al. (2015)
Solanum lycopersicum	*Agrobacterium*-mediated transformation	*Blade-on-petiole (SlBOP)*	Abnormalities in the development of the inflorescence	Xu et al. (2016)
Solanum lycopersicum	*Agrobacterium*-mediated transformation	*Auxin Response Factor Aux/IAA9 (SlIAA9)*	Variations in leaf morphology and production of seedless fruits	Ueta et al. (2017)
Solanum lycopersicum	*Agrobacterium*-mediated transformation	*SlAGAMOUS-LIKE 6 (SlAGL6)*	Seedless fruit production with normal weight and shape under heat stress conditions	Klap et al. (2017)
Solanum lycopersicum	*Agrobacterium*-mediated transformation	*Mildew resistant locus (SlMlo1)*	Resistance to powdery mildew	Nekrasov et al. (2017)
Solanum lycopersicum	*Agrobacterium*-mediated transformation	*Alcobaça (SLALC)*	Long shelf life	Yu et al. (2017)

(continued)

Table 2 (continued)

Plant	CRISPR/Cas delivery method	Target gene	Plant phenotype/target trait	References
Solanum lycopersicum	*Agrobacterium*-mediated transformation	*SlMYC2*	Increased number of flowers, reduced fruit setting rate, and changed fruit shape. Reduced number of seeds, or even seedless fruits, susceptibility to *Botrytis cinerea*	Shu et al. (2020)
Solanum lycopersicum	*Agrobacterium*-mediated transformation	*Glutamate decarboxylase (SlGAD2, SlGAD3)*	Plants carrying a truncated C-terminal in *SlGAD2* exhibited elevated GABA levels but experienced reduced plant size, decreased flowering, and lower fruit yield, while plants with a truncated C-terminal in *SlGAD3* showed enhanced GABA accumulation in fruits without impacting plant size or flowering	Nonaka et al. (2017)
Solanum lycopersicum	*Agrobacterium*-mediated transformation	*Alcohol dehydrogenase (ALC)*	Decrease in the production of ethanol	Yu et al. (2017)
Solanum lycopersicum	*Agrobacterium*-mediated transformation	*Brassinosteroid receptor 1 (BZR1)*	Overexpressing BZR1 resulted in improved heat stress tolerance	Yin et al. (2018)
Solanum tuberosum	PEG-mediated protoplast transformation	*Granule-bound starch synthase (StGBSS)*	Mutants lacked GBSS gene. Lower amylose content in starch, increased amylopectin-amylose ratio	Andersson et al. (2017)
Solanum tuberosum	*Agrobacterium rhizogenes*-mediated transformation of in vitro-grown shoots	*16α-hydroxylation (St16DOX)*	Steroidal glycoalkaloids (SGAs) free potato	Nakayasu et al. (2018)

4 Genome Editing in Ornamental Crops

Ornamental crops are primarily cultivated for their aesthetic value and are used for landscaping, decoration, and overall beautification. The ornamental horticulture industry is a significant sector that generates substantial economic activity. It includes plant nurseries, flower shops, landscape design, maintenance services, and the production and sale of ornamental plants, contributing to job creation and economic growth. The main areas of focus in genome editing for ornamental crops include altering flower color, fragrance, shape, size, and overall plant architecture. The genetic transformation of ornamental crops allows for the development of visually appealing flowers and the creation of plants with unique colors and shapes. However, similar to fruits and vegetables, certain ornamental cultivars pose challenges in terms of transformation using *Agrobacterium*. Most of the initial studies in ornamental crops are focused on targeting the *PDS* gene, to conduct a proof-of-concept study before editing specific traits. Researchers have achieved successful editing of the *PDS* gene in rose (Wang et al. 2023), lily (Yan et al. 2019), petunia (Zhang et al. 2016), chrysanthemum (Liu et al. 2023), and *Hieracium (*Henderson et al. 2020*)* demonstrating the feasibility of gene modification in these ornamental plants. Gene editing techniques in flower research are mainly done to explore and unravel the molecular pathways involved in flower color determination, facilitating the development of novel flower varieties. In poinsettia, the wishbone flower, and *Petunia*, researchers targeted the *flavanone 3-hydroxylase (F3H)* gene to change the flower color. The *F3H* gene is also a great indicator to determine the success of gene editing in flowers apart from the *PDS* gene. Table 3 provides a comprehensive overview of CRISPR/Cas-mediated gene editing in ornamental plants.

Table 3 Recent progress in CRISPR/Cas genome editing in ornamental crops

Plant	CRISPR/Cas delivery system, vectors	Target gene	Plant phenotype/target trait	References
Chrysanthemum indicum	*Agrobacterium*-mediated transformation	*CiPDS*	Albino phenotype	Liu et al. (2023)
Chrysanthemum morifolium	*Agrobacterium*-mediated transformation	*Yellowish-green fluorescent (CpYGFP)*	Disruption of fluorescence protein	Kishi-Kaboshi et al. (2017)
Euphorbia pulcherrima	*Agrobacterium*-mediated transformation	*Flavanone 3-hydroxylase gene (F3H)*	Change of bract color from vivid red (RHS 45B) to vivid reddish-orange (RHS 33A), accompanied by a decreased cyanidin to pelargonidin ratio	Nitarska et al. (2021)
Hieracium piloselloides	*Agrobacterium*-mediated transformation	*PDS*	Albino phenotype	Henderson et al. (2020)

(continued)

Table 3 (continued)

Plant	CRISPR/Cas delivery system, vectors	Target gene	Plant phenotype/target trait	References
Ipomoea nil	*Agrobacterium*-mediated transformation	*Carotenoid cleavage dioxygenase (CCD)*	Change in carotenoid accumulation	Watanabe et al. (2018)
Lilium longiflorum	*Agrobacterium*-mediated transformation	*LpPDS*	Albino phenotype	Yan et al. (2019)
Lilium pumilum	*Agrobacterium*-mediated transformation	*LpPDS*	Albino phenotype	Yan et al. (2019)
Petunia	PEG-mediated protoplast transformation	*PhACO genes*	Extended flower longevity compared with the wild-type (WT) plants due to a significant reduction in ethylene production in the petals and pistils	Xu et al. (2020)
Petunia	PEG-mediated protoplast transformation	*F3H*	Change in flower color	Yu et al. (2021)
Petunia	PEG-mediated protoplast transformation	*Nitrate reductase (PhNR)*	Nitrate assimilation deficiency	Subburaj et al. (2016)
Petunia hybrid	*Agrobacterium*-mediated transformation	*PDS*	Albino phenotype	Zhang et al. (2016)
Phalaenopsis orchid	*Agrobacterium*-mediated transformation	*MADS*	Change in floral initiation and development	Tong et al. (2020)
Rosa hybrida	*Agrobacterium*-mediated transformation	*RhPDS (PHYTOENE DESATURASE)*	Photobleached leaf phenotype resulting from PDS gene loss	Wang et al. (2023)
Rosa hybrida	*Agrobacterium*-mediated transformation	*RhEIN2*	Flower opening completely blocked due to compromised ethylene response	Wang et al. (2023)
Torenia fournieri	*Agrobacterium*-mediated transformation	*Flavanone 3-hydroxylase gene (F3H)*	Change in flower color	Nishihara et al. (2018)

5 Challenges and Future Perspectives for the Improvement of Horticulture Crops Through Genome Editing

The improvement of horticulture crops through CRISPR/Cas genome editing faces several challenges. In contrast to animals, the process of genome editing in plants involves plant transformation, which includes the delivery of editing reagents into plant cells, the selection of edited cells, and the regeneration of complete plants with the desired edits. Despite significant technological advancements spanning over three decades, the steps of transformation and regeneration continue to pose challenges in most crop species (Zhang et al. 2014). Many horticultural crops lack established protocols for gene editing, making the process more time-consuming and resource-intensive. More effort should be made to develop efficient and species-specific gene editing protocols (Bernabé-Orts et al. 2019). One significant concern in genome editing is the possibility of off-target effects, which can lead to unintended modifications in the genome with potential consequences. Ensuring the specificity and precision of genome editing techniques is crucial to minimize off-target effects and their potential impacts on the organism. Designing specific single guide RNAs (sgRNAs) with minimal predicted off-targets is a common strategy to prevent off-target effects in most cases (Manghwar et al. 2020). However, further research is necessary to determine the prevalence of unintended on-target changes, such as large chromosomal insertions, deletions, or inversions, induced by CRISPR/Cas in plants (Hahn and Nekrasov 2019). Using deep learning and predictive tools, the cleavage efficiency and off-target effects of the gRNA can be predicted, which can help in making gene editing in plants more efficient (Konstantakos et al. 2022).

Plant cells contain two subcellular compartments, mitochondria and chloroplasts; each of these organelles has its genome, known as mitochondrial DNA (mtDNA) and chloroplast DNA (cpDNA), respectively, containing essential genes responsible for respiration and photosynthesis (Møller et al. 2021; Sugita and Sugiura 1996). While CRISPR/Cas9 is more efficient for nuclear genome editing due to its RNA-guided specificity, one major challenge is the delivery of the sgRNA and Cas9 endonuclease into the organelles (Glass et al. 2018). *Agrobacterium*-mediated transformation is the most commonly used method for delivering the CRISPR/Cas components. It raises concerns related to genetically modified organisms (GMOs). Once the genome editing process is complete, the CRISPR/Cas9 and sgRNA constructs become redundant and can be removed from the plant. This separation results in transgene-free crop plants that are virtually identical to naturally occurring variations. In many countries, these edited plants are not categorized as GMOs, allowing them to be grown without the customary regulations imposed on GMOs. However, achieving transgene-free plants through genetic segregation can be a demanding and time-consuming task, especially for crops with extensive polyploid genomes. The development and optimization of direct delivery methods offer a promising solution for producing non-transgenic crops across a wider range of plant species.

The application of CRISPR/Cas has already yielded significant advancements in crop breeding, and it is clear that we have only scratched the surface of its potential. As we continue to explore and harness these powerful tools, we can expect even more remarkable breakthroughs in the field of crop improvement, offering immense possibilities for enhancing horticultural productivity, sustainability, and resilience. The ongoing research and innovation in gene editing hold great promise for the future of horticulture, opening up new avenues for addressing global challenges such as food security, climate change, and environmental sustainability.

References

Ahmed T et al (2021) Nanoparticle-based amelioration of drought stress and cadmium toxicity in rice via triggering the stress responsive genetic mechanisms and nutrient acquisition. Ecotoxicol Environ Saf 209:111829

Ali Z et al (2015) Efficient virus-mediated genome editing in plants using the CRISPR/Cas9 system. Mol Plant 8(8):1288–1291

Andersson M et al (2017) Efficient targeted multiallelic mutagenesis in tetraploid potato (*Solanum tuberosum*) by transient CRISPR-Cas9 expression in protoplasts. Plant Cell Rep 36:117–128

Ariga H, Toki S, Ishibashi K (2020) Potato virus X vector-mediated DNA-free genome editing in plants. Plant Cell Physiol 61(11):1946–1953

Bekkaoui F et al (1987) The isolation and culture of protoplasts from an embryogenic cell suspension culture of *Picea glauca* (moench) voss. Plant Cell Rep 6(6):476–479

Benedé S et al (2016) The rise of food allergy: environmental factors and emerging treatments. EBioMedicine 7:27–34

Bernabé-Orts JM et al (2019) Assessment of Cas12a-mediated gene editing efficiency in plants. Plant Biotechnol J 17(10):1971–1984

Biswas S et al (2022) Optimization of protoplast isolation and transformation for a pilot study of genome editing in peanut by targeting the allergen gene Ara h 2. Int J Mol Sci 23(2):837

Brookes G, Barfoot P (2018) Farm income and production impacts of using GM crop technology 1996–2016. GM Crops Food 9(2):59–89

Brooks C et al (2014) Efficient gene editing in tomato in the first generation using the clustered regularly interspaced short palindromic repeats/CRISPR-associated9 system. Plant Physiol 166(3):1292–1297

Chandrasekaran J et al (2016) Development of broad virus resistance in non-transgenic cucumber using CRISPR/Cas9 technology. Mol Plant Pathol 17(7):1140–1153

Cocking EC (1974) The isolation of plant protoplasts. Methods Enzymol 31:578–583

Cody WB, Scholthof HB, Mirkov TE (2017) Multiplexed gene editing and protein overexpression using a tobacco mosaic virus viral vector. Plant Physiol 175(1):23–35

Crossway A et al (1986) Integration of foreign DNA following microinjection of tobacco mesophyll protoplasts. Mol Gen Genet MGG 202:179–185

Demirer GS et al (2020) Carbon nanocarriers deliver siRNA to intact plant cells for efficient gene knockdown. Sci Adv 6(26):eaaz0495

Demirer GS et al (2021) Nanotechnology to advance CRISPR–Cas genetic engineering of plants. Nat Nanotechnol 16(3):243–250

Deng W et al (2009) A novel method for induction of plant regeneration via somatic embryogenesis. Plant Sci 177(1):43–48

Dou S et al (2021) Generation of novel self-incompatible Brassica napus by CRISPR/Cas9. Plant Biotechnol J 19(5):875

Doudna JA, Charpentier E (2014) The new frontier of genome engineering with CRISPR-Cas9. Science 346(6213):1258096
Fagerlund RD, Staals RH, Fineran PC (2015) The Cpf1 CRISPR-Cas protein expands genome-editing tools. Genome Biol 16(1):1–3
García-Murillo L et al (2023) CRISPRa-mediated transcriptional activation of the SlPR-1 gene in edited tomato plants. Plant Sci 329:111617
Ghogare R et al (2021) Genome editing reagent delivery in plants. Transgenic Res 30:321–335
Glass Z et al (2018) Engineering the delivery system for CRISPR-based genome editing. Trends Biotechnol 36(2):173–185
Gleditzsch D et al (2019) PAM identification by CRISPR-Cas effector complexes: diversified mechanisms and structures. RNA Biol 16(4):504–517
Hahn F, Nekrasov V (2019) CRISPR/Cas precision: do we need to worry about off-targeting in plants? Plant Cell Rep 38(4):437–441
Hamada H et al (2017) An in planta biolistic method for stable wheat transformation. Sci Rep 7(1):11443
Hamada H et al (2018) Biolistic-delivery-based transient CRISPR/Cas9 expression enables in planta genome editing in wheat. Sci Rep 8(1):1–7
Hauptmann R et al (1987) Transient expression of electroporated DNA in monocotyledonous and dicotyledonous species. Plant Cell Rep 6:265–270
Henderson SW et al (2020) Efficient crispr/cas9-mediated knockout of an endogenous phytoene desaturase gene in t1 progeny of apomictic hieracium enables new strategies for apomixis gene identification. Genes 11(9):1064
Herath D et al (2022) CRISPR-Cas9-mediated mutagenesis of kiwifruit BFT genes results in an evergrowing but not early flowering phenotype. Plant Biotechnol J 20(11):2064–2076
Holme IB, Wendt T, Holm PB (2013) Intragenesis and cisgenesis as alternatives to transgenic crop development. Plant Biotechnol J 11(4):395–407
Hu C-H et al (2013) An efficient protocol for the production of chit42 transgenic Furenzhi banana (Musa spp. AA group) resistant to *Fusarium oxysporum*. In Vitro Cell Dev Biol Plant 49:584–592
Huang T-K, Puchta H (2021) Novel CRISPR/Cas applications in plants: from prime editing to chromosome engineering. Transgenic Res 30:529–549
Ito Y et al (2015) CRISPR/Cas9-mediated mutagenesis of the RIN locus that regulates tomato fruit ripening. Biochem Biophys Res Commun 467(1):76–82
Kato-Inui T et al (2018) Clustered regularly interspaced short palindromic repeats (CRISPR)/CRISPR-associated protein 9 with improved proof-reading enhances homology-directed repair. Nucleic Acids Res 46(9):4677–4688
Kishi-Kaboshi M, Aida R, Sasaki K (2017) Generation of gene-edited *Chrysanthemum morifolium* using multicopy transgenes as targets and markers. Plant Cell Physiol 58(2):216–226
Klap C et al (2017) Tomato facultative parthenocarpy results from Sl AGAMOUS-LIKE 6 loss of function. Plant Biotechnol J 15(5):634–647
Kleinstiver BP et al (2019) Engineered CRISPR–Cas12a variants with increased activities and improved targeting ranges for gene, epigenetic and base editing. Nat Biotechnol 37(3):276–282
Klimek-Chodacka M et al (2018) Efficient CRISPR/Cas9-based genome editing in carrot cells. Plant Cell Rep 37:575–586
Komarnytsky S et al (2022) Gains and losses of agricultural food production: implications for the twenty-first century. Annu Rev Food Sci Technol 13:239–261
Konstantakos V et al (2022) CRISPR–Cas9 gRNA efficiency prediction: an overview of predictive tools and the role of deep learning. Nucleic Acids Res 50(7):3616–3637
Krens F et al (1982) In vitro transformation of plant protoplasts with Ti-plasmid DNA. Nature 296(5852):72–74
Kujur S, Senthil-Kumar M, Kumar R (2021) Plant viral vectors: expanding the possibilities of precise gene editing in plant genomes. Plant Cell Rep 40:931–934

Lacroix B, Citovsky V (2020) Biolistic approach for transient gene expression studies in plants. Methods Mol Biol 2124:125–139

Lawrenson T et al (2015) Induction of targeted, heritable mutations in barley and *Brassica oleracea* using RNA-guided Cas9 nuclease. Genome Biol 16:1–13

Lazzeri PA (1995) Stable transformation of barley via direct DNA uptake: electroporation-and PEG-mediated protoplast transformation. In: Plant gene transfer and expression protocols. Springer, Berlin, pp 95–106

Lee X, Tan J, Cheng L (2022) Gamma aminobutyric acid (GABA) enrichment in plant-based food—a mini review. Food Rev Intl:1–22

Liu X et al (2022) NS encodes an auxin transporter that regulates the 'numerous spines' trait in cucumber (*Cucumis sativus*) fruit. Plant J 110(2):325–336

Liu L et al (2023) Developing a UV–visible reporter-assisted CRISPR/Cas9 gene editing system to alter flowering time in *Chrysanthemum indicum*. Plant Biotechnol J 21(8):1519–1521

Lowe K et al (2016) Morphogenic regulators Baby boom and Wuschel improve monocot transformation. Plant Cell 28(9):1998–2015

Luo Y et al (2021) Development of a Csy4-processed guide RNA delivery system with soybean-infecting virus ALSV for genome editing. BMC Plant Biol 21(1):1–12

Maher MF et al (2020) Plant gene editing through de novo induction of meristems. Nat Biotechnol 38(1):84–89

Maity A et al (2018) Influence of metal nanoparticles (NPs) on germination and yield of oat (*Avena sativa*) and berseem (*Trifolium alexandrinum*). Proc Natl Acad Sci India Sect B Biol Sci 88:595–607

Malnoy M et al (2016) DNA-free genetically edited grapevine and apple protoplast using CRISPR/Cas9 ribonucleoproteins. Front Plant Sci 7:1904

Manghwar H et al (2020) CRISPR/Cas systems in genome editing: methodologies and tools for sgRNA design, off-target evaluation, and strategies to mitigate off-target effects. Adv Sci 7(6):1902312

Marzban G et al (2005) Localisation and distribution of the major allergens in apple fruits. Plant Sci 169(2):387–394

Mishra R et al (2021) A single transcript CRISPR/Cas9 mediated mutagenesis of CaERF28 confers anthracnose resistance in chilli pepper (*Capsicum annuum L.*). Planta 254(1):5

Molla KA, Yang Y (2019) CRISPR/Cas-mediated base editing: technical considerations and practical applications. Trends Biotechnol 37(10):1121–1142

Møller IM, Rasmusson AG, Van Aken O (2021) Plant mitochondria—past, present and future. Plant J 108(4):912–959

Monsur MB et al (2020) Base editing: the ever expanding clustered regularly interspaced short palindromic repeats (CRISPR) tool kit for precise genome editing in plants. Genes 11(4):466

Murovec J et al (2018) DNA-free genome editing of *Brassica oleracea* and *B. rapa* protoplasts using CRISPR-Cas9 ribonucleoprotein complexes. Front Plant Sci 9:1594

Nakayasu M et al (2018) Generation of α-solanine-free hairy roots of potato by CRISPR/Cas9 mediated genome editing of the St16DOX gene. Plant Physiol Biochem 131:70–77

Nekrasov V et al (2017) Rapid generation of a transgene-free powdery mildew resistant tomato by genome deletion. Sci Rep 7(1):1–6

Nishihara M et al (2018) Application of the CRISPR/Cas9 system for modification of flower color in *Torenia fournieri*. BMC Plant Biol 18(1):331

Nitarska D et al (2021) First genome edited poinsettias: targeted mutagenesis of flavonoid 3′-hydroxylase using CRISPR/Cas9 results in a colour shift. Plant Cell Tissue Organ Cult 147(1):49–60

Nonaka S et al (2017) Efficient increase of ɣ-aminobutyric acid (GABA) content in tomato fruits by targeted mutagenesis. Sci Rep 7(1):1–14

Nonaka S, Ito M, Ezura H (2023) Targeted modification of CmACO1 by CRISPR/Cas9 extends the shelf-life of *Cucumis melo* var. reticulatus melon. Front Genome Ed 5:1176125

O'Connell MR (2019) Molecular mechanisms of RNA targeting by Cas13-containing type VI CRISPR–Cas systems. J Mol Biol 431(1):66–87
Ooi KH et al (2021) An engineered CRISPR-Cas12a variant and DNA-RNA hybrid guides enable robust and rapid COVID-19 testing. Nat Commun 12(1):1739
Pandey AK et al (2019) An improved enzyme-linked immunosorbent assay (ELISA) based protocol using seeds for detection of five major peanut allergens Ara h 1, Ara h 2, Ara h 3, Ara h 6, and Ara h 8. Front Nutr 6:68
Porterfield H et al (2009) Effector activity of peanut allergens: a critical role for Ara h 2, Ara h 6, and their variants. Clin Exp Allergy 39(7):1099–1108
Rasheed A et al (2022) The role of nanoparticles in plant biochemical, physiological, and molecular responses under drought stress: a review. Front Plant Sci 13:976179
Sánchez-Monge R et al (1999) Lipid-transfer proteins are relevant allergens in fruit allergy. J Allergy Clin Immunol 103(3):514–519
Sandhya D et al (2020) The present and potential future methods for delivering CRISPR/Cas9 components in plants. J Genet Eng Biotechnol 18:1–11
Santos A, Van Ree R (2011) Profilins: mimickers of allergy or relevant allergens? Int Arch Allergy Immunol 155(3):191–204
Sharangi AB, Datta S (2015) Value addition of horticultural crops: recent trends and future directions. Springer, Berlin
Shen J et al (2014) Isolation, culture, and transient transformation of plant protoplasts. Curr Protoc Cell Biol 63(1):2.8.1–2.8.17
Shu P et al (2020) CRISPR/Cas9-mediated SlMYC2 mutagenesis adverse to tomato plant growth and MeJA-induced fruit resistance to *Botrytis cinerea*. J Agric Food Chem 68(20):5529–5538
Songstad DD et al (2017) Genome editing of plants. Crit Rev Plant Sci 36(1):1–23
Subburaj S et al (2016) Site-directed mutagenesis in Petunia × hybrida protoplast system using direct delivery of purified recombinant Cas9 ribonucleoproteins. Plant Cell Rep 35(7):1535–1544
Sugita M, Sugiura M (1996) Regulation of gene expression in chloroplasts of higher plants. Plant Mol Biol 32:315–326
Sun Y, Li J, Xia L (2016) Precise genome modification via sequence-specific nucleases-mediated gene targeting for crop improvement. Front Plant Sci 7:1928
Taha EA, Lee J, Hotta A (2022) Delivery of CRISPR-Cas tools for in vivo genome editing therapy: trends and challenges. J Control Release 342:345–361
Tian S et al (2017) Efficient CRISPR/Cas9-based gene knockout in watermelon. Plant Cell Rep 36:399–406
Tong CG et al (2020) High-efficiency CRISPR/Cas-based editing of Phalaenopsis orchid MADS genes. Plant Biotechnol J 18(4):889
Torney F et al (2007) Mesoporous silica nanoparticles deliver DNA and chemicals into plants. Nat Nanotechnol 2(5):295–300
Ueta R et al (2017) Rapid breeding of parthenocarpic tomato plants using CRISPR/Cas9. Sci Rep 7(1):507
Uranga M et al (2021) CRISPR-Cas12a genome editing at the whole-plant level using two compatible RNA virus vectors. CRISPR J 4(5):761–769
Vieths S (2020) Allergens in fruits and vegetables. In: Handbook of plant and fungal toxicants. CRC Press, Boca Raton, pp 157–174
Wang C et al (2023) An efficient CRISPR/Cas9 platform for targeted genome editing in rose (Rosa hybrida). J Integr Plant Biol 65(4):895–899
Watanabe K et al (2018) Alteration of flower colour in *Ipomoea nil* through CRISPR/Cas9-mediated mutagenesis of carotenoid cleavage dioxygenase 4. Transgenic Res 27:25–38
Wu S et al (2020) Establishment of a PEG-mediated protoplast transformation system based on DNA and CRISPR/Cas9 ribonucleoprotein complexes for banana. BMC Plant Biol 20:1–10
Xu C et al (2016) Control of inflorescence architecture in tomato by BTB/POZ transcriptional regulators. Genes Dev 30(18):2048–2061

Xu J et al (2020) CRISPR/Cas9-mediated editing of 1-aminocyclopropane-1-carboxylate oxidase1 enhances Petunia flower longevity. Plant Biotechnol J 18(1):287–297
Yan R et al (2019) Establishment of efficient genetic transformation systems and application of CRISPR/Cas9 genome editing technology in *Lilium pumilum* DC. Fisch and *Lilium longiflorum* White Heaven International. J Mol Sci 20(12):2920
Yin Y et al (2018) BZR1 transcription factor regulates heat stress tolerance through FERONIA receptor-like kinase-mediated reactive oxygen species signaling in tomato. Plant Cell Physiol 59(11):2239–2254
Yu Q-h et al (2017) CRISPR/Cas9-induced targeted mutagenesis and gene replacement to generate long-shelf life tomato lines. Sci Rep 7(1):11874
Yu J et al (2021) Simultaneous targeting of duplicated genes in Petunia protoplasts for flower color modification via CRISPR-Cas9 ribonucleoproteins. Plant Cell Rep 40(6):1037–1045
Zhang F, Wen Y, Guo X (2014) CRISPR/Cas9 for genome editing: progress, implications and challenges. Hum Mol Genet 23(R1):R40–R46
Zhang B et al (2016) Exploiting the CRISPR/Cas9 system for targeted genome mutagenesis in petunia. Sci Rep 6(1):20315
Zhang C et al (2022) Virus-induced gene editing and its applications in plants. Int J Mol Sci 23(18):10202

CRISPR/Cas Genome Editing in Fruit Crops: Recent Advances, Challenges, and Future Prospects

Jayachandran Halka, Nandakumar Vidya, Packiaraj Gurusaravanan, Annamalai Sivaranjini, Arumugam Vijaya Anand, and Muthukrishnan Arun

Abstract Fruit crop offers a wide variety of beneficial metabolites and nutrients that favor human health. Increasing concerns over food and nutritional instability caused by significant climatic changes have posed challenges to crop development, quality, and yield in fruit crops. Conventional farming methods are time-consuming and labor-intensive and are not able to provide long-term solutions to meet the current challenges in fruit cultivation. In this era, newly developed technology like clustered regulatory interspaced short palindromic repeats (CRISPR/Cas)-based genome editing could be a promising approach for trait improvement in fruit crops. This technology can offer efficient means to modify targeted genes that lead to desirable features such as improving fruit quality traits, increasing fruit yield, changing plant architecture that favors fruit development, enhancing nutritional levels, improving shelf life, knocking out genes producing anti-nutrient compounds, improving tolerance to abiotic stress, and reducing fruit disease susceptibility. Furthermore, its simple operation and high mutation efficiency have encouraged researchers working in fruit crops to introduce this technology to generate new germplasm via gene-directed mutation. It is possible to swiftly create newly improved varieties for the development of crucial agronomic traits by precisely editing key genes/transcription factors in fruit crops using CRISPR/Cas. In this chapter,

J. Halka · N. Vidya · M. Arun (✉)
Department of Biotechnology, Bharathiar University, Coimbatore, Tamil Nadu, India
e-mail: arun@buc.edu.in

P. Gurusaravanan
Department of Botany, Bharathiar University, Coimbatore, Tamil Nadu, India

A. Sivaranjini
Department of Biotechnology, Dwaraka Doss Goverdhan Doss Vaishnav College, Chennai, Tamil Nadu, India

A. V. Anand
Department of Human Genetics and Molecular Biology, Bharathiar University, Coimbatore, Tamil Nadu, India
e-mail: avahgmb@buc.edu.in

J.-T. Chen, S. Ahmar (eds.), *Plant Genome Editing Technologies*, Interdisciplinary Biotechnological Advances,
https://doi.org/10.1007/978-981-99-9338-3_10

we summarized the mechanism and applications of CRISPR/Cas as a possible technology in improving agronomical traits that benefit fruit crop breeding.

Keywords CRISPR/Cas · Fruits · Genome editing · Nutritional trait improvement

1 Introduction

Today, in the era of nutritional insecurity, fruits serve as a fundamental component in agriculture production acting as one of the primary reservoirs of vital nutrients for human consumption. In general, fruits are a rich source of vitamins (A, B, C, D, and E), fiber, antioxidants, and minerals (iron, zinc, iodine, and calcium) (Newell-McGloughlin 2008). For centuries, fruit crops have had huge demand and are been extensively cultivated (Dalla et al. 2017). However, fruit cultivation is hampered by biological and environmental factors such as genetically diverse genomes, adverse climate conditions, prolonged life cycles, scarcity of fertile terrain, disease assault, animal-pest attacks, and viral diseases (Leisner 2020). On the other hand, to date, continuous efforts have been made to improve the agronomical traits in fruit crops. Such efforts include achieving higher yields by increasing productivity, obtaining biotic and abiotic-resistant crops, reducing the cost of production, shortening the period for seeds sowing and harvesting, improving nutritional content, and improving wider adaptability (Leisner 2020).

It is imperative to enhance the productivity and sustainability of fruit crops through scientific discoveries and apply time-to-time technological improvements in plant breeding. Advancements in plant breeding have resulted in the adoption of our strategies, namely, conventional breeding (through cross-breeding and selection), mutation-based, transgene-based, and genome editing-based (Hickey et al. 2019). Since conventional breeding and traditional breeding are time-consuming and labor-intensive in nature, researchers are relying on alternative methods for improving the characteristic features of fruit crops (Ashkani et al. 2015). The alternative methods include clustered regularly interspersed short palindromic repeats (CRISPR)/Cas, transcription activator-like effector nucleases (TALENs), and zinc-finger nucleases (ZFNs) (Voytas and Gao 2014). Since TALENS and ZFNs have significant drawbacks, such as inducing recessive random mutations and causing polyploidy in crops while obscuring phenotypic effects, CRISPR/Cas is considered a more effective technique.

CRISPR/Cas genome editing system was adapted from the adaptive immunity of bacteria and archaea as a defense mechanism against viruses and plasmids (Wiedenheft et al. 2012). Thus, CRISPR/Cas 9 is used as a promising tool for altering the genes in fruit crops as it offers a wide range of benefits, including simplified target design, increased efficacy, enhanced precision, ability for multiplexing, absence of off-target effects, capacity to target multiple alleles, cost-effectiveness, ease of delivery and execution, and availability of in silico methods for designing and assessing the designed single-guide RNA (sgRNA) (Yin et al. 2017). The recent advancements in CRISPR/Cas, such as base editor and prime editor, have expanded their possibilities in modifying genomic DNA without the use of double-stranded breaks (DSBs) surpassing traditional DSB-mediated genome editing.

The chapter primarily focuses on discussing the applications of CRISPR/Cas-based genome editing in various fruit crops with special emphasis on nutritional improvements, postharvest loss, biotic and abiotic stresses, and plant growth and development. In addition, this chapter also discusses the future challenges associated with desirable trait improvement in fruit crops using CRISPR/Cas technology.

2 Components and Mechanisms of Action

The CRISPR/Cas system consists of two major components: the Cas9 protein, which acts as a single endonuclease, and synthetic sgRNA (Barrangou et al. 2007). The guide RNA is made of CRISPR RNA (crRNA specifically targets DNA by pairing with target sequence) and trans-activating CRISPR RNA (trac RNA serves as a binding scaffold for Cas9 nuclease). The sgRNA directs Cas9 and recognizes the target sequence from the gene of interest through its 5′crRNA complementary base pair component. Further, gRNA binds to the Cas9 and generates changes in the protein (inactive Cas9 to active form). The sgRNA and Cas 9 form a complex and this complex recognizes the protospacer-adjacent motif (PAM) sequence at 5′-NGG-3′ in creating DSB and triggers cellular DNA repair mechanisms (Chen et al. 2019). Subsequently, these DSBs can be repaired by endogenous DNA repair pathways either via homology-dependent repair (HDR) or by nonhomologous end-joining pathways. The nonhomologous end joining (NHEJ) repair mechanism typically results in small insertions, deletions, or substitutions and does not require a homologous repair template, ultimately resulting in genome modification and loss of gene function. On the other hand, the homology direct repair (HDR) mechanism relies on the homologous DNA template to introduce insertions, mutations, or substitutions of DNA segments (Komor et al. 2016; Ahmar et al. 2020) (Figs. 1 and 2).

3 Application of CRISPR/Cas in Fruit Crops

The remarkable precision and efficiency of CRISPR/Cas technology for genome editing have inspired researchers to utilize this system for the improvement of fruit crops (Biswas et al. 2021; Tiwari et al. 2021). In this chapter, we summarized a comprehensive overview of the diverse applications of CRISPR/Cas in fruit crops (Fig. 3 and Table 1).

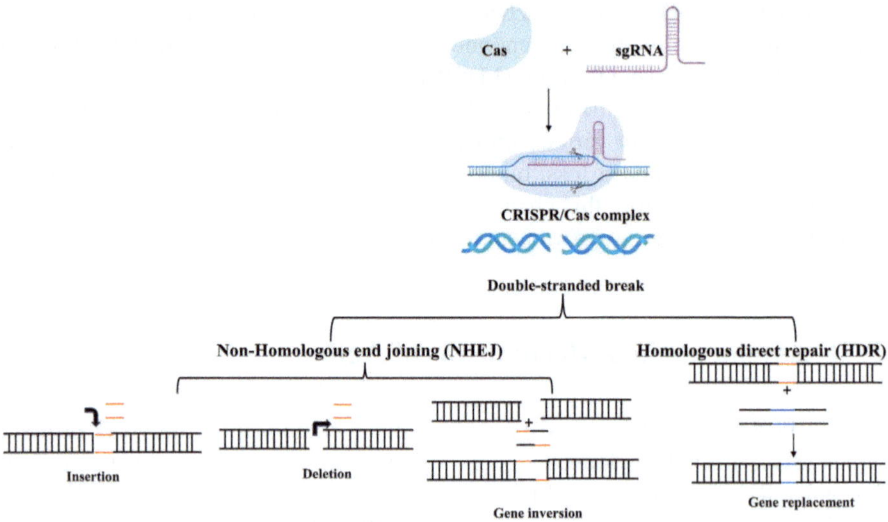

Fig. 1 Mechanism involved in CRISPR/Cas 9 technique

3.1 *Nutritional Improvements*

With a tremendous increase in population, developing nutritional-rich fruits by altering micronutrients that are found in lower quantities using CRISPR/Cas system could be the available approach. Furthermore, this technique opens up the possibilities of enhancing the quality of fruits and investigating the essential heritages of biosynthetic pathways leading to improvements in vitamins, minerals, flavors, and antioxidants (Ren et al. 2016). Ultimately, this technique could be a promising tool for eradicating micronutrient deficiencies in fruit crops (Ren et al. 2016). The *L-idonate dehydrogenase* (*IdnDH*) was reported to play a significant role in the metabolic process of tartaric acid production (Dutt et al. 2020). It was observed that the knockdown of *IdnDH* in grapes using CRISPR/Cas improved the tartaric acid (83.53 mg/g) content with a mutation frequency of 100% by the deletion of 1–3 nucleotides (Dutt et al. 2020). *Phytoene desaturase* (*PDS*) plays an important role in improving the carotenoid content by converting colorless 15 cis phytoene into bright red color. Knocking out *PDS1* and *PDS2* in citrus by CRISPR/Cas9 ameliorates both carotenoid content and chlorophyll (Nakajima et al. 2017). In another study, the researchers showcased the effectiveness of CRISPR/Cas 9 technology in enhancing carotenoid content in grapes by targeting *Vitis vinifera PDS* (*VvPDS*) (Kaur et al. 2020). A similar attempt was made in bananas to enrich carotenoid content in edited lines (24 μg/g) compared to control plants (4 μg/g) by knocking out *Lycopene epsilon cyclase* (*LYCε* is a branching functional gene that modulates the metabolic flux for converting α-carotene to β-carotene in banana) using CRISPR/Cas 9 technology (Wang et al. 2021a). *β-glucosidases* are some of the key functional genes that contribute toward the enhancement of abscisic acid (ABA) content

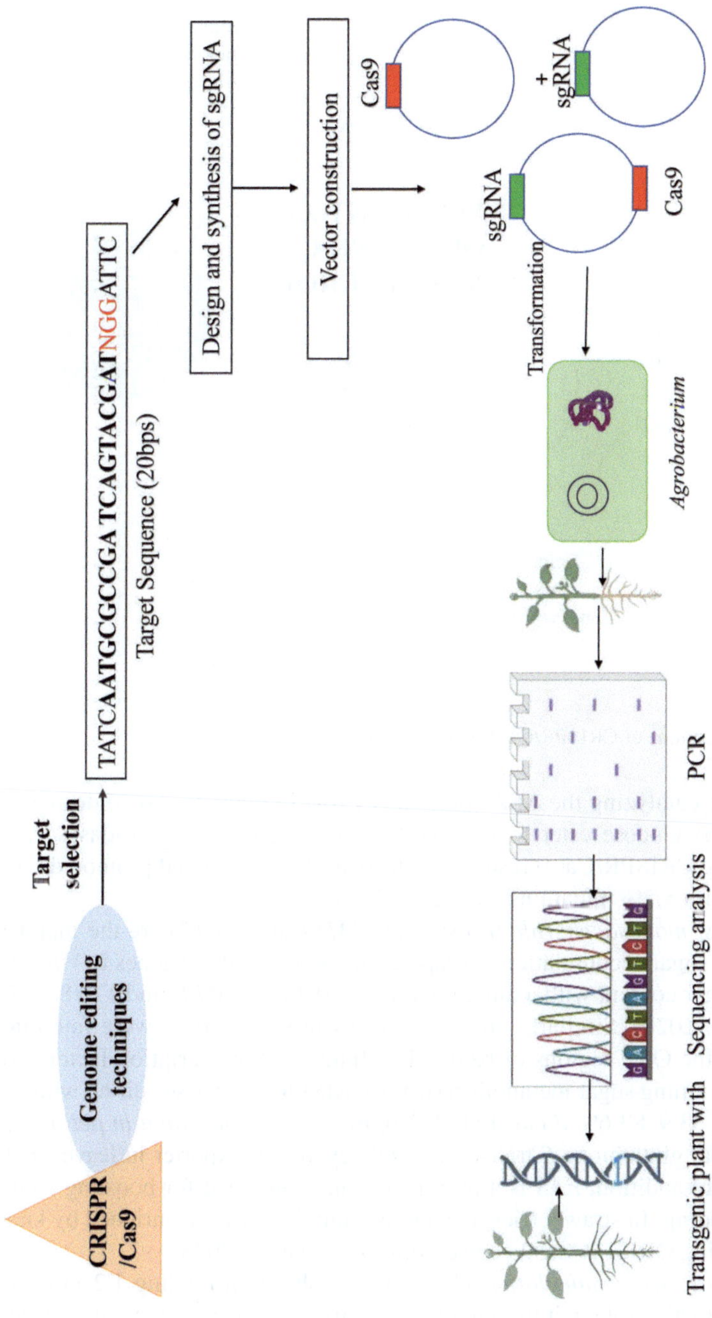

Fig. 2 Systematic scheme describing the various steps involved in CRISPR/Cas9-based genome editing

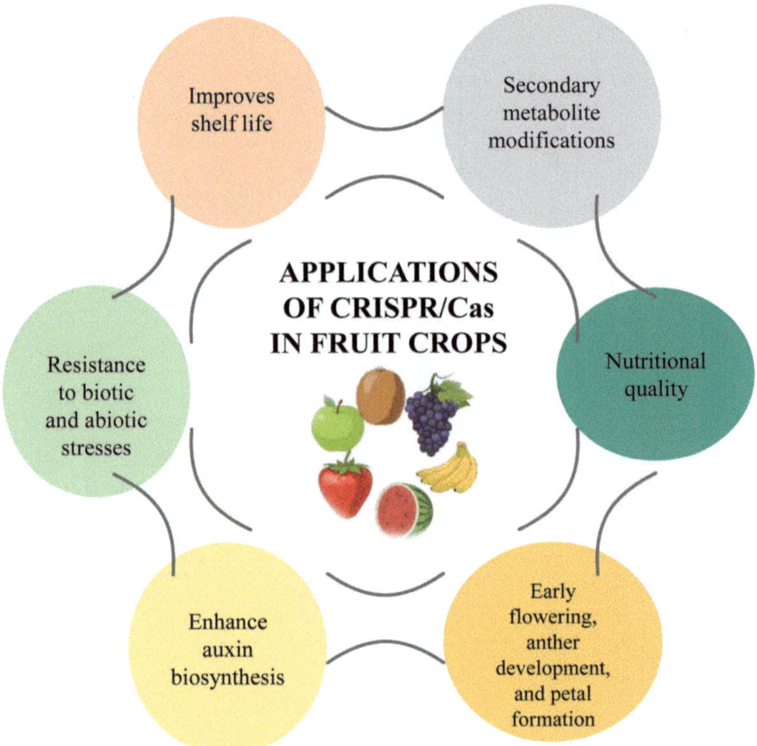

Fig. 3 Applications of CRISPR/Cas 9 in fruit crops

in seeds, by catalyzing the ABA (plays a key role in regulating seed dormancy and germination) glucose ester to free ABA. Knocking out β-glucosidases in watermelon using CRISPR/Cas 9 resulted in decreased seed size and promoted seed germination with 5.3% mutations (Ren et al. 2021).

Tonoplast monosaccharide transporter (*TMT1* and *TMT2*) are the major genes involved in sugar accumulation in grapes. Knocking out these genes in *Vitis vinifera* reduced sugar content with a mutation rate of 10.38% (*TMT1*) and 17.78% (*TMT2*) (Xing et al. 2020). Similarly, the sugar levels in strawberries were fine-tuned by modifying the ORF regions of bZIPs11.1 (functional transcription factor involved in reprogramming sugar metabolism) using CRISPR/Cas9 base editing with a mutation rate of 33.9–83.6% (Gao et al. 2020). *Reduced anthocyanin in petioles* (*RAP*) is a principal glutathione *S*-transferase anthocyanin in exporter in leaves, petioles, and fruits. In addition, *RAS* is a promising candidate gene for boosting fruit color during breeding. In strawberries, the anthocyanin levels were enriched by knocking out *RAS* using CRISPR/Cas9 gene editing (Zhang et al. 2023).

Phosphate overaccumulation (*PHO2*) is a ubiquitin-binding E2 enzyme that plays a distinctive role in improving fruit quality. This enzyme regulates the absorption and transportation of phosphor (P) in plants enhancing plant growth and fruit

Table 1 Summary of recent studies using CRISPR/Cas 9 genome editing in economically important fruit crop

S. no.	Fruit crop	Cultivar	Target gene	Explant and transformation	Improved trait	Mutation rate (%)	References
Nutritional improvement							
1	*Vitis vinifera* (grapes)	Chardonnay	*IdnDH*	Suspension cells; *Agrobacterium*-mediated transformation	Tartaric acid biosynthesis	100	Dutt et al. (2020)
2	*Citrus sinensis* (citrus)	EV2	*PDS1 and PDS2*	Embryonic calli cells; *Agrobacterium*-mediated transformation	Increased carotenoid and chlorophyll content	100	Nakajima et al. (2017)
3	*Vitis vinifera* (grapes)	Neo Muscat	*PDS*	Embryogenic calli *Agrobacterium-mediated* transformation	Increased carotenoid biosynthesis	–	Kaur et al. (2020)
4	*Musa acuminata* (banana)	Grand Naine	*LCYε*	Immature male flower; embryogenic cell suspension	Enhancement of β-carotene	–	Wang et al. (2021a)
5	*Citrullus lanatus* (watermelon)	ZXJM	*BG1*	Seeds; *Agrobacterium*-mediated transformation	Decreased seed size and promoted seed germination	5.3	Ren et al. (2021)
6	*Vitis vinifera* (grapes)	Pinot noir	*TMT1 and TMT2*	Leaves: *Agrobacterium* -mediated transformation	Decreases sugar accumulation in root stock	10.38 and 17.78	Xing et al. (2020)
7	*Fragaria vesca* (strawberry)		*bZIPs1.1*	Leaf; *Agrobacterium*-mediated transformation	Improving sugar content	33.9–83.6	Gao et al. (2020)
8	*Fragaria × Ananassa* (strawberry)	Yellow wonder, Ningu	*RAP*	Calli; *Agrobacterium*-mediated transformation	Fruit color improvement	–	Zhang et al. (2023)
9	*Fragaria vesca* (strawberry)	Ruegen	*PHO2*	Seeds; *Agrobacterium*-mediated transformation	Phosphorous content and improves fruit quality	11–89	Idah et al. (2007)
Post harvest loss							

(continued)

Table 1 (continued)

S. no.	Fruit crop	Cultivar	Target gene	Explant and transformation	Improved trait	Mutation rate (%)	References
10	*Musa acuminata* (banana)	Grand Naine	*MADS1 and MADS2*	Immature male flower; embryogenic cell suspension	Delay ripening and increase shelf life	–	Hu et al. (2021)
11	*Musa acuminata* (banana)	Brazilian	*ACO1*	*Agrobacterium* -mediated transformation	Improves shelf life	–	López-Casado et al. (2023)
12	*Fragaria* × *ananassa Duch.*	Chandler	*PG1*	Leaf disc from micropropagated plant; *Agrobacterium*-mediated transformation	Improves fruit firmness	Small base deletions	Locco et al. (2001)
Increase resistance to disease							
Biotic stress							
13	*Vitis vinifera* (grapes)		*MLO7*	Protoplast transformation	Increases resistance to powdery mildew	0.1	Jia et al. (2017)
14	*Citrus paradisi* (citrus)	Duncan grape fruit	*LOB1*	Epicotyl; agroinfiltration and *Agrobacterium*-mediated transformation	Citrus canker resistance	30–50	Peng et al. (2017)
15	*Citrus sinensis* (citrus)		*LOB1 promoter*	Epicotyl; *Agrobacterium*-mediated transformation	Enhance citrus canker resistance	11.5–64.7	Wang et al. (2019)
16	*Citrus sinensis (citrus)*	Wanjinchen orange	*WRKY22*	Somatic embryos; *Agrobacterium*-mediated transformation	Resistance to citrus canker	68–85.7	Tripathi et al. (2019)
17	*Musa* sp. (banana)	Gonja Manjaya	*eBSV*	Seeds; *Agrobacterium*-mediated transformation	Banana streak virus	95	Zhang et al. (2020)
18	*Citrullus lanatus* (water melon)	Sumi1	*psk1*	*Seeds; Agrobacterium*-mediated transformation	Resistance to *Fusarium oxysporum*	92	Sharma and Gayen (2021)
Abiotic stress							
19	*Vitis vinifera* (grapes)	Muscat Hamburg	*PAT1*	Micropropagated plant; embryogenic calli suspension	Reduced cold tolerance	50	Wang et al. (2021b)
Plant growth and development							

S. no.	Fruit crop	Cultivar	Target gene	Explant and transformation	Improved trait	Mutation rate (%)	References
20	*Fragaria vesca* (strawberry)	Yellow wonder 5AF7	*TAA1 and ARF8*	Protoplast transformation	Controlling sensitivity toward plant hormone auxin	49–75	Zhou et al. (2018)
21	*Actinidia chinensis* (kiwi fruit)	Hort 16A	*CEN*	Male plants; *Agrobacterium*-mediated transformation	Axillary inflorescence to terminal flowers	75%	Varkonyi-Gasic et al. (2019)
22	*Fragaria vesca* (strawberry)	Camarosa	*TM6*	*Agrobacterium*-stable transformation	Anther development and petal formation	–	Martín-Pizarro et al. (2019)
23	*Pyrus communis* (pears)	Conference	*TFL 1.1*	Young leaves; *Agrobacterium*-mediated transformation	Early flowering	9	Charrier et al. (2019)
24	*Vitis vinifera* (grapes)	–	*CCD7&8*	Embryogenic grape calli	Shoot branching	–	Ren et al. (2020)
25	*Actinidia chinensis* (kiwi fruit)	Hort16-A	*BFT*	Seeds; *Agrobacterium*-mediated transformation	Abridged dormancy and early bud break	High mutation rate	Herath et al. (2022)
26	*Fragaria Vesca* (strawberry)	Yellow wonder 5AF7	*AGL62*	Seeds; *Agrobacterium*-mediated transformation	Endosperm development	–	Guo et al. (2022)
27	*Fragaria vesca* (strawberry)	Ruegen	*SEP3*	Seeds; stable transformation	Fruit development	–	Pi et al. (2021)
28	*Musa accuminata* (banana)	Gros michel	*GA20OX2*	Embryogenic cell suspension	Semi dwarf mutants	–	Shao et al. (2020)
Producing albino phenotype							
29	*Mallus* × *domestica* (apple)	Gala	*PDS*	Young leaves; *Agrobacterium*-mediated transformation	Albino phenotype	85	Naing et al. (2019)
30	*Citrullus lanatus* (watermelon)	–	*PDS*	Protoplast; *Agrobacterium*-mediated transformation	Albino type plants	100	Tian et al. (2017)

(continued)

Table 1 (continued)

S. no.	Fruit crop	Cultivar	Target gene	Explant and transformation	Improved trait	Mutation rate (%)	References
31	*Actinidia chinensis* (kiwi fruit)	Hongyang	*PDS*	Leaf; *Agrobacterium*-mediated transformation	Albino and dwarf phenotype	7.4–91.67	Wang et al. (2018a)
32	*Musa species* (banana)	Sukalindiizi	*PDS*	Embryogenic cell suspension; *Agrobacterium*-mediated transformation	Albinism and variegated phenotype	100	Naim et al. (2018)
33	*Musa species* Cavendish (banana)	Williams	*PDS*	Immature male flower; *Agrobacterium*-mediated transformation	Albinism and dwarfism	100	Zhang et al. (2022a)
34	*Fragaria vesca* (strawberry)	Hawaii 4	*PDS*	Leaf and petiole; *Agrobacterium*-mediated transformation	Albino plants	50	Wilson et al. (2019)

Note: *LCYε*, *Lycopene epsilon cyclase*; *PDS*, *phytoene desaturase*; *ACO1*, *aminocyclopropane-1-carboxylate oxidase* 1; *MLO 7*, *mild dew resistance locus* 7; LOB1, *Lateral organ Boundary* 1; *TFL1*, *Terminal flower* 1; *GA20OX2*, *Gibberellin 20 oxidase* 2; *IdnDH*, *L-idonate dehydrogenase*; *TMT*, *Tonoplast monosaccharide transporter*; *RAP*, *Reduced anthocyanin in petioles*; *CEN*, *CENTRORADIALS*; *BFT*, *BROTHER OF FT AND TFL1*; TAA1, *Tryptophan aminotransferase of Arabidopsis*; *ARF8*, *Auxin response factor 8*; *BG1*, *Beta-glucosidases*; *PAT1*, *Phytochrome A signal transduction*; *PHO2*, *Phosphate over accumulator* (*ubiquitin-binding E2 enzyme*); *PG1*, *Polygalacturonase*; *SEP*, *SEPALLATAE 3*; *Psk1*, *Phytosulfokine*; *eBSV*, *Banana streak virus*; *CCD7* and *CCD8*, *Carotenoid cleavage oxygenase 7/8* quality. Recently, it is stated that editing *PHO2* in strawberries using CRISPR/Cas 9 increased phosphorus (40–64%), anthocyanin (30%), and soluble solids content (26–32%) with 11–89% mutation efficiency (Idah et al. 2007).

3.2 *Postharvest Loss*

Postharvest losses and degradation of fruit quality are negatively impacted by pests, microbial infections, the natural ripening process, and environmental factors such as drought, heat, improper harvesting, storing, packing, and transportation (Olayemi et al. 2010; Mrema and Rolle 2002; Hailu and Derbew 2015). Furthermore, postharvest deterioration can be controlled by implementing various measures, such as reducing the storage temperature; utilizing harvesting factors like hydro cooling, room cooling, and forced air cooling; using ethylene inhibitors; applying calcium treatment for reducing firmness; and implementing hot water treatment for reducing the spread of microbes (Gallagher and Mahajan 2011). CRISPR/Cas 9 technique is an alternative tool for solving the above issues and can extend the shelf life of fruits (Elitzur et al. 2016). It is observed that knocking out of *MADS1* and *MADS2* (which are two of the major genes that are involved in fruit ripening) using CRISPR/Cas 9 delayed (6 days) the ripening in edited bananas compared to the control (Hu et al. 2021). Ethylene is one of the gaseous hormones regulated by major genes such as *1-aminocyclopropane-1-carboxylic acid oxidas*e (*ACO*) and *1-aminocyclopropane-1-carboxylic acid* synthase (*ACS*). In the ethylene pathway, the *S*-adenosyl-I-methionine catalyzes the transformation of 1-aminocyclopropane-1-carboxylic acid into ethylene through a reduction process (López-Casado et al. 2023). It is proclaimed that homologous *ACO1* is known to have a significant impact on fruit ripening. Further, knocking down *ACO1* in bananas using CRISPR/Cas results in reduced endogenous ethylene production and even delayed ethylene signal and improved the shelf life of fruits (1–2 days) compared to non-edited bananas (López-Casado et al. 2023). Recently, it is noted that knocking down of *Polygalacturonase 1* (*PG1* that plays a crucial role in maintaining fruit firmness by breaking down pectin found in the plant cell wall) in strawberries improved fruit firmness (30–70%), reduced fungal susceptibility, and decreased transpiration water loss compared to control plant (Locco et al. 2001).

3.3 *Biotic Stress*

The alteration in climatic conditions in fruits alters the quality, quantity, and yield of fruit plants. Biotic stresses such as insects, viruses, fungi, and bacteria can attack plants and cause a great threat to food security (Makarova and Koonin 2015). Powdery mild dew is a fungal disease generally found in fruits. *Mild dew resistance locus O-7* (*MLO-7*) has seven transmembrane domains, and loss of function of this gene results in broad resistance to powdery mild dew (Malnoy et al. 2016). The study revealed that knocking *MLO-7* significantly enhances the resistance to powdery mildew disease with a mutation rate of 0.1–6.9% (Jia et al. 2017). In another study, it is observed that citrus canker-resistant plants were obtained by suppressing the *LOB1* (gene that helps in promoting bacterial growth and pustule formation in citrus) using CRISPR/Cas9 with a mutation rate of 30–50% (Peng et al. 2017). A similar study was conducted on citrus with a different cultivar (Wanjincheng orange) by targeting *LOB1* which showed greater resistance to citrus canker disease with a mutation rate ranging from 11.5% to 64.7% (Wang et al. 2019).

In grapes, a study was conducted using CRISPR/Cas9 to create a mutant *WRKY22* (which is a transcription factor and also facilitates its role as a pathogen-triggered immunity marker). This genetic alteration conferred the resistance of *Xanthomonas citri* subsp. *citri* with 68–85.7% mutation frequency (Tripathi et al. 2019). *Banana streak virus* is an endogenous plant pathogenic *Badnavirus* that affects banana production. In order to control viral pathogenesis in bananas, a daring attempt by knocking out the endogenous banana streak virus (EBSV) using CRISPR/Cas 9 with 95% mutations was conducted (Zhang et al. 2020). *Phytosulfokine 1* (*PSK1*), a pentapeptide plant hormone, has a distinct role in improving plant growth, plant immunity, and plant-pathogen interactions. Knocking *PSK1* in watermelon by CRISPR/Cas 9 conferred resistance to *Fusarium oxysporum* resulting in a 92% mutation rate (Sharma and Gayen 2021).

3.4 Abiotic Stress

Abiotic stresses such as dehydration, soil salinity, and cold and high temperatures have serious risks to fruit crop improvement, yield, and quality (Bressan et al. 2009). Moreover, the adverse effect of abiotic stress negatively influences plant survival and productivity. Generally, multiple genes are responsible for controlling stress responses, which can be studied through physiological and molecular studies using CRISPR/Cas technology (Zhang et al. 2022a). *Phytochrome A signal transduction 1 (PAT1)* is a transcription factor with GRAS domain and *LOX3,* an enzyme crucial for jasmonate biosynthetic aids in regulating jasmonic acid content in fruit crops under cold stress. Knocking down the *PAT1* using CRISPR/Cas9 technique alters *lipoxygenase* (*LOX3*) expression that could significantly lower the cold tolerance in grapes with 50% mutation (Wang et al. 2021b).

3.5 Plant Growth and Development

The growth and development of plants under diverse climatic conditions determine agricultural productivity. Phytohormones, particularly auxin, play a crucial role in directing various processes in plant growth, development, and senescence (Zhang et al. 2022b). Additionally, auxin also modulate photosynthesis and nutrient remobilization. However, fruit crop face hindrances due to various biological and environmental issues, necessitating the exploration of alternative approaches to solve these challenges. It is noted that in strawberries, *tryptophan aminotransferase 1* (*TAA1* prevents the accumulation of indole 3 pyruvic acid) and *Auxin response factor 8* (*ARF8* is a repressor of auxin signaling pathway) are responsible for auxin biosynthesis. In this two-step enzymatic process, indole 3 pyruvic acid is activated by L-tryptophan which is a major component in auxin synthesis, and *YUC* converts indole 3 pyruvic acid to indole acetic acid. Further, *YUC* inhibits auxin biosynthesis and regulates endogenous indole-3-acetic acid (IAA) levels. Knocking out these 2 genes in strawberries using the CRISPR/Cas 9 technique control indole 3 pyruvic acid productions with 49–75% of mutation (Zhou et al. 2018).

CENTRORADICALS (*CEN*) is a key regulator for flowering and inflorescence architecture in plants. A study conducted using male plants as explants to knock down the *CEN* gene by CRISPR/Cas9 rapidly transforms axillary inflorescence (early) into a terminal flower and fruit (long juvenility) in kiwi fruit with 75% mutations (Varkonyi-Gasic et al. 2019). On the other hand, knocking *Tomato MADS-box 6* (*TM6* is responsible for petal formation and stamen development) using CRISPR/Cas9 enhanced the flowering and fruit development stages in strawberries (Martín-Pizarro et al. 2019). Knocking out two genes such as *PDS* and *TFL1* (*Terminal flower 1*) is responsible for regulating flowering time and even plays a special role in regulating the vegetative growth phase of apple and pear. Mutations of these genes by CRISPR/Cas9 culminate in early flowering in apples with a 93% and pears with a 9% mutation rate (Charrier et al. 2019).

In another study, it is reported that knocking two *carotenoid cleavage dioxygenase 7* and *8* (*CCD7* and *CCD8*) helps in improving the shoot branching in grapes using CRISPR/Cas 9 technique (Ren et al. 2020). Later, a subsequent study was conducted in knocking down the *BROTHER OF FT AND TFL1* (*BFT* that delays flowering and improves axillary inflorescences) using CRISPR/Cas9 in kiwifruits to improve the seed dormancy and early bud breakage (Herath et al. 2022). *AGL62* is a functional gene identified as a regulator for endosperm cellularization and regulates expression of the auxin transport from endosperm to seed coat. Knocking *AGL62* using CRISPR/Cas9 finetuned the developmental stages of strawberries (Guo et al. 2022). Deletion of *SEPPALLATE 3* (*SEP3* is a crucial gene necessary for fruit and flower development) in strawberries using the CRISPR/Cas 9 system helps in improving various parthenocarpic phases to facilitate floral organogenesis (Pi et al. 2021). *Gibberellin 20 oxidase 2 (GA20ox2)* regulates seed germination and is also involved in various developmental processes in plants. Semi-dwarf banana mutants were produced by editing the *GA20ox2* using CRISPR/Cas technique (Shao et al. 2020).

3.6 Other Traits

PDS is one of the vital genes that are involved in the synthesis of chlorophyll and carotenoid content. Furthermore, the disruption of *PDS* leads to albinism and dwarfism. Knocking out apple *PDS* using CRISPR/Cas 9 produced an albino phenotype with a mutation rate of 31.8% (Naing et al. 2019). In a separate study, it is proclaimed that knocking the *PDS* in sweet orange by CRISPR/Cas9 resulted in an albino phenotype with a mutation rate of 3.2–3.9% (Idah et al. 2007). Similar to the above study, an albino-type watermelon was developed by knocking *PDS* with 100% mutation efficiency using CRISPR/Cas9 technique (Tian et al. 2017). On the other hand, *PDS* in kiwi were knocked down using CRISPR/Cas9 to generate albino and dwarf phenotypes with a mutation efficiency of 91.67% (Wang et al. 2018a). In bananas, using CRISPR/Cas9 technique, the *PDS* is knocked down to produce 100% mutation with 1–105 nt insertions and 1–55 nt deletions (Naim et al. 2018). A similar investigation was made by knocking down the *PDS* using CRISPR/Cas in bananas which showed 100% indel mutation efficiency (Zhang et al. 2022a). Additionally, efforts were made in producing biallelic mutants at high frequency in strawberries by editing the *PDS* using CRISPR/Cas9 technology (Wilson et al. 2019). As a whole *PDS* is employed as a convenient indicator for CRISPR/Cas9-mediated induced knockouts.

4 Challenges

The applications of using CRISPR/Cas9-based genome editing in fruit crops for increasing the nutritional value, improving the levels of secondary metabolites, and increasing disease resistance in fruits over the past 10 years have posed precise methodological challenges. However, various challenges have been encountered by CRISPR/Cas9 genome editing in fruits. The lack of efficient transformation methods in fruits is one of the bottlenecks that resist the applications of CRISPR in fruits such as grape, strawberry, apple, pear, melon, and banana. Additionally, some fruit cultivars show low susceptibility to *Agrobacterium*-mediated transformation resulting in a very low mutation rate (Mezzetti et al. 2002). Moreover, this technique is laborious and inefficient to discriminate between edited and non-edited lines from suspension cultures (Najafi et al. 2022; Kim et al. 2014; Wang et al. 2018b). In addition, chemical modifications of gRNA and variants of Cas engineering were precisely reported to improve the specificity for about 25,000 folds in fruits (Cromwell et al. 2018; Jin et al. 2020; Zhang et al. 2019). For instance, grapevine has a long juvenility, making it challenging to obtain regenerated plants in a very short period. On the other hand, it is difficult to evaluate and optimize enzyme activity and the protein expression cascade in different fruits. More importantly, obtaining trans-free plants in the next generation is a very tedious process and is one of the major challenges of CRISPR/Cas 9 in fruit crops.

5 Future Perspectives and Conclusion

Among different genome editing technologies, the CRISPR/Cas technique is widely accepted and rapidly expanding and offers various advantages over the existing plant breeding methods. However, some problems remained unsolved which could be addressed by CRISPR technology in fruit crops. The development of next-generation sequencing high throughput helps in generating the genomic data of several unidentified fruit crops (Dongariyal et al. 2023). Moreover, the satisfaction of global demands in fruit crops is revolutionized by enhancing the nutritional attributes of fruits either by improving the yield, producing germplasm resistant to biotic and abiotic stress, improving the post-harvest host, and improving the plant growth and development using CRISPR/Cas technique. Additionally, researchers were able to remove some antinutrients that are harmful to human health in fruit crops using CRISPR/Cas 9 either by altering the genome sequence or by targeting the genes that code for their function. However, fruit crops that are parthenocarpic (banana, orange, and grapefruit) and those with long juvenility (mango, apple, pear) are difficult to edit using CRISPR/Cas technique. Despite these limitations, CRISPR/Cas technique serves as a promising technology for improving the traits and expanding this technique in other fruit crops for the benefit of society.

Acknowledgments The corresponding author is thankful to The Tamil Nadu State Council for Higher Education-Research Grant Project (TANSCHE-RGP), Government of Tamil Nadu, (RGP/2019-20/BU/HECP-0018, Dt.27.04.2021) for providing financial support to Dr. Muthukrishnan Arun.

References

Ahmar S, Saeed S, Khan MHU, Ullah Khan S, Mora-Poblete F, Kamran M, Faheem A, Maqsood A, Rauf M, Saleem S, Hong WJ (2020) A revolution toward gene-editing technology and its application to crop improvement. Int J Mol Sci 21(16):5665

Ashkani S, Rafii MY, Shabanimofrad M, Miah G, Sahebi M, Azizi P, Tanweer FA, Akhtar MS, Nasehi A (2015) Molecular breeding strategy and challenges towards improvement of blast disease resistance in rice crop. Front Plant Sci 6:886

Barrangou R, Fremaux C, Deveau H, Richards M, Boyaval P, Moineau S, Romero DA, Horvath P (2007) CRISPR provides acquired resistance against viruses in prokaryotes. Science 315(5819):1709–1712

Biswas S, Zhang D, Shi J (2021) CRISPR/Cas systems: opportunities and challenges for crop breeding. Plant Cell Rep 40(6):979–998

Bressan R, Bohnert H, Zhu JK (2009) Abiotic stress tolerance: from gene discovery in model organisms to crop improvement. Mol Plant 2(1):1–2

Charrier A, Vergne E, Dousset N, Richer A, Petiteau A, Chevreau E (2019) Efficient targeted mutagenesis in apple and first time edition of pear using the CRISPR-Cas9 system. Front Plant Sci 10:40

Chen K, Wang Y, Zhang R, Zhang H, Gao C (2019) CRISPR/Cas genome editing and precision plant breeding in agriculture. Annu Rev Plant Biol 70:667–697

Cromwell CR, Sung K, Park J, Krysler AR, Jovel J, Kim SK, Hubbard BP (2018) Incorporation of bridged nucleic acids into CRISPR RNAs improves Cas9 endonuclease specificity. Nat Commun 9(1):1448

Dalla CL, Malnoy M, Gribaudo I (2017) Breeding next generation tree fruits: technical and legal challenges. Hortic Res 4:17067

Dongariyal A, Chandra AK, Dongriyal A, Kumar A, Sharma P (2023) Tending genome editing via CRISPR/Cas9-induced mutagenesis: opportunity and challenges for yield, quality and nutritional improvement of fruit crops. Sci Hortic 311:111790

Dutt M, Mou Z, Zhang X, Tanwir SE, Grosser JW (2020) Efficient CRISPR/Cas9 genome editing with citrus embryogenic cell cultures. BMC Biotechnol 20(1):1–7

Elitzur T, Yakir E, Quansah L, Zhangjun F, Vrebalov J, Khayat E, Giovannoni JJ, Friedman H (2016) Banana MaMADS transcription factors are necessary for fruit ripening and molecular tools to promote shelf-life and food security. Plant Physiol 171(1):380–391

Gallagher MS, Mahajan PV (2011) The stability and shelf life of fruit and vegetables. In: Food and beverage stability and shelf life. Elsevier, Amsterdam, pp 641–656

Gao Q, Luo H, Li Y, Liu Z, Kang C (2020) Genetic modulation of RAP alters fruit coloration in both wild and cultivated strawberry. Plant Biotechnol J 18(7):1550–1561

Guo L, Luo X, Li M, Joldersma D, Plunkert M, Liu Z (2022) Mechanism of fertilization-induced auxin synthesis in the endosperm for seed and fruit development. Nat Commun 13:3985

Hailu G, Derbew B (2015) Extent, causes and reduction strategies of postharvest losses of fresh fruits and vegetables—a review. J Biol Agric Healthcare 5(5):49–64

Herath D, Voogd C, Mayo-Smith M, Yang B, Allan AC, Putterill J, Varkonyi-Gasic E (2022) CRISPR-Cas9-mediated mutagenesis of kiwifruit BFT genes results in an evergrowing but not early flowering phenotype. Plant Biotechnol J 20(11):2064–2076

Hickey LT, Hafeez AN, Robinson H, Jackson SA, Leal-Bertioli SCM, Tester M, Gao C, Godwin ID, Hayes BJ, Wulff BBH (2019) Breeding crops to feed 10 billion. Nat Biotechnol 37:744–754

Hu C, Sheng O, Deng G, He W, Dong T, Yang Q, Dou T, Li C, Gao H, Liu S, Yi F (2021) CRISPR/Cas9-mediated genome editing of MaACO1 (aminocyclopropane-1-carboxylate oxidase 1) promotes the shelf life of banana fruit. Plant Biotechnol J 19(4):654

Idah PA, Ajisegiri ES, Yisa MG (2007) Fruits and vegetables handling and transportation in Nigeria. AU J Technol 10(3):175–183

Jia H, Zhang Y, Orbović V, Xu J, White FF, Jones JB, Wang N (2017) Genome editing of the disease susceptibility gene Cs LOB 1 in citrus confers resistance to citrus canker. Plant Biotechnol J 15(7):817–823

Jin S, Fei H, Zhu Z, Luo Y, Liu J, Gao S, Zhag F, Chen YH, Wang Y, Gao C (2020) Rationally designed APOBEC3B cytosine base editors with improved specificity. Mol Cell 79(5):728–740

Kaur N, Alok A, Kumar P, Kaur N, Awasthi P, Chaturvedi S, Pandey P, Pandey AK, Tiwari S (2020) CRISPR/Cas9 directed editing of lycopene epsilon-cyclase modulates metabolic flux for β-carotene biosynthesis in banana fruit. Metab Eng 59:76–86

Kim S, Kim D, Cho SW, Kim J, Kim JS (2014) Highly efficient RNA-guided genome editing in human cells via delivery of purified Cas9 ribonucleoproteins. Genome Res 24:1012–1019

Komor AC, Kim YB, Packer MS, Zuris JA, Liu DR (2016) Programmable editing of a target base in genomic DNA without double-stranded DNA cleavage. Nature 533(7603):420–424

Leisner CP (2020) Climate change impacts on food security-focus on perennial cropping systems and nutritional value. Plant Sci 293:110412

Locco P, Franks T, Thomas MR (2001) Genetic transformation of major wine grape cultivars of Vitis vinifera L. Transgenic Res 10(2):105

López-Casado G, Sánchez-Raya C, Ric-Varas PD, Paniagua C, Blanco-Portales R, Muñoz-Blanco J, Pose S, Matas AJ, Mercado JA (2023) CRISPR/Cas9 editing of the polygalacturonase FaPG1 gene improves strawberry fruit firmness. Hortic Res 10(3):uhad011

Makarova KS, Koonin EV (2015) Annotation and classification of CRISPR-Cas systems. In: CRISPR: methods and protocols. Springer, Berlin, pp 47–75

Malnoy M, Viola R, Jung MH, Koo OJ, Kim S, Kim JS, Velasco R, Nagamangala Kanchiswamy C (2016) DNA-free genetically edited grapevine and apple protoplast using CRISPR/Cas9 ribonucleoproteins. Front Plant Sci 7:1904

Martín-Pizarro C, Triviño JC, Posé D (2019) Functional analysis of the TM6 MADS-box gene in the octoploid strawberry by CRISPR/Cas9-directed mutagenesis. J Exp Bot 70(3):885–895

Mezzetti B, Pandolfini T, Navacchi O, Landi L (2002) Genetic transformation of Vitis vinifera via organogenesis. BMC Biotechnol 27(2):18

Mrema CG, Rolle SR (2002) Status of the postharvest sector and its contribution to agricultural development and economic growth. In: 9th JIRCAS international symposium-value addition to agricultural product, pp 13–20

Naim F, Dugdale B, Kleidon J, Brinin A, Shand K, Waterhouse P, Dale J (2018) Gene editing the phytoene desaturase alleles of Cavendish banana using CRISPR/Cas9. Transgenic Res 27:451–460

Naing AH, Kyu SY, Pe PPW, Park KI, Lee JM, Lim KB, Kim CK (2019) Silencing of the phytoene desaturase (*PDS*) gene affects the expression of fruit-ripening genes in tomatoes. Plant Methods 15:110

Najafi S, Bertini E, D'Incà E, Fasoli M, Zenoni S (2022) DNA-free genome editing in grapevine using CRISPR/Cas9 ribonucleoprotein complexes followed by protoplast regeneration. Hortic Res 10(1):uhac240

Nakajima I, Ban Y, Azuma A, Onoue N, Moriguchi T, Yamamoto T, Toki S, Endo M (2017) CRISPR/Cas9-mediated targeted mutagenesis in grape. PLoS One 12(5):e0177966

Newell-McGloughlin M (2008) Nutritionally improved agricultural crops. Plant Physiol 147(3):939–953

Olayemi FF, Adegbola JA, Bamishaiye EI, Daura AM (2010) Assessment of post-harvest challenges of small scale farm holders of tomotoes, bell and hot pepper in some local government areas of Kano State, Nigeria. Bayero J Pure Appl Sci *3*(2):39–42

Peng A, Chen S, Lei T, Xu L, He Y, Wu L, Yao L, Zou X (2017) Engineering canker-resistant plants through CRISPR/Cas9-targeted editing of the susceptibility gene Cs LOB 1 promoter in citrus. Plant Biotechnol J 15(12):1509–1519

Pi M, Hu S, Cheng L, Zhong R, Cai Z, Liu Z, Yao JL, Kang C (2021) The MADS-box gene FveSEP3 plays essential roles in flower organogenesis and fruit development in woodland strawberry. Hortic Res 8:247

Ren C, Liu X, Zhang Z, Wang Y, Duan W, Li S, Liang Z (2016) CRISPR/Cas9-mediated efficient targeted mutagenesis in Chardonnay (Vitis vinifera L.). Sci Rep 6(1):32289

Ren C, Guo Y, Kong J, Lecourieux F, Dai Z, Li S, Liang Z (2020) Knockout of VvCCD8 gene in grapevine affects shoot branching. BMC Plant Biol 20(1):1–8

Ren C, Liu Y, Guo Y, Duan W, Fan P, Li S, Liang Z (2021) Optimizing the CRISPR/Cas9 system for genome editing in grape by using grape promoters. Hortic Res 8(1):52

Shao X, Wu S, Dou T, Zhu H, Hu C, Huo H, He W, Deng G, Sheg O, Bi F, Gao H (2020) Using CRISPR/Cas9 genome editing system to create MaGA20ox2 gene-modified semi-dwarf banana. Plant Biotechnol J 18(1):17

Sharma P, Gayen D (2021) Plant protease as regulator and signaling molecule for enhancing environmental stress-tolerance. Plant Cell Rep 40(11):2081–2095

Tian S, Jiang L, Gao Q, Zhang J, Zong M, Zhang H, Ren Y, Guo S, Gong G, Liu F, Xu Y (2017) Efficient CRISPR/Cas9-based gene knockout in watermelon. Plant Cell Rep 36:399–406

Tiwari M, Kumar Trivedi P, Pandey A (2021) Emerging tools and paradigm shift of gene editing in cereals, fruits, and horticultural crops for enhancing nutritional value and food security. Food Energy Secur 10(1):e258

Tripathi JN, Ntui VO, Ron M, Muiruri SK, Britt A, Tripathi L (2019) CRISPR/Cas9 editing of endogenous banana streak virus in the B genome of Musa spp. overcomes a major challenge in banana breeding. Commun Biol 2(1):46

Varkonyi-Gasic E, Wang T, Voogd C, Jeon S, Drummond RS, Gleave AP, Allan AC (2019) Mutagenesis of kiwifruit CENTRORADIALIS-like genes transforms a climbing woody

perennial with long juvenility and axillary flowering into a compact plant with rapid terminal flowering. Plant Biotechnol J 17(5):869–880
Voytas DF, Gao C (2014) Precision genome engineering and agriculture: opportunities and regulatory challenges. PLoS Biol 12(6):e1001877
Wang Z, Wang S, Li D, Zhang Q, Li L, Zhong C, Liu Y, Huang H (2018a) Optimized paired-sgRNA/Cas9 cloning and expression cassette triggers high-efficiency multiplex genome editing in kiwifruit. Plant Biotechnol J 16(8):1424–1433
Wang X, Tu M, Wang D, Liu J, Li Y, Li Z, Wang Y, Wang X (2018b) CRISPR/Cas9-mediated efficient targeted mutagenesis in grape in the first generation. Plant Biotechnol J 16(4):844–855
Wang L, Chen S, Peng A, Xie Z, He Y, Zou X (2019) CRISPR/Cas9-mediated editing of CsWRKY22 reduces susceptibility to Xanthomonas citri subsp. citri in Wanjincheng orange (Citrus sinensis (L.) Osbeck). Plant Biotechnol Rep 13:501–510
Wang Y, Wang J, Guo S, Tian S, Zhang J, Ren Y, Li M, Gong G, Zhang H, Xu Y (2021a) CRISPR/Cas9-mediated mutagenesis of ClBG1 decreased seed size and promoted seed germination in watermelon. Hortic Res 8(1):70
Wang Z, Wong DCJ, Wang Y, Xu G, Ren C, Liu Y, Kuang Y, Fan P, Li S, Xin H, Liang Z (2021b) GRAS-domain transcription factor PAT1 regulates jasmonic acid biosynthesis in grape cold stress response. Plant Physiol 186(3):1660–1678
Wiedenheft B, Sternberg SH, Doudna JA (2012) RNA-guided genetic silencing systems in bacteria and archaea. Nature 482(7385):331–338
Wilson FM, Harrison K, Armitage AD, Simkin AJ, Harrison RJ (2019) CRISPR/Cas9-mediated mutagenesis of phytoene desaturase in diploid and octoploid strawberry. Plant Methods 15:1–13
Xing S, Chen K, Zhu H, Zhang R, Zhang H, Li B, Gao C (2020) Fine-tuning sugar content in strawberry. Genome Biol 21(1):1–14
Yin K, Gao C, Qiu JL (2017) Progress and prospects in plant genome editing. Nat Plants 3(8):1–6
Zhang Y, Malzahn AA, Sretenovic S, Qi Y (2019) The emerging and uncultivated potential of CRISPR technology in plant science. Nat Plants 5(8):778–794
Zhang M, Liu Q, Yang X, Xu J, Liu G, Yao X, Ren R, Xu J, Lou L (2020) CRISPR/Cas9-mediated mutagenesis of Clpsk1 in watermelon to confer resistance to Fusarium oxysporum f. sp. niveum. Plant Cell Rep 39:589–595
Zhang Q, Gong M, Xu X, Li H, Deng W (2022a) Roles of auxin in the growth, development, and stress tolerance of horticultural plants. Cells 11(17):2761
Zhang Y, Yu J, Xu X, Wang R, Liu Y, Huang S, Wei H, Wei Z (2022b) Molecular mechanisms of diverse auxin responses during plant growth and development. Int J Mol Sci 23(20):12495
Zhang C, Liu Y, Wang B, Li H, Zhang J, Ma Y, Dai H, Wang Y, Zhang Z (2023) CRISPR/Cas9 targeted knockout FvPHO2 can increase phosphorus content and improve fruit quality of woodland strawberry. Sci Hortic 317:112078
Zhou J, Wang G, Liu Z (2018) Efficient genome editing of wild strawberry genes, vector development and validation. Plant Biotechnol J 16(11):1868–1877

Plant Tissue Culture: A Boon or Enigma in Gene Editing for Plants Using CRISPR/Cas System

Shampa Purkaystha, Biswajit Pramanik, Anamika Das, Sushmita Kumari, and Sandip Debnath

Abstract This chapter delves into the exploration of metabolic pathways, unraveling cellular activities, genetic enhancement, and the production of stress-resistant cell lines, all of which contribute to the development of crops with superior characteristics. It presents how these objectives are attainable through the integration of tissue culture and molecular biology techniques. The potential of gene editing in studying gene function and precise crop breeding is enormous, particularly through its capacity to specifically alter plant structures. The prevalent use of CRISPR/Cas9 technology in gene editing is due to its high precision, simplicity, and efficacy. It enables the simultaneous addition of beneficial alleles and the removal of undesired ones within a single operation. However, the current application of this technology, while effective, is laborious, expensive, and time-consuming and is only achievable in a select few plant species. This limitation presents a considerable hurdle in the plant gene editing field. The chapter also explores various innovative strategies for the administration of CRISPR agents to plants, bypassing the need for tissue culture and regeneration processes. These strategies include viral vector delivery, de novo meristem induction, and possible applications of nanotechnology. A novel method is also introduced that enables gene-edited plant production without the need for tissue culture, called the fast-TrACC technique. This streamlines the process of

---

S. Purkaystha
Department of Genetics and Plant Breeding and Seed Science and Technology, Centurion University of Technology and Management, Paralekhamundi, Odisha, India

B. Pramanik · S. Debnath (✉)
Department of Genetics and Plant Breeding, Palli Siksha Bhavana (Institute of Agriculture), Visva-Bharati, Sriniketan, West Bengal, India
e-mail: sandip.debnath@visva-bharati.ac.in

A. Das
Department of Genetics and Plant Breeding, Bidhan Chandra Krishi Viswavidyalaya, Mohanpur, Nadia, West Bengal, India

S. Kumari
Department of Environmental Science, Institute of Science, Banaras Hindu University, Varanasi, Uttar Pradesh, India

J.-T. Chen, S. Ahmar (eds.), *Plant Genome Editing Technologies*, Interdisciplinary Biotechnological Advances,
https://doi.org/10.1007/978-981-99-9338-3_11

creating gene-edited plants, significantly contributing to the era of precision crop breeding.

Keywords CRISPR/Cas9 · Gene editing · Tissue culture · Molecular biology · Fast-TrACC

1 Introduction

Reverse genetics techniques often attempt to produce null mutants to explore the function of genes, particularly in fundamental science. However, advantageous features in agriculture are frequently caused by gain or modification of function, much like many other mutations that result in phenotypic alterations. As a result, agriculture has to develop molecular tools for precise gene changes. Although many editing methods have been demonstrated to function effectively in some species, changes in cellular or environmental factors may make them inapplicable to use in plants. The selective elimination of harmful or advantageous features from agricultural plants is the principal pre-requisite for the production of single or multiple mutations at the required loci of the plant genome. To create new types, plant breeders began using artificial mutation induction in the middle of the twentieth century. This was accomplished by using genotoxic substances, such as ionizing radiation, which causes many genomic double-strand breaks (DSBs) at random (Stadler 1928). Nonhomologous end joining (NHEJ) is the predominant DSB repair process in plants, and it frequently causes mutations at the break site (Puchta 2005). Site-specific nucleases (SSNs) were made practical to utilize in multicellular eukaryotes (Puchta et al. 1993) and the enzymatic induction of single genomic. It became possible to induce single genomic DSBs enzymatically. The latest developments in CRISPR/Cas technology will aid agriculture in addressing the problems of the twenty-first century connected to pollution, global warming, and the ensuing food crisis (Schaeffer and Nakata 2015).

Due to their enormous potential for use in medicine and all branches of biology and science, CRISPR/Cas-derived genome engineering technologies have progressed extraordinarily quickly in recent years. Many methods are worth examining for their potential use in plants. The development of CRISPR/Cas-mediated genome editing technologies has advanced remarkably in recent years (Zhang et al. 2015). There are several characterized natural CRISPR/Cas nuclease variations. We will assess the potential and practical use of these various technologies in plants. In order to reduce temperature sensitivity, off-side effects, and PAM constraints, engineered Cas proteins have been created. Both types of enzymes have now been used extensively and effectively in several plant species to multiplexed produce one or multiple mutations at the appropriate loci. Specifically created CRISPR/Cas systems provide more precise gene editing in addition to DSB-induced mutagenesis, resulting in both random mutations and preset modifications. Even plants can use the Cas9 nuclease to induce heritable chromosomal rearrangements like inversions and translocations. This method will enable the controlled modification of genetic connections and bring a new degree of genome engineering to plant breeding techniques (Bao et al. 2019).

Apart from CRISPR/Cas systems, several other genome editing techniques are available for more precisely targeting genes of interest. To target DSB distinct sites in the genome, several artificial nucleases, such as zinc-finger nucleases (ZFNs) and transcription activator-like effector nucleases (TALENs), have been created (Voytas 2013). Inducing error-prone NHEJ DSB repair in plants is theoretically the most effective way to employ SSNs for mutagenesis. They can also be utilized to significantly boost the amount of gene targeting (GT) mediated by homologous recombination (HR) (Puchta et al. 1996).

Agrobacterium-mediated transformation of a T-DNA with Wus2, STM, and ipt while editing was induced simultaneously recently represented a significant advancement. Thus, soil-grown tobacco might be used to regrow genome-altered shoots without the need for a tissue culture stage (Maher et al. 2020). With this technique, it may be less expensive and time-consuming to produce gene-edited crops than at the time of the first infection. Additionally, it was proven that the mutation was passed down through the germline to the next generations. Sonchus yellow net rhabdovirus is used as a vector to produce the nuclease Cas9, and the gRNA is another technique for DNA-free editing of somatic plant cells (Ma et al. 2016). The utilization of the virus has a limited host range; use in crops necessitates the employment of RNA viruses that are able to carry the few kbs of extra genetic material.

Additionally, it would be beneficial to take into account additional strategies like grafting for DNA-free editing. Thus, there may be an increase in the number of alternatives for DNA-free plant genome editing over the next years. The use of an RNA viral vector derived from the tobacco rattle virus is another new strategy to pass tissue culture. The virus was altered in a way that led to the production of a sgRNA fused to the mRNA of FT, a flowering factor known to be capable of traveling great distances from cell to cell through the phloem and even across grafting junctions between species. For viral infection, Cas9-expressing transgenic strains of tobacco plants were employed. For gene editing in the infected leaves, the upper regions of the plants than at the time of initial infection are useful. Additionally, passing the mutation down the germline additionally, it was proven that the mutation was passed down through the germline to the next generations (Song et al. 2022).

Genome editing technologies are developing quickly, and current research is focused on using the same precise technology to produce nontransgenic organisms. Various technical developments are there, which may offer a comprehensive understanding of the necessity for genome editing devoid of transgenes. In order to remove the barrier of transgenic regulation, researchers worldwide are devoting consistent effort to this area. However, this strategy has serious ramifications that are genuine for reducing the tension between the acceptability, dependability, and generosity of gene editing technology and sustainably addressing the issues associated with feeding the growing world population. However, this technique has encountered a number of difficulties (Doudna and Charpentier 2014). Major obstacles preventing it from achieving its potential include target specificity, efficiency, usage in a nontransgenic manner, use in perennial crops, optimum delivery mechanism, etc. Transgene-free editing is necessary to apply in crops, but effective gene containment is difficult to achieve. Productions of foreign DNA-free agriproducts are the current improvising area for researchers. DNA-free genome editing can be defined as altering genomic sequences without keeping a trace of foreign DNA footprint in the target organism (Metje-Sprink et al. 2019).

The transformation procedure is one of the constraints of genome editing in plants. Agrobacterium-mediated transformation, biolistic transformation, and PEG-mediated transformation are the three main plant transformation techniques. The transformation of the majority of crops needs tissue culture in order to produce viable plants from somatic cells, except for *Arabidopsis* and its near relatives, which are susceptible to floral dip transformation. This process takes a long time and causes a significant bottleneck for many agricultural plants. Hence, to limit these constraints, techniques for genome editing without using tissue culture were discovered, which might be useful in overcoming the transformation limits in many crops. Voytas and Gao (2014) recently devised two methods, either by de novo induction of the meristem or by mobile gRNAs to edit meristem, both of which are capable of producing genome-modified crops while bypassing the tissue culture stage.

2 Genome Editing Techniques with Special Reference to CRISPR

Since the last decade of the twentieth century, genome modification technologies have surfaced to develop several crops with desirable features. There are a lot of fundamental as well as social misconceptions raised regarding the use of these genetically modified (GM) crops, whether it be the adverse effects on the juxtapositioned genes' activities or public uneasiness due to the conviction of introgression of foreign genes. Despite the endless skeptical notions toward this newly developed technique, several researchers have appointed it to be ecologically safe for humans just like the conventionally raised crops (Das et al. 2023). Hence, taking the developmental expenses and regulatory prerequisites into account, the government decided to limit the applications of transgenesis within a number of commodity crops such as soybean, cotton, maize, etc. Although a bunch of enhancement has been achieved in the approaches of DNA manipulations, all of them come with their imperfections, which can be surpassed by this genome editing. The utilization of modern-day perceptions in these gene manipulation procedures offers more precise sequence alterations in DNA, which eventually leads to the expression of the genes of interest along with lesser cellular toxicity and improved reproducibility (Voytas and Gao 2014). These functionalities have significantly been employed to develop numerous stress-resistant or nutritionally rich crops. Moreover, the application can be witnessed in those lands where cultivation remains a troublesome job to pursue.

Principally, two enzymes, namely, recombinases and nucleases, respectively, are involved in the entire process of genome editing, or more precisely genome engineering. Furthermore, the two strategies are named after these two enzymes, i.e., site-specific recombinases (SSRs), led by the former, and site-specific nucleases (SSNs), led by the latter one.

2.1 SSRs-Mediated Genome Engineering

Generally, permanent DNA sequence alterations are conducted by a number of recombinases in various plant systems in order to manipulate the gene expressions associated with these. The journey of recombinases for three decades progressed through scores of events such as the introgression of complex transgenes combinedly in a single copy number, removal of redundant DNA sequences, explicit insertion of DNA into the target region, and many more. A research study with tobacco has also demonstrated the exemplary capabilities of SSRs to modulate DNA sequences in the extranuclear genome of plastids. Moreover, the multipurpose utility of both insertion of transgenes and the removal of undesirable selectable markers has been well-documented by the study of Wang et al. (2011).

The key reaction followed by recombinases is a recombination of the DNA base pairs, which is immediately preceded by the recognition of the desired sequence. The external SSRs are mainly used in genome modifications of higher organisms as it is difficult to detect their recognition sites. The target sites inside a genome may either be identical or opposite, where recombination reaction takes place, and based on the orientation of the respective sites and recombinase activities, gene products are formed. Notably, a tag and exchange methodology is deployed by SSRs for gene swapping (Fig. 1). The extensive use of SSRs is witnessed in knock-in or knock-out studies in the eukaryotic genome. The exertion of recombinases irrespective of the generation of double-strand breaks (DSBs) and intracellular repair machineries has offered this gene editing tool an added advantage along with the simplicity in site recognition due to the integration of donor DNA. On the other hand, off-target outcomes and tremendously expressed cellular toxicity act as the major hindrances in its implementation (Anastassiadis et al. 2009).

2.2 SSNs-Operated Genome Editing

SSNs, with the DNA- or RNA-binding domains within them, generally cleave the pre-decided sequences in the genome. The cleavage is, after that, succeeded by DNA repair, which creates the fundamental difference from the aforesaid procedure. Compared to the former, these editing methodologies are easier, inexpensive, and efficiently active. Although four different types of nuclease-operated systems are available to be used in genome modifications, namely, meganucleases,

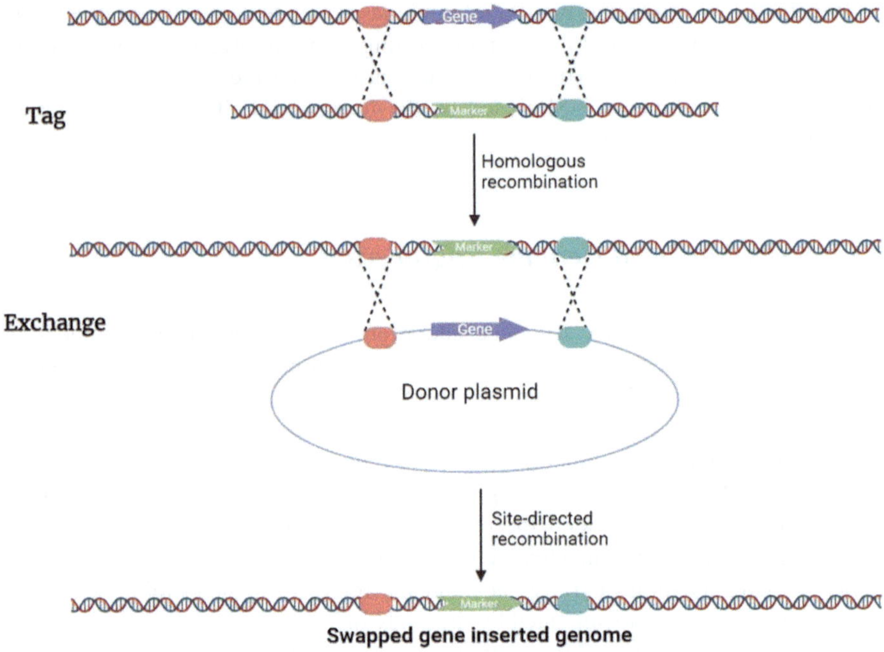

Fig. 1 Diagrammatic representation of gene swapping in SSRs-mediated genome editing. (Created in BioRender: https://biorender.com) (Adapted from Abdallah et al. 2015)

zinc-finger nucleases (ZFNs), transcription activator-like effector nucleases (TALENs), and clustered regularly interspaced short palindromic repeat and its associated protein 9 (CRISPR/Cas9), the last one can be considered to be well-supported and recommended by most groups of researchers (Čermák et al. 2015).

2.2.1 Meganucleases-Associated Editing

Meganucleases belong to a group of endonucleases, which got its breakthrough in the last of the 1980s. These naturally abundant nucleases possess the ability to cleave larger DNA sequences of up to 40 bases, which provides this tool a perfection toward genetic engineering (Fig. 2a). In addition, the creation of a lesser toxic cellular environment gifts a supplementary advantage to its worth. On the contrary, the availability of fewer numbers of meganucleases with insufficient coverage to the entire genome gives backlash to its productivity. The aforesaid drawback is also accompanied by soaring expenses and time consumption during the construction of sequence-directed endonucleases when compared to the rest of the gene-modifying machinery. Besides, meganucleases pose technical challenges while implementing owing to the customization procedures required during the preliminary protein modification techniques. Furthermore, conflicts regarding patent issuance continue to be a hindrance to its execution (Smith et al. 2006).

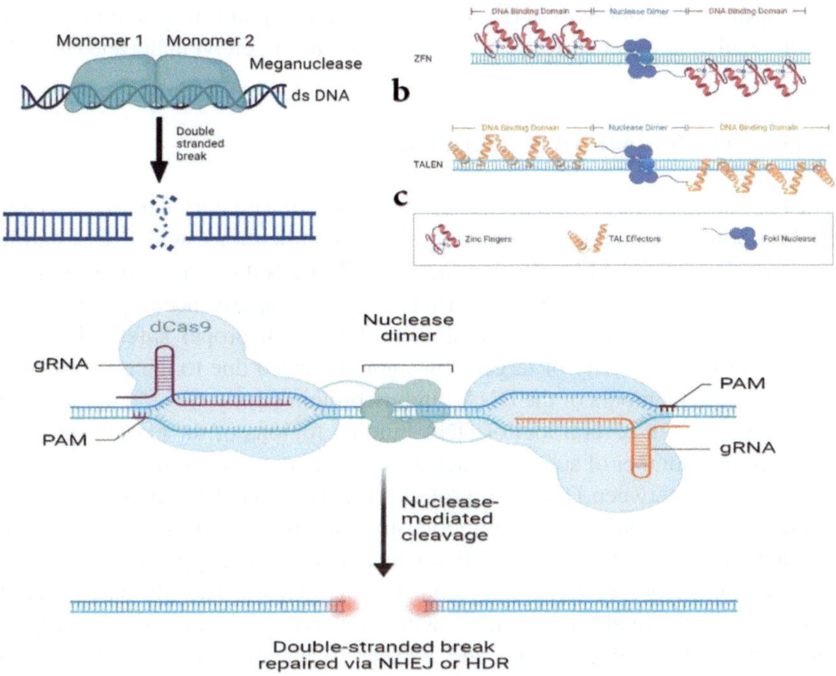

Fig. 2 Schematic representation of SSRs-based genome modifications using (**a**). Meganucleases (**b**). ZFNs (**c**). TALENs (**d**). CRISPR/Cas9. (Created in BioRender: https://biorender.com)

2.2.2 ZFNs-Mediated Editing

Since its discovery in 1996, ZFN also acts as a transcription factor. A very short sequence of DNA ranging up to three base pairs (bp) is detected by the carboxyl (–COOH) terminus of each finger. Thus, almost seven fingers with candidate recognition sequences are required to form the engineered protein sequence that can bind with a DNA of 20 bp length. These specifically designed ZFNs-sourced proteins, after merging with the catalytic domain of endonucleases, get activated to induce DSBs (Fig. 2b). These protein sequences are actually used as gene-modifying tools. A couple of proteins assembled by the association of ZFN and endonuclease usually detect a pair of DNA sequences some bases apart from each other. This event empowers the endonucleases to form a dimer, which eventually generates DSBs (Abdallah et al. 2015).

Like the previous systems, ZFNs-mediated genetic engineering also coexists with a number of disadvantages. This method is practically much more complex than the previous one, and even the management expenses during the construction of protein domains are to some extent higher too. Moreover, the probability of off-target impacts due to inappropriate cleavage between the desired DNA sequences and the cytotoxic nature of nucleases continue to remain as barriers against its efficiency. Therefore, an array of modifications needs to be implemented for developing its productivity, viz., proper selection of plant tissues, minimization of cytological toxicity, occurrence of off-target mutations, etc. (Puchta and Hohn 2010).

2.2.3 TALENs-Induced Modifications

The beginning of the second decade of the twenty-first century witnessed a new start in the domain of nuclease-mediated genetic modifications, which was entitled as transcription activator-like effector nucleases (TALENs) system. This *Xanthomonas*-originated genome engineering tool has exhibited an advanced approach with its much higher efficiency and accessibility along with proper safety (Urnov et al. 2010). Here, the activity of target genes got enhanced due to the effector proteins, which are the members of the DNA-binding protein family. Several tandemly repetitive sequences in several loci of DNA binding regions of the effector proteins are involved in a number of activities such as sequence recognition, improved efficiency of attachment between DNA and proteins, etc. Both the catalytic domain of endonuclease and its befitting monomeric structure formed from the DNA binding domain collectively fabricate TALEN, which is basically a type of artificially developed endonuclease to target the sequence of interest. The ultimate structure of TALEN is collectively incorporated with a catalytic domain, an artificially manufactured DNA binding region, and nuclear localization signal (NLS). Artificial nucleases get attached to the carboxyl terminus, which results in DSBs by the process of dimerization (Fig. 2c) (Abdallah et al. 2015).

The principal obstacle associated with the application of TALENs is the formation of suitable vectors with proper DNA-binding ability. Although the time taken to construct TALENs posed a significant disadvantage, various commercial platforms play a major role in this manufacturing procedure at a large scale, in recent times.

2.2.4 CRISPR/Cas9-Based Gene Editing

Among all the SSNs-mediated genome editing systems, CRISPR/Cas9 evolves to be the most recent one, which originated in the year of 2013. This genome modifying tool is developed with two components, e.g., one is the nuclease part, namely, CRISPR-associated protein, i.e., Cas9, and the rest is the noncoding RNA complex. The complementarity between the RNA and desired DNA sequence helps in the recognition step, after which the cleaving is well-performed by Cas9 (Fig. 2d) (Sar et al. 2022). Moreover, the proper mode of action of this system depends on three fundamental prerequisites, viz., CRISPR mRNA, Cas9 nuclease or Cas9 nickase, and tracrRNA. Cas9 nuclease or wild-type nuclease differs from the Cas9 nickase in its nature of cleavage or cut. The former creates a DSB in DNA, while the latter develops a single-stranded cut or break (SSB) in the targeted sequence. Factually, Cas9, after undergoing a range of mutational events, forms nickase. Further indicated, nickase-mediated genome editing requires two guide RNAs to minimize the off-target reactions. On the other hand, tracrRNA is the transactive RNA that forms an active complex after binding with the crRNA, and this complex provides the breakthrough of target detection as well as cleavage much more efficiently. This complex construct can replace a stretch of 30 bp sequence of DNA with the artificially developed one (Nemudryi et al. 2014).

Basically, its emergence took place for resisting virus attacks in bacteria in order to protect them. Three fundamental steps are involved in this protection mechanism, viz., adaptation or spacer acquisition, transcription or expression of cr RNA, and interference or targeting (Fig. 3). The first stage involves the integration of foreign viral DNA sequences into the host genome. In this step, little segments of the viral genome, in the form of specially designed repeats, get inserted into the CRISPR locus inside the host DNA isolated by short palindromic sequences, i.e., spacer. Cas9 is actually an endonuclease, which is juxtaposed by CRISPR itself. This CRISPR-associated protein also possesses helicase activity (Haft et al. 2005). The next step intrinsically enfolds the process of transcription, where the aforementioned locus gets converted into a stretch of pre-CRISPR RNA or pre-crRNA, which in due course gets matured into CRISPR or crRNA, embracing a single spacer. The prevailing assembly of Cas9 with crRNA-tracrRNA complex at the 3′ end evolves in the last step. Cas-operated DNA interference is caused by tracrRNA immediately after the recognition of complementary sequences to the protospacer by crRNA. These cumulative events ultimately result in the degradation of foreign DNA following the cleavage of the matching sequences. Noteworthy to indicate, another significant element in viral DNA named as protospacer adjacent motif or

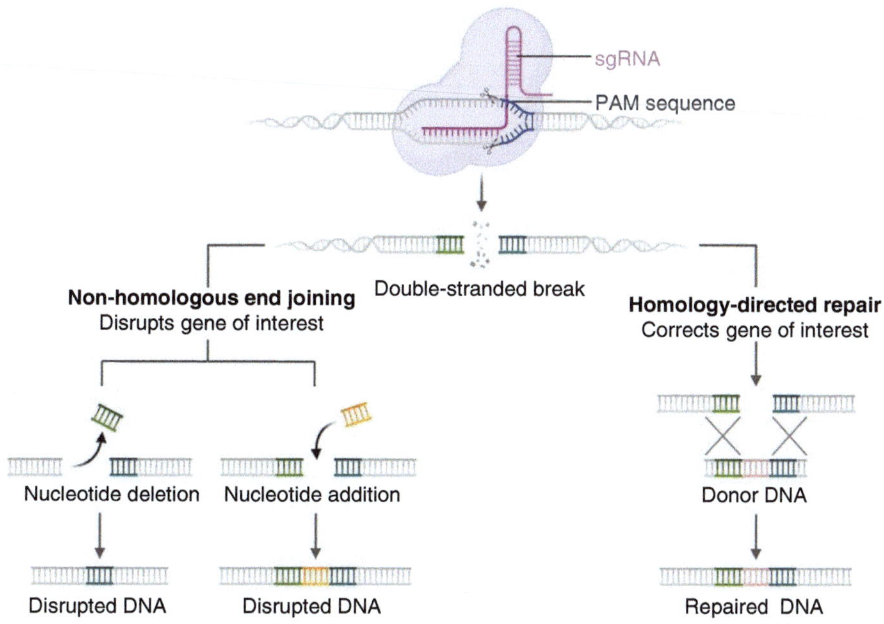

Fig. 3 DNA repair mechanisms followed by CRISPR/Cas9. (Created in: https://biorender.com)

PAM acts as the distinguishing factor between foreign and artificially designed genomes despite their identical sequences.

CRISPR-mediated editing generally goes through DNA repairing while generating DSBs in DNA. Effectively, it follows two distinct pathways for DNA repair, one is homologous end joining (HDR), and the rest is known as nonhomologous end joining (NHEJ) (Fig. 4). The rudimentary difference between these pathways can be recognized through their principles of action. The former follows insertion or substitution policy in accordance with sequence homology, while the latter is able only to insert irrespective of the base sequences. Moreover, NHEJ is independent of the cell cycle, whereas synthesis (S) and second growth or gap (G_2) phases are recommended to be the best for the rest. Notable to mention, although nonhomologous end joining is usually proposed for gene knockout programs, it lacks precision due to the dearth of sequence similarities (Sar et al. 2022).

A cascade of modifications has been acquired while redesigning CRISPR as an RNA-based gene alteration tool in eukaryotes, which comprises of standardization of codons for enhancing Cas9 activities in order to improve transcription inside the cell and construction of guide RNA or sgRNA replacing the expression of two engaged noncoding RNAs, i.e., pre-crRNA and tracrRNA.

CRISPR modifications are recruited with a higher-level precision along with much lesser chances of cytologic toxicity and erroneous events as compared to the rest. Notably, meganucleases, ZFNs, and TALENs act on the basis of protein-DNA interaction. CRISPR/Cas system outpaces this backlash with its progressive mode

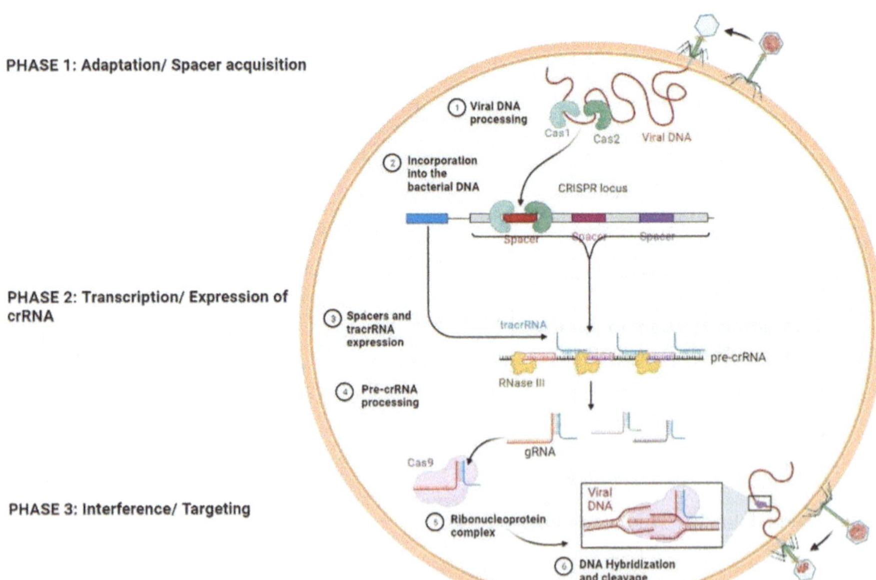

Fig. 4 CRISPR-associated stages of immunization of bacteria against viral genome. (Created in BioRender: https://biorender.com)

of action implying RNA-DNA interaction, which excels the provinces of accuracy. In brief, CRISPR/Cas9 is an economic and effortlessly operated mechanism for accurate detection, editing, rearrangement, and modifications of loci of interest throughout the genome of the living world. Thus, this innovative platform has opened up a revolutionary approach to the world of genetic modifications in the realm of applied biology.

3 PTC (Plant Tissue Culture) and CRISPR/Cas9

PTC has been at the forefront of progress ever since the CRISPR/Cas9 technology was first used in plants. Transgenic plants that have been modified using CRISPR/Cas9 have been regenerated using PTC systems. The CRISPR-Cas system combined with plant tissue culture can directly introduce desired genetic modifications into genotypes that are important for commercial use. Compared to conventional breeding techniques, this technology can integrate genetic modifications into the appropriate lines more quickly and precisely.

Among many PTC systems, some are extensively used for the regeneration of the modified plants by CRISPR/Cas9 such as protoplast culture, callus culture, suspension culture, hairy root (HR) culture, etc. Protoplast culture has been employed for the transformation and regeneration of transgenic plants in monocots like rice, wheat, maize, and barley as well as dicot plants including *Arabidopsis*, tobacco, tomato, etc. The efficiency of the transformation may be calculated with the help of protoplasts transformed, which varies from species to species and even within a species. For instance, it was noted that 18 h after transformation, over half of the protoplasts in *Oryza sativa* produced the marker gene green fluorescence protein, whereas after 36–72 h following the change, this percentage rose to almost 90% (Xie and Yang 2013). After 72 h, one-fourth of the phytoene desaturase gene in *O. sativa* protoplasts had been successfully targeted by gRNA (Shan et al. 2013). In regenerated lines, targeted mutation of *O. sativa* miRNA genes by CRISPR-Cas9 varied from 48% to 89% at all investigated miRNA target locations (Zhou et al. 2017). Various factors can affect how effective a transformation is. For instance, in *Triticum aestivum* transformation efficiency has been reported to range from 3.8% to 60–80% of protoplasts transformed with a construct containing the GFP. In *Zea mays*, 10.67% of the targeted genes were mutated (Loyola-Vargas and Avilez-Montalvo 2018).

Although callus culture is employed less frequently than protoplast culture, they are nonetheless incredibly helpful in the production of transgenic plants in various species. The average mutation frequency for a single-gene targeted sgRNA in the T0 generation ranged from 27.6% to 96.6% in *B. napus* (Yang et al. 2017), whereas a high frequency of mutation (up to 83.56%) was observed in *S. lycopersicum* (Pan et al. 2016). However, using callus cultures, an effective transformation was recorded as a result of which the glutamate decarboxylase of *S. lycopersicum* underwent a mutation eventually enhancing the GABA buildup 7–15-fold and altering fruit size and production (Nonaka et al. 2017).

Transgenic plants transformed by the CRISPR/Cas9 technology have also been successfully regenerated using the shoot culture system in many species such as tomato (*S. lycopersicum*), potato (*S. tuberosum*), Brassica (*B. napus*), etc. It's quite intriguing that only dicotyledonous plants have been able to regenerate from shoots. Dicots are simpler to regenerate from calli than monocots; therefore, this is presumably a reflex. Six lines of Duncan grapefruit were developed as a result of the CRISPR/Cas9 alteration of the canker susceptibility gene CsLOB1 in the fruit. When infected with the pathogen *Xanthomonas citri* subsp. citri (Xcc), those lines with less than 40% mutation rate had canker symptoms resembling those of wild-type grapefruit. After 4 days of Xcc inoculation, the lines with mutation rates greater than 40% did not exhibit any canker signs (Jia et al. 2017).

Hairy root cultures have been extensively utilized to examine secondary metabolites. This technique has been used to regenerate plants with CRISPR/Cas9 modifications in species like *S. lycopersicum*, *B. carinata*, and *Glycine max*. This technique has proven to be quite effective as DNA alterations were found in 95% of the 88 hairy root transgenic events examined in *G. max* (Jacobs et al. 2015). Similarly, suspension cultures and somatic embryogenesis were employed for the transformation in *T. aestivum* and *Ipomoea*, respectively (Loyola-Vargas and Avilez-Montalvo 2018).

For the majority of plant species, tissue culture is necessary after transgene integration into recipient cells in order to produce transgenic plants from the transformed cells. However, it is time-consuming and often requires the growing medium to be supplemented with costly and ecologically hazardous agents (such as antibiotics) in order to select transgenic plants. Furthermore, efficient genetic transformation techniques have yet to be created for many plants, and the efficacy of genetic transformation is frequently genotype-dependent. For some plant species, such as *Arabidopsis*, researchers have developed planta transformation (the transformation of intact plants or plant tissues without callus culture or regeneration), which is efficient at delivering T-DNA into reproductive cells and produces transgenic offspring without involving tissue culture processes (Altpeter et al. 2016). However, major food, oil, and fiber crops have not yet been included in this method. The tissue culture stage in CRISPR-mediated gene editing has been made simpler in a few recent investigations by skipping the selection reagents (Chen et al. 2019), but the complete transforming procedure is still necessary. It is obvious that it is desirable to skip the tissue culture-based genetic transformation process for the practical and effective implementation of plant gene editing technologies. Recently, a promising gene editing method has been developed that eliminates tissue culture by reprogramming genome-edited somatic cells into meristems via the co-expression of developmental regulators (DRs) and gene editing components, allowing direct regeneration of genome-edited plants from somatic cells (Maher et al. 2020).

4 Conclusion

For sexually reproduced crops, genetic segregation is widely recognized to be extremely beneficial when the desired feature is being introduced, and undesirable characteristics may be readily deleted in subsequent generations. However, the introduction of a characteristic frequently results in the insertion of an unfavorable DNA fragment in crops that are reproduced asexually. Additionally, it would be challenging to delete the transgenes by genetic segregation if the transgenic locus and modified locus are related. Genetic segregation such as the use of the recombinase system may therefore be used to separate transgene-free plants from the offspring of backcrossed populations; although the procedure is time-consuming and labor-intensive, it can be used to eliminate transgenes (Anand et al. 2019).

CRISPR-Cas systems are valuable tools for advancing fundamental plant biology and agricultural development due to their resilience in producing targeted genomic alterations. The increasing variety of viral vectors on the market is assisting in overcoming species- or cultivar-specific obstacles to genome editing. When the whole Cas nuclease-sgRNA complex is transiently produced by an RNA virus, using viral vectors as delivery vehicles for CRISPR-Cas reaction components dramatically enhances editing efficiency and gets rid of the time-consuming backcrossing process. Additionally, several studies have shown that the virus is absent in the offspring of infected plants, unless they are vegetatively reproduced (Ali et al. 2018; Ellison et al. 2020; Uranga et al., 2021). Due to the fact that RNA viruses like SYNV (Sonchus yellow net rhabdovirus) do not integrate into host chromosomes during replication, mutant plants grown from infected tissues are not regarded as transgenic. The present bottleneck results from viral vectors' incapacity to change the germline, which has a negative or nonexistent impact on the recovery of plants with heritable modifications (Uranga et al. 2021).

In contrast to earlier discovered DNA-free delivery technologies that required the separation of specialized plant cells or tissues, SYNV enables the direct administration of CRISPR-Cas9 reagents into whole plants, enabling the regeneration of a variety of plant tissues that are suited for regeneration.

The progeny viruses may also be manually transferred from plant to plant, eliminating the usage of *Agrobacterium* pathogens, in addition to the amazing stability of the vector. The potential for off-target alterations when there is a mismatch of two to five nucleotides, as well as the construct's protracted stability inside the host cell, are significant drawbacks. In general, the host range connected to a particular virus species restricts viral delivery techniques. However, it is getting easier to employ reverse-genetics methods for viruses that are comparable to them, and efforts are also being made to create or find a more practical viral vector system (Cao et al. 2021).

In conclusion, there is immense promise for CRISPR-Cas-mediated genome editing in plants using viral vectors. Future developments in plant functional genomics and agricultural biotechnology will be aided by the widespread use of viral delivery methods, which we anticipate will increase as their popularity grows. It is of utmost importance to expand our vision from a tissue culture based to a tissue cultureless regeneration system which ultimately saves time, cost, and labor during transformation.

References

Abdallah NA, Prakash CS, McHughen AG (2015) Genome editing for crop improvement: challenges and opportunities. GM Crops Food 6(4):183–205

Ali M, Anwar S, Shuja MN, Tripathi RK, Singh J (2018) The genus Luteovirus from infection to disease. Eur J Plant Pathol 151:841–860

Altpeter F, Springer NM, Bartley LE, Blechl AE, Brutnell TP, Citovsky V et al (2016) Advancing crop transformation in the era of genome editing. Plant Cell 28(7):1510–1520

Anand A, Wu E, Li Z, TeRonde S, Arling M, Lenderts B et al (2019) High efficiency Agrobacterium-mediated site-specific gene integration in maize utilizing the FLP-FRT recombination system. Plant Biotechnol J 17(8):1636–1645

Anastassiadis K, Fu J, Patsch C, Hu S, Weidlich S, Duerschke K, Stewart AF (2009) Dre recombinase, like Cre, is a highly efficient site-specific recombinase in E. coli, mammalian cells and mice. Dis Model Mech 2(9–10):508–515

Bao A, Burritt DJ, Chen H, Zhou X, Cao D, Tran LSP (2019) The CRISPR/Cas9 system and its applications in crop genome editing. Crit Rev Biotechnol 39(3):321–336

Cao Y, Zhou H, Zhou X, Li F (2021) Conferring resistance to plant RNA viruses with the CRISPR/CasRx system. Virol Sin 36:1–4

Čermák T, Baltes NJ, Čegan R, Zhang Y, Voytas DF (2015) High-frequency, precise modification of the tomato genome. Genome Biol 16:1–15

Chen K, Wang Y, Zhang R, Zhang H, Gao C (2019) CRISPR/Cas genome editing and precision plant breeding in agriculture. Annu Rev Plant Biol 70:667–697

Das A, Mahanta M, Pramanik B, Purkayastha S (2023) Genetically modified crops and crop species adapted to global warming in dry regions. In: Enhancing resilience of dryland agriculture under changing climate: interdisciplinary and convergence approaches. Springer Nature, Singapore, pp 385–409

Doudna JA, Charpentier E (2014) Genome editing. The new frontier of genome engineering with CRISPR-Cas9. Science 346(6213):1258096

Ellison EE, Nagalakshmi U, Gamo ME, Huang PJ, Dinesh-Kumar S, Voytas DF (2020) Multiplexed heritable gene editing using RNA viruses and mobile single guide RNAs. Nat Plants 6(6):620–624

Haft DH, Selengut J, Mongodin EF, Nelson KE (2005) A guild of 45 CRISPR-associated (Cas) protein families and multiple CRISPR/Cas subtypes exist in prokaryotic genomes. PLoS Comput Biol 1(6):e60

Jacobs TB, LaFayette PR, Schmitz RJ, Parrott WA (2015) Targeted genome modifications in soybean with CRISPR/Cas9. BMC Biotechnol 15:1–10

Jia H, Zhang Y, Orbović V, Xu J, White FF, Jones JB, Wang N (2017) Genome editing of the disease susceptibility gene Cs LOB 1 in citrus confers resistance to citrus canker. Plant Biotechnol J 15(7):817–823

Loyola-Vargas VM, Avilez-Montalvo RN (2018) Plant tissue culture: a battle horse in the genome editing using CRISPR/Cas9. In: Loyola-Vargas V, Ochoa-Alejo N (eds) Plant cell culture protocols. Methods in molecular biology, vol 1815. Humana Press, New York, pp 131–148

Ma X, Zhu Q, Chen Y, Liu Y-G (2016) CRISPR/Cas9 platforms for genome editing in plants: developments and applications. Mol Plant 9(7):961–974

Maher MF, Nasti RA, Vollbrecht M, Starker CG, Clark MD, Voytas DF (2020) Plant gene editing through de novo induction of meristems. Nat Biotechnol 38(1):84–89

Metje-Sprink J, Menz J, Modrzejewski D, Sprink T (2019) DNA-free genome editing: past, present and future. Front Plant Sci 9:1957
Nemudryi AA, Valetdinova KR, Medvedev SP, Zakian SM (2014) TALEN and CRISPR/Cas genome editing systems: tools of discovery. Acta Nat 6(3):19–40
Nonaka S, Arai C, Takayama M, Matsukura C, Ezura H (2017) Efficient increase of ɣ-aminobutyric acid (GABA) content in tomato fruits by targeted mutagenesis. Sci Rep 7(1):1–14
Pan C, Ye L, Qin L, Liu X, He Y, Wang J et al (2016) CRISPR/Cas9-mediated efficient and heritable targeted mutagenesis in tomato plants in the first and later generations. Sci Rep 6(1):24765
Puchta H (2005) The repair of double-strand breaks in plants: mechanisms and consequences for genome evolution. J Exp Bot 56(409):1–14
Puchta H, Hohn B (2010) Breaking news: plants mutate right on target. Proc Natl Acad Sci 107(26):11657–11658
Puchta H, Dujon B, Hohn B (1993) Homologous recombination in plant cells is enhanced by *in vivo* induction of double strand breaks into DNA by a site-specific endonuclease. Nucleic Acids Res 21:5034–5040
Puchta H, Dujon B, Hohn B (1996) Two different but related mechanisms are used in plants for the repair of genomic double-strand breaks by homologous recombination. Proc Natl Acad Sci U S A 93:5055–5060
Sar P, Pramanik B, Debnath S (2022) CRISPR/Cas: a new horizon in crop improvement. In: Debnath S, Manjula IK, Shah KR, Bhattacharjee S (eds) Futuristic trends in biotechnology. IIP International Publishers, Chikkamagaluru, pp 9–27
Schaeffer SM, Nakata PA (2015) CRISPR/Cas9-mediated genome editing and gene replacement in plants: transitioning from lab to field. Plant Sci 240:130–142
Shan Q, Wang Y, Li J, Zhang Y, Chen K, Liang Z, Gao C (2013) Targeted genome modification of crop plants using a CRISPR-Cas system. Nat Biotechnol 31(8):686–688
Smith J, Grizot S, Arnould S, Duclert A, Epinat JC, Chames P, Duchateau P (2006) A combinatorial approach to create artificial homing endonucleases cleaving chosen sequences. Nucleic Acids Res 34(22):e149
Song GQ, Urban G, Ryner JT, Zhong GY (2022) Gene editing profiles in 94 CRISPR-Cas9 expressing T0 transgenic tobacco lines reveal high frequencies of chimeric editing of the target gene. Plants 11(24):3494
Stadler LJ (1928) Genetic effect of X-rays in maize. Proc Natl Acad Sci U S A 14:69–75
Uranga M, Vazquez-Vilar M, Orzáez D, Daròs JA (2021) CRISPR-Cas12a genome editing at the whole-plant level using two compatible RNA virus vectors. CRISPR J 4(5):761–769
Urnov FD, Rebar EJ, Holmes MC, Zhang HS, Gregory PD (2010) Genome editing with engineered zinc finger nucleases. Nat Rev Genet 11:636–646
Voytas DF (2013) Plant genome engineering with sequence-specific nucleases. Annu Rev Plant Biol 64:327–350
Voytas DF, Gao C (2014) Precision genome engineering and agriculture: opportunities and regulatory challenges. PLoS Biol 12(6):e1001877
Wang Y, Yau YY, Perkins-Balding D, Thomson JG (2011) Recombinase technology: applications and possibilities. Plant Cell Rep 30:267–285
Xie K, Yang Y (2013) RNA-guided genome editing in plants using a CRISPR–Cas system. Mol Plant 6(6):1975–1983
Yang H, Wu JJ, Tang T, Liu KD, Dai C (2017) CRISPR/Cas9-mediated genome editing efficiently creates specific mutations at multiple loci using one sgRNA in Brassica napus. Sci Rep 7(1):7489
Zhang F, Puchta H, Thomson JG (2015) Advances in new technology for targeted modification of plant genomes. Springer, New York
Zhou J, Deng K, Cheng Y, Zhong Z, Tian L, Tang X et al (2017) CRISPR-Cas9 based genome editing reveals new insights into microRNA function and regulation in rice. Front Plant Sci 8:1598

The Use of Gene Editing Technology to Introduce Targeted Modifications in Woody Plants

Samim Dullah, Rahul Gogoi, Anshu, Priyadarshini Deka, Amarjeet Singh Bhogal, Jugabrata Das, and Sudipta Sankar Bora

Abstract Genome editing technologies help in precise target genome modifications. Various powerful editing tools such as transcription activator-like effector nucleases (TALENs), zinc-finger nucleases (ZFNs), clustered regularly interspaced short palindromic repeats-associated protein systems (CRISPR/Cas), etc. have been used to modify various target genomes efficiently. CRISPR/Cas is the most recent technology used to target gene modifications in almost all kingdoms of life, including plants. One of the best uses of this technology can be observed in woody plants as these plants show a complexity in their genome, viz., a huge genome size, ploidy nature, and a long life span. The species-independent nature of CRISPR/Cas helps in controlling the gene modifications in woody plants. CRISPR/Cas has been used for studying the growth and development, wood properties of these plants, as well as their responses to biotic and abiotic stresses. This

S. Dullah
Nanda Nath College, Titabar, Assam, India

R. Gogoi
Department of Agricultural Biotechnology, Assam Agricultural University, Jorhat, Assam, India

Anshu
All India Institute of Medical Sciences, New Delhi, India

P. Deka
Hereditary Biosciences, Guwahati, Assam, India

A. S. Bhogal
Department of Plant Breeding and Genetics, Assam Agricultural University, Jorhat, Assam, India

J. Das
College of Horticulture and Farming System Research, Assam Agricultural University, Nalbari, Assam, India

S. S. Bora (✉)
DBT-North East Centre for Agricultural Biotechnology (DBT-NECAB), Assam Agricultural University, Jorhat, Assam, India

J.-T. Chen, S. Ahmar (eds.), *Plant Genome Editing Technologies*, Interdisciplinary Biotechnological Advances,
https://doi.org/10.1007/978-981-99-9338-3_12

book chapter highlights different genome editing tools and recent advances for the alteration of genes in woody plants as well as current challenges encountered during the use of this technology for the improvement of traits, products, and germplasm improvement.

Keywords Genome editing · ZF nucleases · TAL effectors · CRISPR/Cas · Woody plants

1 Introduction

Gene diversification is an age-old phenomenon that has been practiced for years for the development of plants with novel traits. Several techniques have been used in order to develop plant varieties that have improved genetic architecture, such as the conventional method of plant breeding and random mutagenesis (either physicochemical or biological) (Ahmar et al. 2020). Though these methods have been vividly used for a long time, the development of modern tools has been the need of the hour, to obtain plants with better varieties rapidly and precisely. Due to this reason, genome editing technology was developed in order to produce plants with better resistance to biotic/abiotic stress and better crop production (Gu et al. 2021).

Genome editing is a technique developed recently for the manipulation of targeted genes. It has become a revolutionary approach in breeding crops along with focusing on the progression of plant science (Nerkar et al. 2022). There are different mechanisms involved in obtaining genome-edited plants, namely, (SDNs) site-directed nucleases, including transcription activator-like effector nucleases (TALENs), meganucleases, zinc-finger nucleases (ZFN), and clustered regularly interspaced short palindromic repeats-associated protein system (CRISPR/Cas) (Kim et al. 2021). Of all the mechanisms mentioned above, the CRISPR/Cas system is widely used nowadays because it is a simple and easy-to-use technology for the development of plants with an edited genome. The CRISPR/Cas system needs two components. One of the components is a Cas nuclease (e.g., Cas9, Cpf1). The second component is a guide RNA (gRNA) (Xu and Li 2020). In the target site, the function of the Cas nuclease is to make a double-strand break (DSB), mostly performed through a nonhomologous end joining (NHEJ) pathway. On the other hand, the gRNA, the second component of the CRISPR/Cas system, is programmed to attach to the target DNA that has to be edited during the genome-editing process (Schubert et al. 2021).

One of the prime reasons for genome editing is to obtain crops with better quality and yields. The other significant use of genome editing is to produce plants that can tolerate biotic and abiotic stress and plants that are resistant to herbicides (Pixley et al. 2022). A very important reason for developing genome-edited plants is that due to minor indels and short sequence modifications by homologous recombination (HDR), very minute changes are obtained in the plant genome through this technology making the edited plant breed almost indistinguishable from the natural genome variant (Gu et al. 2021). Due to this reason, genome-edited plants in certain countries are not considered genetically modified organisms (GMO); as a result, they are exempted from GMO regulation (Schmidt et al. 2020). The progress of genome editing technology has magnificently helped in the progression of plant biology, crop yield, and quality due to its precision causing very minute changes in the plant genome. This chapter focuses on the requirement of genome editing in woody plants, along with the various mechanisms involved in the genome modification of woody plants, emphasizing particularly the CRISPR/Cas system by highlighting its significance. The difficulties encountered when aiming gene editing toward woody plants are also covered in this chapter.

2 Need for Genome Modification in Woody Plants

Woody plants described as bushes or sub-shrubs are plants having slightly complex and upright stems with high lignin along with numerous lignified cells (Crivellaro et al. 2022). These long-term macrophanerophytes are beneficial in various fields such as pharmaceuticals, medicine, and the construction of buildings, industry, and food. Macro-phanerophytes are crucial in maintaining biodiversity and the global climate (Niklas 2008). Therefore, it is of utmost importance to focus on the improvement of the traits of woody plants to increase the production of valuable resources and develop environmental resistance. This can be achieved by genome editing technology which would facilitate the cultivation of new higher-quality woody plant varieties (Min et al. 2022). Unlike annual herbs, woody plants demonstrate some phenomenal natural attributes which makes the application of genome editing tools in these plants more essential (Osakabe et al. 2016). Some of the interesting characteristics are a better life span, increased growth period, and a complex genome. Initially, cross-breeding was the only process through which modification of woody plants was done; but the problem with the process was that the period was long leading to a lengthened production cycle (Wu et al. 2016). Now with the use of modern technologies that enable the manipulation of DNA, the production and breeding cycles are shortened. This made the breeding process more efficient, making it suitable for research and improvement of the woody plants (Osakabe et al. 2016). A lot of emphasis has been given to the genetic modifications of woody plants in order to develop stress (biotic as well abiotic)-resistant varieties and also focus on the improvement of traits (Osakabe et al. 2011). The transgenic process, being the core of conventional genetic engineering for the development of new plant breeds, has led to the improvement of germplasm. Moreover, the expression of targeted plant genes can be regulated by this technology (Basso et al. 2020).

3 Mechanism of Genomic Editing Tools

3.1 Comparison Among Genome Editing Tools

Genome editing technologies hold the ability to modify the genome by taking advantage of elimination, genomic recombination, and site-driven mutagenesis at stipulated locations. Furthermore, this process is performed by inducing (DSBs) double-stranded breaks in the DNA of cytogenetic regions targeted by sequence-specific nucleases (SSNs). These (DSBs) double-stranded breaks are successively repaired either using the homology-directed repair (HDR) pathway that directs DNA sequence substitutions in the sites using the homologous template or via the error-prone nonhomologous end joining (NHEJ) pathway triggering nucleotide insertions or else deletions in absence of homologous template (Khalil 2020). An error that occurs during the genetic repair process in the coding site of the DNA may result in frameshift mutations or codon mutations in the DNA. Indeed, genome editing approaches, in addition, have a greater probability of causing mutations and are additionally more convenient and effective than spontaneous mutagenesis (Zhang et al. 2017). The three primary sequence-specific nucleases (SSNs) used in genome editing technologies are the clustered regularly interspaced short palindromic repeats/CRISPR-associated protein 9 (CRISPR/Cas9) system, transcription activator-like effector nucleases (TALENs), and zinc finger nucleases (ZFNs). ZFNs and TALENs are both dimers of chimeric nucleases consisting of a FokI cleavage domain and a DNA-binding domain. ZFNs were the first SSNs to be utilized to modify plant genomes (Gaj et al. 2016). ZFN constructions, on the other hand, are very expensive and complex to develop and can recognize nonspecific DNA sequences. Later, the TALENs produced from *Xanthomonas* bacteria were created and employed for gene editing in plants (Joung and Sander 2013). However, the intricate tandem repeat characteristic in the TALEN protein's DNA binding domains makes TALEN gene synthesis time-consuming and difficult. The CRISPR/Cas9 system is a new genome editing technology that is quickly replacing older methods. CRISPR/Cas9-based genome editing tools have also emerged as a valuable system in plant science research due to their selectivity and efficiency (Nerkar et al. 2022).

3.2 Mechanism of the CRISPR/Cas9 Nuclease System

A single-guide RNA (sgRNA) and the Cas9 protein are the two main parts of the CRISPR/Cas9 system (multiplex genome engineering using CRISPR/Cas systems) (Fig. 1). A particular RNA sequence called a guide RNA can accurately link with a target location in the genome's DNA through DNA hybridization. A protospacer neighboring motif must be immediately upstream of the target location. The two nuclease domains of Cas9 are activated as a result of this precise binding, triggering a DSB in the target sequence (Khalil 2020). Cas9 is made up of two nuclease

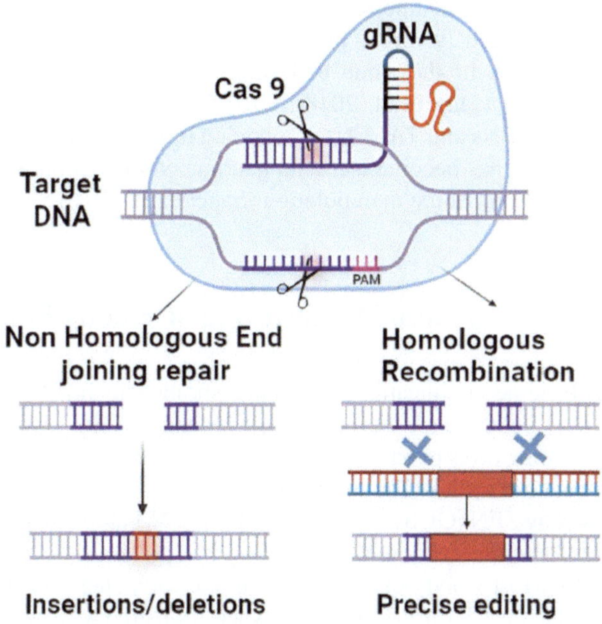

Fig. 1 Mechanism of CRISPR/Cas9-based genome editing tool

domains: RucV and HNH. Each nuclease domain cleaves one of the target DNA strands, resulting in a blunt-ended DSB (the double-edged sword of CRISPR/Cas systems) produced simultaneously. The sgRNA is a fusion of two distinct RNAs required for CRISPR activity: the protospacer-matching crRNA (crRNA) as well as the transactivating crRNA (tracrRNA) (Xu and Li 2020). The sgRNA structure closely resembles its original crRNA-tracrRNA duplex structure, but because it is encoded by a single gene, it substantially simplifies CRISPR/Cas9 system modification. The target recognition sequence, which is positioned at the 5′ end of the sgRNA, is in charge of accurate hybridization to the chromosomal DNA target site (multiplex genome engineering using CRISPR/Cas systems) (Jiang and Doudna 2017).

Cas9 nuclease and sgRNA combine to generate a Cas9-ribonucleoprotein complex. The hybridization ability of sgRNA aids in the specificity of this complex to the target DNA sequence (Zhang et al. 2017). However, the protospacer adjacent motif-PAM (5′-NGG-3′) sequence must be present next to the target gDNA for the Cas9/sgRNA complex to bind to it (Ma et al. 2016). After locating the target site, the Cas9/sgRNA complex divides the target DNA strands to accelerate sgRNA pairing with the target complementary strand via its 5crRNA complementary base pair component. The RucV and HNH domains of the Cas9 complex then cut one of the two DNA strands containing three nucleotides upstream of PAM, resulting in a blunt-ended DSB. The DSBs can then be repaired by either NHEJ or HDR, resulting in alterations in the targeted location (Konstantakos et al. 2022).

Multiple sgRNAs with distinct targeting sequences can bind to separate Cas9 proteins and simultaneously direct the multiple Cas9/sgRNA complexes to the respective target genes in the same cell (multiplex genome engineering using CRISPR/Cas systems) (Jang et al. 2018). CRISPR/Cas9 is easier to modify and more effective than ZFNs and TALENs at inducing targeted alterations. Indeed, the CRISPR/Cas9 system has become a crucial genome editing technology, representing a major advance in genome manipulation (Sauer et al. 2016).

4 Challenges for Genome Engineering of Woody Plants

The practice of genome modification in trees and shrubs remains the subject of multiple investigations, yet there are currently a few instances of this practice in varieties of trees. Implementing tools to edit genomes for tree species involves a limited number of experimentation cycles. The uidA transgene, which encodes b-glucuronidase (GUS), is targeted by ZFN (GUS) and was used, and ZFN-based site-driven mutations in apple (*Malus domestica*) and fig (*Ficus carica*) had previously been demonstrated (Peer et al. 2015). Most woody plants are still in the early stages of building a gene editing system using the CRISPR/Cas system when compared with herbaceous plants. There are still numerous obstacles and challenges in adapting the CRISPR/Cas system to woody plants. First, complete genome sequences of only a few forest species have been available to date. Second, preliminary analysis has revealed these genomes to be enormous, ploidy, repetitive, and heterozygous which specifically poses challenges while designing gRNA. Third and most importantly, an efficient and stable transformation protocol for most of the woody plants is yet to be optimized. The low transformation rate, poor in vitro regeneration ability, and slow growth of explants frequently inhibit genome-editing technologies (Miladinovic et al. 2021).

The CRISPR/Cas system has been used to study a variety of important behaviors in forestry breeding, trait development, and quality enhancement. Off-target mutations are an issue with the CRISPR/Cas9 technology that will need to be addressed in plant genome editing. A solution was however shown in a recent study using a double-nicking CRISPR/Cas9 system (Ran et al. 2013). The approach could also be beneficial to woody plant species. Another system including a *Fok* I-based Cas9 nuclease system (Tsai et al. 2014) as well as a novel gRNA design approach with a target sequence of 17–18 nucleotides (Fu et al. 2014) have also been demonstrated to significantly reduce off-target mutation efficiency.

5 Significance of CRISPR in Genome Editing of Woody Plants

In woody plants, with better durability and out-crossing mating systems, editing their genome is a challenging game. However, a recent study has used the CRISPR/Cas mechanism to study the functionality of genes involved in *Eucalyptus* hairy root development (Dai et al. 2020). Amid difficulties, there are about ten woody plant species that have been successfully edited using CRISPR/Cas9 technology: apple (Nishitani et al. 2016), citrus (Jia et al. 2019), grape (Nakajima et al. 2017), cassava (Odipio et al. 2017), cacao (Fister et al. 2018), coffee (Breitler et al. 2018), kiwifruit (Wang et al. 2018), *Parasponia andersonii* (van Zeijl et al. 2018), pomegranate (Chang et al. 2019), and poplar (Fan et al. 2015). The applicability of CRISPR/Cas9 as an important tool in gene editing technology woody plants is briefly presented in Table 1. In the case of *Populus*, the CRISPR/Cas9 system was used for knocking out two genes, namely, 4CL1 and 4CL2, which are responsible for lignin production. Postmutation, the plants showed that the amount of lignin was greatly reduced to about 23% reduction, resulting in a decrease in the concentration of lignin mono-phenol with a reddish-brown *Populus* (Zhou et al. 2015). A similar study also confirmed the role of *cse* gene for the wood property of poplar plants. CRISPR/Cas was used to knockout the *CSE1* and *CSE2* genes to generate *cse1-cse2* double mutants. These mutants had 35% less lignin content and also showed growth restriction (de Vries et al. 2021). The two repressed TF (transcription factor) genes, namely, B3 heat-shock TF (HSFB3-1) and MYB092, were recognized when CRISPR/Cas9-mediated tension wood (TW) induction was performed in *P. trichocarpa* stem. Further knockout of the genes PtrHSFB3-1 and PtrMYB092 was conducted which showed condensation of lignin and high cellulose content in the mutants. Thus, in the process of production of cell wall constituents, while TW synthesis takes place, the TFs are found to have a regulatory role (Liu et al. 2021). Another study depicted the role of the CRISPR/Cas9 system

Table 1 A summary of the use of CRISPR/Cas9 genome editing technology in several woody plants

Woody plants with altered genomes	Gene of interest	Performative trait	References
Populus	*4CL1, 4CL2*	Decreased lignin content, discoloration of stems	Zhou et al. (2015)
Populus tomentosa	*PtoMYB156*	Negatively influencing the development of secondary walls	Yang et al. (2017)
Populus tomentosa	*PtoDWF4*	Decreased the xylem development	Shen et al. (2018)
Apple	*MdTFL1.1*	Bloom early	Charrier et al. (2019)
Populus trichocarpa	*PtrADA2b-3*	Resistance to drought	Li et al. (2019)
Cassava	*ncbp-1, ncbp-2*	Anti-cassava brown spot virus defense	Gomez et al. (2019)
Populus trichocarpa	*PtrWRKY18, PtrWRKY35*	Resistance to *Melampsora*	Jiang et al. (2017)
Populus	*PdNF-YB21*	Drought resistance	Zhou et al. (2020)
Populus trichocarpa	*PtrHSFB3-1 PtrMYB092*	Lower lignin and higher cellulose	Liu et al. (2021)
Populus alba var. pyramidalis	*PalWRKY77*	Resistance to salt	Jiang et al. (2021)

in knocking out the TFL1 gene of pear and apple which is an antiflowering pigment. As per the reports, 93% of apple mutant lines and 9% of pear mutant lines showed early flowering (Charrier et al. 2019). Moreover, a protein called TERMINAL FLOWER1 (TFL1) has been observed to inhibit flowering in woody and flowering plants, thus acting as an antiflowering compound. The gene editing system was also successful in decreasing the cytokinin content in *Jatropha curcas* leaves by knocking out a gene called JcCYP735A which is related to cytokine synthesis (Cai et al. 2018). Another gene called IdnDH which is related to the synthesis of tartaric acid (TA) was knock out in grapes which resulted in a decrease in the concentration of tartaric acid in the mutant, confirming the usefulness of gene editing technology in grapes (Ren et al. 2016). Similarly, *Populus tomentosa* PtoDWF4 mutants generated through CRISPR/Cas9 had 8.79–11.67% lesser lignin content compared to controls that demonstrated the role of PtowDWF4 gene in positively regulating xylem development during wood formation (Shen et al. 2018). The CRISPR/Cas system was also useful in a simultaneous edition of two genes of the Xinjiang wild apple (Zhang et al. 2021). The favorable outcome of this study helped in stretching the diversity of plant species whose genome can be modified by the CRISPR/Cas9 system, thus helping in providing technical aid for the production of altered germplasm.

6 Conclusion and Prospects

Gene editing technologies, especially CRISPR/CAS systems, are growing rapidly and have found their operations in a wide variety of living organisms comprising agriculturally important trees and crops. The recent surge in publications, reports, novel traits, etc. is a testament to the high scientific and commercial popularity of such technologies. Despite the widespread adoption of genome editing technologies, serious practical as well as ethical trials remain to be addressed. Technical challenges include precise delivery of the editing tools to the target genome and differentiating between the edited genomes from natural and mutagen-induced genomes, while the ethical challenges are social responsibility and environment, such as policies, regulations, and public acceptance. It is believed that technical innovations should focus more on the development of efficient viral systems with transitory editing and inducible recombinase systems for timely elimination. On the other front, ethical innovations should comparatively assess the risk/benefit ratio of the developed products/cultivars to that of conventionally bred products. Keeping an eye on the changing global climate and its possible impact on the forest ecosystem, it is opined that important challenges remain ahead for woody plants, including complete genome sequencing, plant-specific gene editing systems, and effective methods for knock-ins and substitutions.DeclarationFigure 1 was locally created using the software BioRender and does not involve any copyright infringement.

References

Ahmar S, Gill RA, Jung K-H et al (2020) Conventional and molecular techniques from simple breeding to speed breeding in crop plants: recent advances and future outlook. Int J Mol Sci 21:2590. https://doi.org/10.3390/ijms21072590

Basso MF, Arraes FBM, Grossi-de-Sa M et al (2020) Insights into genetic and molecular elements for transgenic crop development. Front Plant Sci 11:509. https://doi.org/10.3389/fpls.2020.00509

Breitler J-C, Dechamp E, Campa C et al (2018) CRISPR/Cas9-mediated efficient targeted mutagenesis has the potential to accelerate the domestication of Coffea canephora. Plant Cell Tissue Organ Cult 134:383–394. https://doi.org/10.1007/s11240-018-1429-2

Cai L, Zhang L, Fu Q, Xu Z-F (2018) Identification and expression analysis of cytokinin metabolic genes IPTs , CYP735A and CKXs in the biofuel plant Jatropha curcas. PeerJ 6:e4812. https://doi.org/10.7717/peerj.4812

Chang L, Wu S, Tian L (2019) Effective genome editing and identification of a regiospecific gallic acid 4-O-glycosyltransferase in pomegranate (Punica granatum L.). Hortic Res 6:123. https://doi.org/10.1038/s41438-019-0206-7

Charrier A, Vergne E, Dousset N et al (2019) Efficient targeted mutagenesis in apple and first time edition of pear using the CRISPR-Cas9 system. Front Plant Sci 10:40. https://doi.org/10.3389/fpls.2019.00040

Crivellaro A, Piermattei A, Dolezal J et al (2022) Biogeographic implication of temperature-induced plant cell wall lignification. Commun Biol 5:767. https://doi.org/10.1038/s42003-022-03732-y

Dai Y, Hu G, Dupas A et al (2020) Implementing the CRISPR/Cas9 technology in eucalyptus hairy roots using wood-related genes. Int J Mol Sci 21:3408. https://doi.org/10.3390/ijms21103408

de Vries L, Brouckaert M, Chanoca A et al (2021) CRISPR-Cas9 editing of CAFFEOYL SHIKIMATE ESTERASE 1 and 2 shows their importance and partial redundancy in lignification in Populus tremula × P. alba. Plant Biotechnol J 19:2221–2234. https://doi.org/10.1111/pbi.13651

Fan D, Liu T, Li C et al (2015) Efficient CRISPR/Cas9-mediated targeted mutagenesis in Populus in the first generation. Sci Rep 5:12217. https://doi.org/10.1038/srep12217

Fister AS, Landherr L, Maximova SN, Guiltinan MJ (2018) Transient expression of CRISPR/Cas9 machinery targeting TcNPR3 enhances defense response in theobroma cacao. Front Plant Sci 9:268. https://doi.org/10.3389/fpls.2018.00268

Fu Y, Sander JD, Reyon D et al (2014) Improving CRISPR-Cas nuclease specificity using truncated guide RNAs. Nat Biotechnol 32:279–284. https://doi.org/10.1038/nbt.2808

Gaj T, Sirk SJ, Shui S-L, Liu J (2016) Genome-editing technologies: principles and applications. Cold Spring Harb Perspect Biol 8:a023754. https://doi.org/10.1101/cshperspect.a023754

Gomez MA, Lin ZD, Moll T et al (2019) Simultaneous CRISPR/Cas9-mediated editing of cassava eIF4E isoforms nCBP-1 and nCBP-2 reduces cassava brown streak disease symptom severity and incidence. Plant Biotechnol J 17:421–434. https://doi.org/10.1111/pbi.12987

Gu X, Liu L, Zhang H (2021) Transgene-free genome editing in plants. Front Genome Ed 3:805317. https://doi.org/10.3389/fgeed.2021.805317

Jang DE, Lee JY, Lee JH et al (2018) Multiple sgRNAs with overlapping sequences enhance CRISPR/Cas9-mediated knock-in efficiency. Exp Mol Med 50:1–9. https://doi.org/10.1038/s12276-018-0037-x

Jia H, Orbović V, Wang N (2019) CRISPR-LbCas12a-mediated modification of citrus. Plant Biotechnol J 17:1928–1937. https://doi.org/10.1111/pbi.13109

Jiang F, Doudna JA (2017) CRISPR-Cas9 structures and mechanisms. Annu Rev Biophys 46:505–529. https://doi.org/10.1146/annurev-biophys-062215-010822

Jiang Y, Guo L, Ma X et al (2017) The WRKY transcription factors PtrWRKY18 and PtrWRKY35 promote melampsora resistance in Populus. Tree Physiol 37:665–675. https://doi.org/10.1093/treephys/tpx008

Jiang Y, Tong S, Chen N et al (2021) The PalWRKY77 transcription factor negatively regulates salt tolerance and abscisic acid signaling in Populus. Plant J 105:1258–1273. https://doi.org/10.1111/tpj.15109

Joung JK, Sander JD (2013) TALENs: a widely applicable technology for targeted genome editing. Nat Rev Mol Cell Biol 14:49–55. https://doi.org/10.1038/nrm3486

Khalil AM (2020) The genome editing revolution: review. J Genet Eng Biotechnol 18:68. https://doi.org/10.1186/s43141-020-00078-y

Kim Y-C, Kang Y, Yang E-Y et al (2021) Applications and major achievements of genome editing in vegetable crops: a review. Front Plant Sci 12:688980. https://doi.org/10.3389/fpls.2021.688980

Konstantakos V, Nentidis A, Krithara A, Paliouras G (2022) CRISPR-Cas9 gRNA efficiency prediction: an overview of predictive tools and the role of deep learning. Nucleic Acids Res 50:3616–3637. https://doi.org/10.1093/nar/gkac192

Li S, Lin Y-CJ, Wang P et al (2019) The AREB1 transcription factor influences histone acetylation to regulate drought responses and tolerance in Populus trichocarpa. Plant Cell 31:663–686. https://doi.org/10.1105/tpc.18.00437

Liu B, Liu J, Yu J et al (2021) Transcriptional reprogramming of xylem cell wall biosynthesis in tension wood. Plant Physiol 186:250–269. https://doi.org/10.1093/plphys/kiab038

Ma X, Zhu Q, Chen Y, Liu Y-G (2016) CRISPR/Cas9 platforms for genome editing in plants: developments and applications. Mol Plant 9:961–974. https://doi.org/10.1016/j.molp.2016.04.009

Miladinovic D, Antunes D, Yildirim K et al (2021) Targeted plant improvement through genome editing: from laboratory to field. Plant Cell Rep 40:935–951. https://doi.org/10.1007/s00299-020-02655-4

Min T, Hwarari D, Li D et al (2022) CRISPR-based genome editing and its applications in woody plants. Int J Mol Sci 23:10175. https://doi.org/10.3390/ijms231710175

Nakajima I, Ban Y, Azuma A et al (2017) CRISPR/Cas9-mediated targeted mutagenesis in grape. PLoS One 12:e0177966. https://doi.org/10.1371/journal.pone.0177966

Nerkar G, Devarumath S, Purankar M et al (2022) Advances in crop breeding through precision genome editing. Front Genet 13:880195. https://doi.org/10.3389/fgene.2022.880195

Niklas KJ (2008) Life forms, plants. In: Jørgensen SE, Fath BD (eds) Encyclopedia of ecology. Academic Press, Oxford, pp 2160–2167

Nishitani C, Hirai N, Komori S et al (2016) Efficient genome editing in apple using a CRISPR/Cas9 system. Sci Rep 6:31481. https://doi.org/10.1038/srep31481

Odipio J, Alicai T, Ingelbrecht I et al (2017) Efficient CRISPR/Cas9 genome editing of phytoene desaturase in Cassava. Front Plant Sci 8:1780. https://doi.org/10.3389/fpls.2017.01780

Osakabe Y, Kajita S, Osakabe K (2011) Genetic engineering of woody plants: current and future targets in a stressful environment. Physiol Plant 142:105–117. https://doi.org/10.1111/j.1399-3054.2011.01451.x

Osakabe Y, Sugano SS, Osakabe K (2016) Genome engineering of woody plants: past, present and future. J Wood Sci 62:217–225. https://doi.org/10.1007/s10086-016-1548-5

Peer R, Rivlin G, Golobovitch S et al (2015) Targeted mutagenesis using zinc-finger nucleases in perennial fruit trees. Planta 241:941–951. https://doi.org/10.1007/s00425-014-2224-x

Pixley KV, Falck-Zepeda JB, Paarlberg RL et al (2022) Genome-edited crops for improved food security of smallholder farmers. Nat Genet 54:364–367. https://doi.org/10.1038/s41588-022-01046-7

Ran FA, Hsu PD, Lin C-Y et al (2013) Double nicking by RNA-guided CRISPR Cas9 for enhanced genome editing specificity. Cell 154:1380–1389. https://doi.org/10.1016/j.cell.2013.08.021

Ren C, Liu X, Zhang Z et al (2016) CRISPR/Cas9-mediated efficient targeted mutagenesis in Chardonnay (Vitis vinifera L.). Sci Rep 6:32289. https://doi.org/10.1038/srep32289

Sauer NJ, Narváez-Vásquez J, Mozoruk J et al (2016) Oligonucleotide-mediated genome editing provides precision and function to engineered nucleases and antibiotics in plants. Plant Physiol 170:1917–1928. https://doi.org/10.1104/pp.15.01696

Schmidt SM, Belisle M, Frommer WB (2020) The evolving landscape around genome editing in agriculture: many countries have exempted or move to exempt forms of genome editing from GMO regulation of crop plants. EMBO Rep 21:e50680. https://doi.org/10.15252/embr.202050680

Schubert MS, Thommandru B, Woodley J et al (2021) Optimized design parameters for CRISPR Cas9 and Cas12a homology-directed repair. Sci Rep 11:19482. https://doi.org/10.1038/s41598-021-98965-y

Shen Y, Li Y, Xu D et al (2018) Molecular cloning and characterization of a brassinosteriod biosynthesis-related gene PtoDWF4 from Populus tomentosa. Tree Physiol 38:1424–1436. https://doi.org/10.1093/treephys/tpy027

Tsai SQ, Wyvekens N, Khayter C et al (2014) Dimeric CRISPR RNA-guided FokI nucleases for highly specific genome editing. Nat Biotechnol 32:569–576. https://doi.org/10.1038/nbt.2908

van Zeijl A, Wardhani TAK, Seifi Kalhor M et al (2018) CRISPR/Cas9-mediated mutagenesis of four putative symbiosis genes of the tropical tree Parasponia andersonii reveals novel phenotypes. Front Plant Sci 9:284. https://doi.org/10.3389/fpls.2018.00284

Wang Z, Wang S, Li D et al (2018) Optimized paired-sgRNA/Cas9 cloning and expression cassette triggers high-efficiency multiplex genome editing in kiwifruit. Plant Biotechnol J 16:1424–1433. https://doi.org/10.1111/pbi.12884

Wu HX, Hallingbäck HR, Sánchez L (2016) Performance of seven tree breeding strategies under conditions of inbreeding depression. G3 (Bethesda) 6:529–540. https://doi.org/10.1534/g3.115.025767

Xu Y, Li Z (2020) CRISPR-Cas systems: overview, innovations and applications in human disease research and gene therapy. Comput Struct Biotechnol J 18:2401–2415. https://doi.org/10.1016/j.csbj.2020.08.031

Yang L, Zhao X, Ran L et al (2017) PtoMYB156 is involved in negative regulation of phenylpropanoid metabolism and secondary cell wall biosynthesis during wood formation in poplar. Sci Rep 7:41209. https://doi.org/10.1038/srep41209

Zhang Y, Ma X, Xie X, Liu Y-G (2017) Chapter 8—CRISPR/Cas9-based genome editing in plants. In: Weeks DP, Yang B (eds) Gene editing in plants (Progress in molecular biology and translational science). Academic Press, New York, pp 133–150

Zhang Y, Zhou P, Bozorov TA, Zhang D (2021) Application of CRISPR/Cas9 technology in wild apple (Malus sieverii) for paired sites gene editing. Plant Methods 17:79. https://doi.org/10.1186/s13007-021-00769-8

Zhou X, Jacobs TB, Xue L et al (2015) Exploiting SNPs for biallelic CRISPR mutations in the outcrossing woody perennial Populus reveals 4-coumarate:CoA ligase specificity and redundancy. New Phytol 208:298–301. https://doi.org/10.1111/nph.13470

Zhou Y, Zhang Y, Wang X et al (2020) Root-specific NF-Y family transcription factor, PdNF-YB21, positively regulates root growth and drought resistance by abscisic acid-mediated indoylacetic acid transport in Populus. New Phytol 227:407–426. https://doi.org/10.1111/nph.16524

Single-Base Editing in the *Arabidopsis* SUMO Conjugating Enzyme by Adenine Base Edition and Screening for a Rare Editing Event

Lilian Nehlin, Vera Schoft, Volodymyr Shubchynskyy, Andreas Sommer, and Andreas Bachmair

Abstract SUMO conjugation is an essential process that regulates development and supports plant adaptation to environmental stress. We previously found that a Lys to Arg change in position 28 of the single-copy *Arabidopsis thaliana* SUMO conjugating enzyme SCE1 has increased in vitro activity, so we wanted to test the effect of this mutation in planta. The change is within reach for A to G base editing (AAG to AGG codon change) by an adenine deaminase-Cas9 fusion protein. We describe how we generated and identified two independent mutants carrying this change. The procedure may be useful for other cases of directed genome changes that occur at low frequency.

Keywords Cas9 · Genome editing · Small ubiquitin-related modifier (SUMMO) · Site-directed mutagenesis

---

Supplementary Information The online version contains supplementary material available at https://doi.org/10.1007/978-981-99-9338-3_13.

L. Nehlin · A. Bachmair (✉)
Max Perutz Labs, Department of Biochemistry and Cell Biology, University of Vienna, Vienna, Austria
e-mail: andreas.bachmair@univie.ac.at

V. Schoft
Vienna BioCenter Core Facilities, Protein Technologies, Vienna, Austria

V. Shubchynskyy · A. Sommer
Vienna BioCenter Core Facilities, Next Generation Sequencing, Vienna, Austria

J.-T. Chen, S. Ahmar (eds.), *Plant Genome Editing Technologies*, Interdisciplinary Biotechnological Advances,
https://doi.org/10.1007/978-981-99-9338-3_13

1 Biological Background

SUMO conjugation is the attachment of a small ubiquitin-related modifier (SUMO) to a substrate protein. It is conducted by SUMO activating enzyme and SUMO conjugating enzyme. SUMO ligases as dedicated substrate recognition components frequently enable or enhance this process (Clark et al. 2022; Vertegaal 2022; Novatchkova et al. 2012). SUMO-specific proteases serve to activate and recycle SUMO protein by deconjugation, and they can increase substrate specificity by fast reversal of inappropriate linkages (Morrell and Sadanandom 2019). Sumoylation is an essential process that increases drastically under environmental stress conditions, e.g., heat, cold, or salt. Mutant and overexpression studies suggest a positive role for SUMO conjugation in plant survival of stressful conditions and adaptation to altered conditions in general (Clark et al. 2022). Substrates of the pathway have been identified by in vivo proteome analysis (Vertegaal 2022; Millar et al. 2019). There is usually a good correlation between in vivo sumoylation with in vitro SUMO attachment to substrates by purified enzymes. Using purified enzymes and identified substrates to study SUMO conjugating enzyme SCE1, we found that, whereas most mutations in this enzyme are either neutral or decrease function, a change of Lys 28 to Arg (K28R) increased in vitro activity (Tomanov et al. 2018). It turned out that this position lies in a potential guide RNA for Cas9, so targeted mutagenesis via adenine deaminase base editing seemed feasible. In order to study the consequence of this change on the whole plant level, we used site-directed in vivo methods to generate the mutant.

2 Constructs for Site-Directed Mutagenesis

Deamination of adenine to generate xanthine generates a position in DNA that base pairs with cytosine instead of thymine. When xanthine-containing DNA is replicated, DNA polymerase inserts cytosine instead of thymine at the complementary position, generating a sequence change to a G·C base pair upon further replication, instead of the previous A·T base pair at this position. Gaudelli et al. (2017) showed that an improved adenine deaminating enzyme, adenine deaminase (ADA; ADA7.10), gets access to some bases of the guide RNA when linked to modified Cas9 (a "nickase" nCas9). The use of this mutant is expected to decrease both the site-specific and ectopic mutagenesis potential of Cas9 because the enzyme has single-strand cleavage (nicking) activity only. Nicks can be precisely repaired by resealing of the DNA strand, as opposed to the nonhomologous end joining usually employed by plant cells to repair double-strand breaks. For site-directed mutagenesis in the model plant *Arabidopsis*, we copied the domain arrangement that was successfully used in animal cells, namely, two ADA domains fused in tandem to Cas9 with loss of the nuclease function of the RuvC I domain via Asp to Ala change in the active site (a D10A change in Cas9). Plant promoter and terminator elements were linked to the ADA-nCas9 cassette, with codon usage optimized for plant

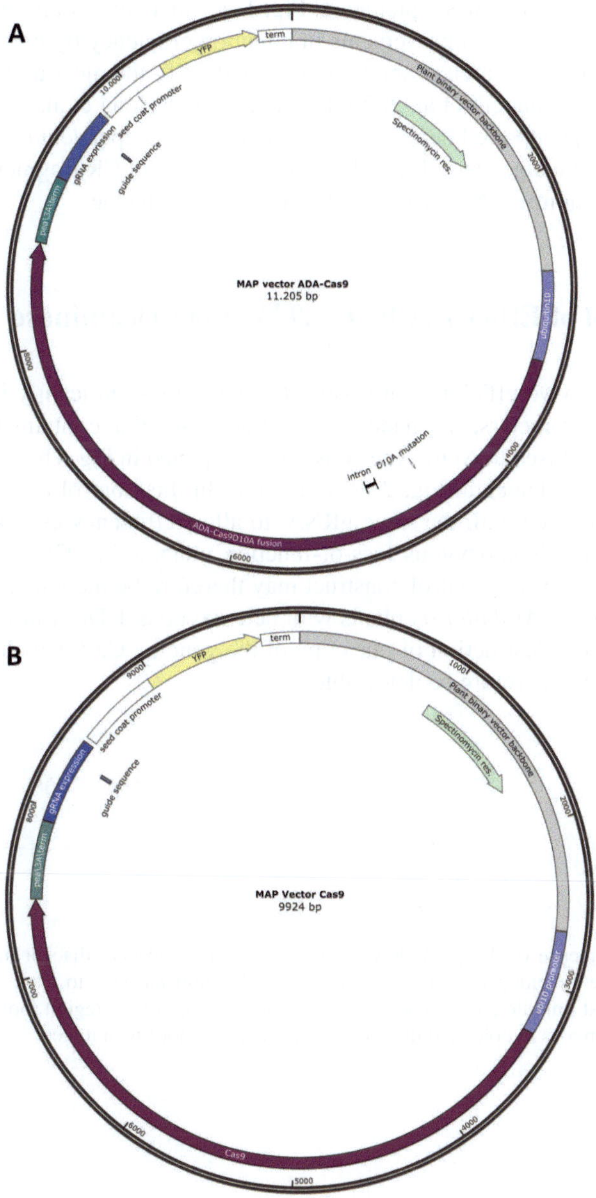

Fig. 1 Maps of plasmids for expression of ADA-nCas9 (panel **a**) and Cas9 (panel **b**) in plants. The vector backbone is pGGZ001 (Addgene #48868). Sequence information can be obtained in Supplemental Figs. 1 and 2

expression. Plant binary vector pGGZ001 (Addgene #48868) was used as the vector backbone, and the cloning procedure followed the protocols by Richter et al. (2018). A schematic drawing of the vector is shown in Fig. 1a. The DNA sequence of the construct can be found in Supplemental Fig. 1 (a ".dna" file). Later work by others documented significant improvements in mutation efficiency by improved vectors, so we recommend a search of the literature to identify the most up-to-date protein constructs (e.g., Gaudelli et al. 2020; Hua et al. 2020; Zong et al. 2022). However, most of the steps reported here to obtain mutants are independent of the exact construct used for mutagenesis. The only condition is that a PCR fragment can be generated from genomic DNA that contains the intended change.

3 Control of Efficiency by Cas9 Without Deaminase

Because the in vivo efficiency of Cas9 gRNAs differs significantly in dependence on unknown parameters, we made a parallel construct that contains enzymatically active Cas9 not fused to ADA. The construct is depicted in Fig. 1b; its sequence can be found in Supplemental Fig. 2 as a .dna file. In this control construct, unfused Cas9 was combined with the same gRNA, to allow efficiency estimates regarding gRNA targeting. Homozygous loss-of-function mutants in SCE1 are lethal. The efficient activity of the control construct may therefore be incompatible with plant growth. However, *Arabidopsis* plants with heterozygous T-DNA insertion in SCE1 grow fine, so the destruction of one copy of the gene by Cas9 should not interfere with growth and therefore be detectable.

GGCAAAGCCGGAGACGGGGC*AGG (NGG PAM motif)*

aaatttagGGTTTTGTGGCA**AAG**CCGGAGACGGGGCAGGATGGAACTGTGAATCTAATG...

(intron)GlyPheValAla**Lys**ProGluThrGlyGlnAspGlyThrValAsnLeuMet...

Fig. 2 The sequence of the guide RNA with adjacent protospacer adjacent motif PAM (top), sequence of the Arabidopsis SCE1 locus (At3g57870) from intron 1 to exon 2 containing the region of interest (middle), and translated amino acid sequence of the region (bottom). The codon targeted for change is in bold, and the protospacer adjacent motif is in italics

4 Assessment of Efficiency: Somatic Mutation Rate and Heat Treatment

Transgenic lines of *Arabidopsis* harboring either the Cas9 control or the ADA-nCas9 construct were generated and tested in parallel. In particular, we assessed the somatic mutation rate. While the somatic mutation rate is expected to increase with increasing enzyme activity and therefore give hints regarding the feasibility of the approach, the measured frequency in somatic cells may not be the same as the frequency of mutagenesis in zygotes, or in L2 layer cells that form the germ line. Mutation in either of these cell types is necessary to generate a heritable change. A key point to consider is whether the used promoter shows activity in the L2 layer and/or in the flower, in particular during meiosis or in early zygotes. The ubiquitin promoter used in these experiments is well-suited and offers high activity in somatic and flower tissue. Generally, promoters that allow efficient generation of heritable mutations by Cas9 should also serve well for site-directed mutagenesis. Figure 2 shows the gRNA sequence (top) above the genomic sequence of the mutated region (intron sequence in small, exon sequence in capitals). As pointed out above, because the loss of two SCE1 gene copies is lethal, we expected somatic frameshift mutations by Cas9 would occur mostly in the heterozygous state, and their occurrence to be an underestimation of the actual mutation rate, due to decreased cell viability in case the second copy is also mutated. Regarding mutations by the ADA-Cas9 construct, we expected no such problems due to the potential increase-of-function in vitro properties of the K28R mutant.

We grew transgenic *Arabidopsis* lines that carry the ADA-nCas9 and the Cas9 constructs. To compensate for varying expression levels of the transgenes in different plant lines, a mixture of leaf material from several independently generated transgenic lines was taken. At the rosette stage, leaf material was collected for DNA preparation using standard methods. The leaf DNA was subjected to PCR amplification. The primers used for amplification, and the amplified sequence including primers, are shown in Fig. 3. The fragments amplified from a mixture of leaf material from different transgenic lines underwent quality control by fragment size analysis and qPCR followed by Illumina (paired-end) sequencing on a MiSeq instrument. A total of 3.5 million reads were obtained from the plant material containing ADA-nCas9. We looked at changes in adenine residues. Adenine residues not at the mutable position served as the standard for mutation errors inherent in the method. The context AAGCCGGAG**A**CG was found to mutate 852 times among the 3.5 million reads. The boldface A is within the gRNA but at a position considered unfavorable for mutagenesis due to proximity to the protospacer adjacent motif (Gaudelli et al. 2017). The context GGATGGA**A**CT contained a mutant A 703 times; the

Primer 1
GTTGATGGTGAAATTTAGGGT

Primer 2
CAGAGGAATAAGAAAACTGAAGAAA

Amplified sequence:

*gttgatggtgaaatttagGGT*TTTGTGGCA**AA**GCCGGAGA
CGGGGCAGGATGGAACTGTGAATCTAATGGTGTGGCATTG
CACTATACCTGGTAAAGCTGGGgtaagtatcaatcgtgaa
attagg*tttcttcagttttcttattcctctg*

Fig. 3 Primers used for PCR to amplify the potentially mutated region (top) and sequence of the amplified region (bottom). Intronic parts of the latter sequence are in small letters and primer binding regions are in italics. Mutagenizable adenine residues are in bold. The fragment was further processed and Illumina-sequenced to assess somatic mutation frequency

neighboring A (context GGATGG**A**ACT) was found mutated in 1416 reads. These values define the frequency of mutant detection that is inherent in the method of PCR amplification and sequencing of these short DNA fragments. In contrast, the two "mutable" As (see Figs. 2 and 3, bold) were found mutated more frequently: Lys AAG codon was found converted to GAG (codon for Glu) 1069 times, which is, however, still close to the frequency of erroneous reads (or spontaneous somatic mutations) of the "negative control" As. Most importantly, the AAG to AGG (Lys to Arg, the intended change) mutation was found 2242 times, clearly above the background error rate. Interestingly, the double change AAG to GGG (Gly codon) was not found at all.

The changes were compared between plants grown at 22 °C and those shifted to 37 °C. *Arabidopsis* ecotypes grow well at 37 °C but have reduced fertility when kept permanently at this temperature. We, therefore, switched back to 22 °C for a day after every 2 days at 37 °C. The heat treatment followed a publication showing that Cas9 without fused ADA polypeptide results in increased mutagenic potential at 37 °C (LeBlanc et al. 2018). We compared ADA-nCas9 with Cas9. Figure 4 shows the summary. Whereas mutations due to Cas9 cleavage by the unmodified enzyme increase with increasing temperature, the frequency of adenine base editing does not change significantly.

CAS9- and ADA-CAs9-induced changes at gRNA positions at K28 of SCE1

Enzyme	Growth condition	Mutagenesis efficiency	Nature of change
Cas9	22 °C	2.8 %	any change
Cas9	37 °C	16 %	any change
ADA-Cas9	22 °C	0.054 %	AAG to A**G**G
ADA-Cas9	37 °C	0.064 %	AAG to A**G**G
ADA-Cas9	22 °C	0.030 %	AAG to **G**AG
ADA-Cas9	37 °C	0.030 %	AAG to **G**AG
ADA-Cas9	22 °C	0.000 %	AAG to **GG**G
ADA-Cas9	37 °C	0.000 %	AAG to **GG**G

Fig. 4 Frequencies of somatic mutations with ADA construct and Cas9 control, under two different growth conditions as defined in the text

5 Screening Scheme: Pooling of Leaf Material in Search of Heritable Changes

We, therefore, concluded that plant growth at 22 °C was optimal for our purpose and that in somatic tissue, a mutation rate of 2000 among 3.5 million cases could be expected, i.e., 0.06%, or one change among 1600 genomes. The background error rate, possibly due to the method (e.g., PCR errors and sequencing artifacts), is similar to one true change in approximately 3000 genomes. An inherent assumption was that heritable changes did occur at a similar frequency as somatic mutations. This was based on the known activity of the ubiquitin promoter used in the construct, which is highly expressed in leaves and during flower and seed development. If our assumptions were correct, 1 out of 1600 progeny plants would be heterozygous for the intended mutation.

We decided to screen 2000 plants of the following generation. We chose to grow seedlings from a mother plant with a high expression level of the transgene, assuming that this choice would support mutagenesis. DNA from the leaves of 2000 plants was collected. Leaf material from ten plants was pooled before DNA isolation so that 200 DNA preparations had to be carried out. As a next step, the pools were subjected to PCR amplification, using primers similar to those used for somatic analysis. If 1 heterozygous plant were present in a pool, 1 in 20 PCR fragments

Primer 1
TCGTCGGCAGCGTC AGATGTGTATAAGAGACAG
GTTGATGGTGAAATTTAGGGT

Primer 2
GTCTCGTGGGCTCGG AGATGTGTATAAGAGACAG
CAGAGGAATAAGAAAACTGAAGAAA

Primer 1var(i5/5XX)
AATGATACGGCGACCACCGAGATCTACAC
NNNNNNNN TCGTCGGCAGCGTC

Primer 2var(i7/7XX)
CAAGCAGAAGACGGCATACGAGAT
NNNNNNNN GTCTCGTGGGCTCGG

Fig. 5 Primers are used to amplify and add sequence tags to different pools of potentially mutated plant lines. The 3′ parts of primers 1 and 2 (in italics) are identical to primers 1 and 2 of Fig. 3. Underlined: Overlap with the second set of primers that served to introduce sequence tags (shown as NNNNNNNN in the figure). Supplemental Figs. 3 and 4 provide tag information

should carry the mutation. Such a frequency can be easily detected by the method described below.

Differing from the efficiency assessment (Sect. 4), the somewhat longer primers 1 and 2 of Fig. 5 were used for the amplification of DNA from each batch. This PCR was followed by a PCR using Primers 1var and 2var (Fig. 5), to introduce unique sequence identifiers. The NNNNNNNN positions were filled with different nucleotides for each of the 200 PCR fragment pools, chosen according to the lists recommended by the manufacturer (Supplemental Figs. 3 and 4; .xlsx files). With these adapted identifier tags, all fragments could be pooled for sequencing in one Illumina lane. After sequence collection, the identifier tags would allow sorting sequences according to the batch number.

The first and second PCR were carried out in microtiter format, to allow robotic handling of the samples and reactions as available in our sequencing facility. In particular, some microtiter wells were kept empty to allow for the robotic deposition of liquids according to the used program. Most importantly, and independent of the exact handling of the samples, a one-on-one assignment between the batch number of the DNA and microtiter well is essential.

6 Sequence Interpretation

Figure 6 shows readouts from the sequencing. Results were grouped according to the original pool, as identified by the added linker barcodes. Panel a shows the result of a negative pool. Eighty percent of all reads were of WT sequence. PolyG was included as an internal reference and occurred in 1.5% of all reads. Other sequences were below 0.6%. Characteristically, the searched sequence change occurred at 0.36%, which is in the order of the previously established somatic mutagenesis rate for this position (see Fig. 4). Two pools contained the AAG to AGG change at a frequency in the range of 5%, which would amount to one heterozygous genomic change per 10 plants (one in 20 amplified ds DNA strands; Fig. 6b). A third pool contained the sequence at 8% abundance.

A

Sequence	Count	Percentage
GTTGATGGTGAAATTTAGGGTTTTGTGGCA**AAG**CCGGAGACGGGGCAGGA	50307	80.84432802481238
GGGGGGGGGGGGGGGGGGGGGGGGGGGGGGGGGGGGGGGGGGGG	973	1.5636299355585197
GTTGAGGGTGAAATTTAGGGTTTTGTGGCAAAGCCGGAGACGGGGCAGGA	329	0.5287094026708663
GTTGATGGTGAAATTTAGGGTTTTGTGGCAAAGCCGGAGACGGGGCAGGAT	297	0.4772847799186848
GTTGAAGGTGAAATTTAGGGTTTTGTGGCAAAGCCGGAGACGGGGCAGGA	227	0.3647934176482877
GTTGATGGTGAAATTTAGGGTTTTGTGGCA**AGG**CCGGAGACGGGGCAGGA	224	0.35997235926527066
GTTGATGGTGAAATTTAGGTTTTTGTGGCAAAGCCGGAGACGGGGCAGGA	207	0.33265302842817424
GTTGATGGTGAAATTTAGGGTTTTGTGGCGAAGCCGGAGACGGGGCAGGA	190	0.3053336975910778
GTTGATGGTGAAATTTAGGGTTTTGTGGCAAAGCCGGAGACGGGGCGGGA	180	0.2892635029810211
GTTGATGGTGAAATTTAGGGTTTTGTGGCAAAGCCGGAGGCGGGGCAGGA	180	0.2892635029810211
GTTGATGGTGAAATTTAGGGTTTTGTGGCAAAGCCGGAGACGGGGCAGGG	178	0.2860494640590097
GTTGATGGTGAAATTTAGGGTTTTGCGGCAAAGCCGGAGACGGGGCAGGA	177	0.2844424445980041
GTTGATGGTGAAATTTAGGGTTTTGTGGCAAAGCCGGGGACGGGGCAGGA	176	0.28283542513699844
GTTGATGGTGAAATTTAGGGTTCTGTGGCAAAGCCGGAGACGGGGCAGGA	169	0.2715862889099587
GTTGATGGTGAAATTTAGGGTTTTGTGGCAGAGCCGGAGACGGGGCAGGA	163	0.2619441721439247
GTTCATGGTGAAATTTAGGGTTTTGTGGCAAAGCCGGAGACGGGGCAGGA	160	0.25712311376090763
GTTGATGGTGAAATTTAGGGTTTCGTGGCAAAGCCGGAGACGGGGCAGGA	158	0.2539090748388963

B

Sequence	Count	Percentage
GTTGATGGTGAAATTTAGGGTTTTGTGGCA**AAG**CCGGAGACGGGGCAGGA	117429	74.36074418369027
GTTGATGGTGAAATTTAGGGTTTTGTGGCA**AGG**CCGGAGACGGGGCAGGA	9323	5.903696855329981
GGGGGGGGGGGGGGGGGGGGGGGGGGGGGGGGGGGGGGGGGGGG	2654	1.6806190554591625
GTTGAGGGTGAAATTTAGGGTTTTGTGGCAAAGCCGGAGACGGGGCAGGA	837	0.5300219101052445
GTTGATGGTGAAATTTAGGGTTTTGTGGCAAAGCCGGAGACGGGGCAGGAT	679	0.42996998442229506
GTTGAAGGTGAAATTTAGGGTTTTGTGGCAAAGCCGGAGACGGGGCAGGA	552	0.34954849985435477
GTTGATGGTGAAATTTAGGTTTTTGTGGCAAAGCCGGAGACGGGGCAGGA	510	0.32295241834369737
GTTGATGGTGAAATTTAGGGTTTCGTGGCAAAGCCGGAGACGGGGCAGGA	464	0.2938233766891678
GTTGATGGTGAAATTTAGGGTTTTGTGGCGAAGCCGGAGACGGGGCAGGA	462	0.2925568966172317
GTTGATGGTGAAATTTAGGGTTTTGTGGCAAAGCCGGAGGCGGGGCAGGA	451	0.2855912562215834
GTTGATGGTGAAATTTAGGGTTTTGTGGCAAAGCCGGGGACGGGGCAGGA	441	0.279258855861903
GTTGATGGTGAAATTTAGGGTTTTGTGGCAAAGCCGGAGACGGGGCGGGA	426	0.2697602553223825
GTTGATGGTGAAATTTAGGGTTTTGTGGCAGAGCCGGAGACGGGGCAGGA	414	0.2621613748907661
GTTCATGGTGAAATTTAGGGTTTTGTGGCAAAGCCGGAGACGGGGCAGGA	404	0.2558289745310857
GTTGATGGTGAAATTTAGGGTCTTGTGGCAAAGCCGGAGACGGGGCAGGA	401	0.25392925442318165
GTTGATGGTGAAATTTAGGGTTCTGTGGCAAAGCCGGAGACGGGGCAGGA	394	0.2494965741714054
GTTGATGGTGAAATTTAGGGTTTTGCGGCAAAGCCGGAGACGGGGCAGGA	393	0.2488633413543738

Fig. 6 An example of sequence read information from a negative pool of plants (panel **a**) and from a pool containing one plant that is heterozygous for the desired mutation (panel **b**). Count means the number of reads obtained from the pool with the sequence shown to the left

7 De-convolution of Positive Pools: Identification of Mutated Plants

Seeds from the three positive pools were sown. Plants were tested individually by DNA isolation, PCR amplification of the potentially mutated region, and Sanger sequencing. Positive plants were identified from two of the three candidate pools. Both independently generated mutants were subjected to four rounds of outcrossing, to remove the ADA-nCas9 transgene and potential additional, unintended mutations. We followed the K28R mutation in a heterozygous state through the generations. The first round of backcrossing replaced 50% of the DNA with non-transformed Col-0 WT DNA, and each following round reduced that remaining DNA from the original mutant plant by half, ending at 6.25%. Selfing of the ensuing plant(s) resulted in progeny that can be expected to be homozygous at loci of such unintended changes with 1.56% the frequency of their original occurrence (1/4 of 6.25% according to Mendel's law). Because we selected for the intended mutation in each round, the DNA around the SCE1 locus was inherited preferentially from the mutant parent, though (generating a linkage disequilibrium). However, assuming that potential mutations genetically linked to SCE1 were distinct in the two independently generated mutants gives additional statistical quality to the experiments. As an additional measure, comparison of siblings from the repeated outcrosses that are either homozygous for the WT allele or for the K28R mutant allele should be more reliable than comparison of the mutant to the WT used for outcrossing. Having this independent genetic material therefore further improves the reliability of the ensuing phenotypic analysis.

8 Summary

We describe a method for the identification of *Arabidopsis* plants with directed mutation. The mutation was obtained at a low frequency. The method can be adapted to any rare mutational event in plants of choice.

Acknowledgments We would like to thank Karoline Hilse-Koller for cloning experiments. Work in the author's lab was supported by the Austrian Science Foundation FWF (grants F7904B and DOC 111B).

References

Clark L, Sue-Ob K, Mukkawar V, Jones AR, Sadanandom A (2022) Understanding SUMO-mediated adaptive responses in plants to improve crop productivity. Essays Biochem 66(2):155–168

Gaudelli NM, Komor AC, Rees HA, Packer MS, Badran AH, Bryson DI, Liu DR (2017) Programmable base editing of A·T to G·C in genomic DNA without DNA cleavage. Nature 551(7681):464–471

Gaudelli NM, Lam DK, Rees HA, Solá-Esteves NM, Barrera LA, Born DA, Edwards A, Gehrke JM, Lee SJ, Liquori AJ, Murray R, Packer MS, Rinaldi C, Slaymaker IM, Yen J, Young LE, Ciaramella G (2020) Directed evolution of adenine base editors with increased activity and therapeutic application. Nat Biotechnol 38(7):892–900

Hua K, Tao X, Liang W, Zhang Z, Gou R, Zhu J-K (2020) Simplified adenine base editors improve adenine base editing efficiency in rice. Plant Biotechnol J 18(3):770–778

LeBlanc C, Zhang F, Mendez J, Lozano Y, Chatpar K, Irish VF, Jacob Y (2018) Increased efficiency of targeted mutagenesis by CRISPR/Cas9 in plants using heat stress. Plant J 93(2):377–386

Millar AH, Heazlewood JL, Giglione C, Holdsworth MJ, Bachmair A, Schulze WX (2019) The scope, functions, and dynamics of posttranslational modifications. Annu Rev Plant Biol 70:119–151

Morrell R, Sadanandom A (2019) Dealing with stress: a review of plant SUMO proteases. Front Plant Sci 10:1122

Novatchkova M, Tomanov K, Hofmann K, Stuible H-P, Bachmair A (2012) Update on sumoylation: defining core components of the plant SUMO conjugation system by phylogenetic comparison. New Phytol 195(1):23–31

Richter J, Watson JM, Stasnik P, Borowska M, Neuhold J, Berger M, Stolt-Bergner P, Schoft V, Hauser MT (2018) Multiplex mutagenesis of four clustered CrRLK1L with CRISPR/Cas9 exposes their growth regulatory roles in response to metal ions. Sci Rep 8(1):12182

Tomanov K, Nehlin L, Ziba I, Bachmair A (2018) SUMO chain formation relies on the amino-terminal region of SUMO conjugating enzyme and has dedicated substrates in plants. Biochem J 475(1):61–68

Vertegaal ACO (2022) Signalling mechanisms and cellular functions of SUMO. Nat Rev Mol Cell Biol 23(11):715–731

Zong Y, Liu Y, Xue C, Li B, Li X, Wang Y, Li J, Liu G, Huang X, Cao X, Gao C (2022) An engineered prime editor with enhanced editing efficiency in plants. Nat Biotechnol 40(9):1394–1402

MIX
Papier aus verantwortungsvollen Quellen
Paper from responsible sources
FSC® C105338

If you have any concerns about our products,
you can contact us on
ProductSafety@springernature.com

In case Publisher is established outside the EU,
the EU authorized representative is:
Springer Nature Customer Service Center GmbH
Europaplatz 3, 69115 Heidelberg, Germany

Printed by Libri Plureos GmbH
in Hamburg, Germany